T0329643

Annals of Mathematics Studies

Number 163

Mathematical Aspects of Nonlinear Dispersive Equations

Jean Bourgain, Carlos E. Kenig, and
S. Klainerman, Editors

PRINCETON UNIVERSITY PRESS

PRINCETON AND OXFORD

2007

Published by Princeton University Press, 41 William Street, Princeton, New Jersey 08540

In the United Kingdom: Princeton University Press, 3 Market Place, Woodstock, Oxfordshire OX20 1SY

Library of Congress Cataloging-in-Publication Data

Mathematical aspects of nonlinear dispersive equations / Jean Bourgain, Carlos E. Kenig, and S. Klainerman, editors.
    p. cm. — (Annals of mathematics studies; no. 163)
    Includes bibliographical references and index.
    ISBN-13: 978-0-691-12860-3 (acid-free paper)
    ISBN-10: 0-691-12860-X (acid-free paper)
    ISBN-13: 978-0-691-12955-6 (pbk.: acid-free paper)
    ISBN-10: 0-691-12955-X (pbk.: acid-free paper)
  1. Differential equations, Nonlinear–Congresses. 2. Nonlinear partial differential operators–Congresses. I. Bourgain, Jean, 1954– II. Kenig, Carlos E., 1953– III. Klainerman, Sergiu, 1950–

QA372.M387 2007
515'.355–dc22   2006050254

British Library Cataloging-in-Publication Data is available

This book has been composed in Times Roman in LATEX

Printed on acid-free paper.∞

press.princeton.edu

Printed in the United States of America

10 9 8 7 6 5 4 3 2 1

Contents

Preface

This book contains the written accounts of a number of lectures given during the CMI/IAS Workshop on mathematical aspects of nonlinear PDEs in the spring of 2004 at the Institute for Advanced Study in Princeton, New Jersey. Several of them have an expository nature, describing the state of the art and research directions.Topics that are discussed in this volume are new developments on Schrödinger operators, non-linear Schrödinger and wave equations, hyperbolic conservation laws, and the Euler and Navier-Stokes equations. There has been intensive activity in recent years in each of these areas, leading in several cases to very significant progress and almost always broadening the subject.

The workshop is the conclusion of a year-long program at IAS centered around the analysis of nonlinear PDEs and the emergence of new analytical techniques. That year is marked by at least two important breakthroughs. The first is in the understanding of the blowup mechanism for the critical focusing Schrödinger equation. The other is a proof of global existence and scattering for the 3D quintic equation for general smooth data. Both cases illustrate in a striking way the role of hard analysis in addition to the more geometric approach and the role of energy estimates. This point of view is also reflected through the material presented in this volume.

The articles are written in varying styles. As mentioned, some are mainly expository and meant for a broader audience. They are not an overall survey but present more focused perspectives by a leader in the field. Others are more technical in nature with an emphasis on a specific problem and the related analysis and are addressed to active researchers. All of them are fully original accounts.

In conclusion, the editors would like to express their thanks to the Clay Mathematical Institute for its involvement and funding of the workshop.

Chapter One

On Strichartz's Inequalities and the Nonlinear Schrödinger Equation on Irrational Tori

J. Bourgain

1.0 INTRODUCTION

Strichartz's inequalities and the Cauchy problem for the nonlinear Schrödinger equation are considerably less understood when the spatial domain is a compact manifold M, compared with the Euclidean situation $M = \mathbb{R}^d$. In the latter case, at least the theory of Strichartz inequalities (i.e., moment inequalities for the linear evolution, of the form $\|e^{it\Delta}\phi\|_{L_{x,t}^p} \leq C\|\phi\|_{L_x^2}$) is basically completely understood and is closely related to the theory of oscillatory integral operators. Let $M = \mathbb{T}^d$ be a flat torus. If M is the usual torus, i.e.,

$$(e^{it\Delta}\phi)(x) = \sum_{n\in\mathbb{Z}^d} \hat\phi(n)e^{2\pi i(nx+|n|^2 t)} \quad (|n|^2 = n_1^2 + \cdots + n_d^2), \qquad (1.0.1)$$

a partial Strichartz theory was developed in [B1], leading to the almost exact counterparts of the Euclidean case for $d = 1, 2$ (the exact analogues of the $p = 6$ inequality for $d = 1$ and $p = 4$ inequality for $d = 2$ are *false* with periodic boundary conditions). Thus, assuming supp $\hat\phi \subset B(0, N)$,

$$\|e^{it\Delta}\phi\|_{L^6_{([0,1]\times[0,1])}} \ll N^\varepsilon \|\phi\|_2 \quad \text{for } d = 1 \qquad (1.0.2)$$

and

$$\|e^{it\Delta}\phi\|_{L^4_{([0,1]^2\times[0,1])}} \ll N^\varepsilon \|\phi\|_2 \quad \text{for } (d = 2). \qquad (1.0.3)$$

For $d = 3$, we have the inequality

$$\|e^{it\Delta}\phi\|_{L^4([0,1]^3\times[0,1])} \ll N^{\frac{1}{4}+\varepsilon}\|\phi\|_2 \quad (d = 3), \qquad (1.0.4)$$

but the issue:

PROBLEM. *Does one have an inequality*

$$\|e^{it\Delta}\phi\|_{L^{10/3}([0,1]^3\times[0,1)} \ll N^\varepsilon\|\phi\|_2 \quad (d = 3)$$

for all $\varepsilon > 0$ and supp $\hat\phi \subset B(0, N)$?

is still unanswered.

There are two kinds of techniques involved in [B1]. The first kind are arithmetical, more specifically the bound

$$\#\{(n_1, n_2) \in \mathbb{Z}^2 \mid |n_1| + |n_2| \le N \quad \text{and} \quad |n_1^2 + n_2^2 - A| \le 1\} \ll N^\varepsilon, \quad (1.0.5)$$

which is a simple consequence of the divisor function bound in the ring of Gaussian integers. Inequalities (1.0.2), (1.0.3), (1.0.4) are derived from that type of result.

The second technique used in [B1] to prove Strichartz inequalities is a combination of the Hardy-Littlewood circle method together with the Fourier-analytical approach from the Euclidean case (a typical example is the proof of the Stein-Tomas L^2-restriction theorem for the sphere). This approach performs better for larger dimension d although the known results at this point still leave a significant gap with the likely truth.

In any event, (1.0.2)–(1.0.4) permit us to recover most of the classical results for NLS

$$iu_t + \Delta u - u|u|^{p-2} = 0,$$

with $u(0) \in H^1(\mathbb{T}^d)$, $d \le 3$ and assuming $p < 6$ (subcriticality) if $d = 3$.

Instead of considering the usual torus, we may define more generally

$$\Delta\phi(x) = \sum_{n \in \mathbb{Z}^d} Q(n)\hat{\phi}(n)e^{2\pi i n.x}, \quad (1.0.6)$$

with $Q(n) = \theta_1 n_1^2 + \cdots \theta_d n_d^2$ and, say, $\frac{1}{C} \le \theta_i < C$ ($1 \le 1 \le d$) arbitrary (what we refer to as "(irrational torus)."

In general, we do not have an analogue of (1.0.5), replacing $n_1^2 + n_2^2$ by $\theta_1 n_1^2 + \theta_2 n_2^2$. It is an interesting question what the optimal bounds are in N for

$$\#\{(n_1, n_2) \in \mathbb{Z}^2 \mid |n_1| + |n_2| \le N \quad \text{and} \quad |\theta_1 n_1^2 + \theta_2 n_2^2 - A| \le 1\} \quad (1.0.7)$$

and

$$\#\{(n_1, n_2, n_3) \in \mathbb{Z}^3 \mid |n_1| + |n_2| + |n_3| \le N$$
$$\text{and} \quad |\theta_1 n_1^2 + \theta_2 n_2^2 + \theta_3 n_3^2 - A| \le 1\} \quad (1.0.8)$$

valid for all $\frac{1}{2} < \theta_i < 2$ and A.

Nontrivial estimates may be derived from geometric methods such as Jarnick's bound (cf. [Ja], [B-P]) for the number of lattice points on a strictly convex curve. Likely stronger results are true, however, and almost certainly better results may be obtained in a certain averaged sense when A ranges in a set of values (which is the relevant situation in the Strichartz problem). Possibly the assumption of specific diophantine properties (or genericity) for the θ_i may be of relevance.

In this paper, we consider the case of space dimension $d = 3$ (the techniques used have a counterpart for $d = 2$ but are not explored here).

Taking $\frac{1}{C} < \theta_i < C$ arbitrary and defining Δ as in (1.0.6), we establish the following:

PROPOSITION 1.1 *Let supp* $\hat{\phi} \subset B(0, N)$. *Then for* $p > \frac{16}{3}$,

$$\|e^{it\Delta}\phi\|_{L_t^p L_x^4} \le CN^{\frac{3}{4} - \frac{2}{p}} \|f\|_2, \quad (1.0.9)$$

where L_t^p *refers to* $L_{[0,1]}^p(dt)$.

PROPOSITION 1.3′. *Let supp* $\hat{\phi} \subset B(0, N)$. *Then*

$$\|e^{it\Delta}\phi\|_{L^4_{x,t}} < C_\varepsilon N^{\frac{1}{3}+\varepsilon} \|\phi\|_2 \ for \ all \ \varepsilon > 0. \tag{1.0.10}$$

The analytical ingredient involved in the proof of (1.0.9) is the well-known inequality for the squares

$$\left\| \sum_{j=1}^N e^{2\pi i j^2 \theta} \right\|_{L^q(\mathbb{T})} < CN^{1-\frac{2}{q}} \quad for \ q > 4. \tag{1.0.11}$$

The proof of (1.0.10) is more involved and relies on a geometrical approach to the lattice point counting problems, in the spirit of Jarnick's estimate mentioned earlier. Some of our analysis may be of independent interest. Let us point out that both (1.0.9), (1.0.10) are weaker than (1.0.4). Thus,

PROBLEM. *Does* (1.0.4) *hold in the context of* (1.0.6)?

Using similar methods as in [B1, 2] (in particular $X_{s,b}$-spaces), the following statements for the Cauchy problem for NLS on a 3D irrational torus may be derived.

PROPOSITION 1.2 *Let* Δ *be as in* (1.0.6). *Then the 3D defocusing NLS*

$$iu_t + \Delta u - u|u|^{p-2} = 0$$

is globally wellposed for $4 \le p < 6$ *and* H^1*-data.*

PROPOSITION 1.4′. *Let* Δ *be as in* (1.0.6). *Then the 3D defocusing cubic NLS*

$$iu_t + \Delta u \pm u|u|^2 = 0$$

is locally wellposed for data $u(0) \in H^s(\mathbb{T}^3)$, $s > \frac{2}{3}$.

This work originates from discussion with P. Gerard (March, 04) and some problems left open in his joint paper [B-G-T] about NLS on general compact manifolds. The issues in the particular case of irrational tori, explored here for the first time, we believe, unquestionably deserve to be studied more. Undoubtedly, further progress can be made on the underlying number theoretic problems.

1.1 AN INEQUALITY IN 3D

$$Q(n) = \theta_1 n_1^2 + \theta_2 n_2^2 + \theta_3 n_3^2, \tag{1.1.1}$$

where the θ_i are arbitrary, θ_i and θ_i^{-1} assumed bounded. Write

$$(e^{it\Delta} f)(x) = \sum_{n \in \mathbb{Z}^3} \hat{f}(n) e^{2\pi i(n.x + Q(n)t)}. \tag{1.1.2}$$

PROPOSITION 1.1 *For $p > \frac{16}{3}$, we have*

$$\|e^{it\Delta} f\|_{L_t^p L_x^4} \leq C_p N^{\frac{3}{4} - \frac{2}{p}} \|f\|_2 \tag{1.1.3}$$

assuming supp $\hat{f} \subset B(0, N)$. *Here L_t^p denotes L_t^p (loc).*

Remark. Taking $f(x) = N^{-3/2} \sum_{|n| < N} e^{inx}$, we see that (1.1.3) is optimal.

Proof of Proposition 1.1.

$$\|e^{it\Delta} f\|_{L_t^p L_x^4}^2 = \|(e^{it\Delta} f)^2\|_{L_t^{p/2} L_x^2}$$

$$= \left\| \left[\sum_{a \in \mathbb{Z}^3} \left| \sum_n \hat{f}(n) \hat{f}(a - n) e^{i[Q(n) + Q(a-n)]t} \right|^2 \right]^{1/2} \right\|_{L_t^{p/2}}$$

$$\leq \left[\sum_{a \in \mathbb{Z}^3} \left\| \sum_n \hat{f}(n) \hat{f}(a - n) e^{i[Q(n) + Q(a-n)]t} \right\|_{L_t^{p/2}}^2 \right]^{1/2} \tag{1.1.4}$$

since $p \geq 4$.

Denote $c_n = |\hat{f}(n)|$. Applying Hausdorff-Young,

$$\| \cdots \|_{L_t^{p/2}} \lesssim \left[\sum_{k \in \mathbb{Z}} \left| \sum_{|Q(n) + Q(a-n) - k| \leq \frac{1}{2}} c_n c_{a-n} \right|^{\frac{p}{p-2}} \right]^{\frac{p-2}{p}}. \tag{1.1.5}$$

Rewrite $|Q(n) + Q(a - n) - k| \leq \frac{1}{2}$ as $|Q(2n - a) + Q(a) - 2k| \leq 1$ and hence

$$2n \in a + \mathfrak{S}_\ell,$$

where

$$\ell = 2k - Q(a) \quad \text{and} \quad \mathfrak{S}_\ell = \{m \in \mathbb{Z}^3 | \ |Q(m) - \ell| \leq 1\}. \tag{1.1.6}$$

Clearly (1.1.5) may be replaced by

$$\left[\sum_{\ell \in \mathbb{Z}} \left| \sum_{2n \in a + \mathfrak{S}_\ell} c_n c_{a-n} \right|^{\frac{p}{p-2}} \right]^{\frac{p-2}{p}} \tag{1.1.6'}$$

and an application of Hölder's inequality yields

$$\left[\sum_\ell |\mathfrak{S}_\ell|^{\frac{p}{2(p-2)}} \left(\sum_{2n \in a + \mathfrak{S}_\ell} c_n^2 c_{a-n}^2 \right)^{\frac{p}{2(p-2)}} \right]^{\frac{p-2}{p}}$$

$$\lesssim \left(\sum_\ell |\mathfrak{S}_\ell|^{\frac{p}{p-4}} \right)^{\frac{p-4}{2p}} \left[\sum_n c_n^2 c_{a-n}^2 \right]^{1/2} \tag{1.1.7}$$

(since the $\mathfrak{S}_\ell$ are essentially disjoint).

Substitution of (1.1.7) in (1.1.4) gives the bound

$$\|e^{it\Delta}f\|_{L_t^p L_x^4} \leq C\left(\sum_\ell |\mathfrak{S}_\ell|^{\frac{p}{p-4}}\right)^{\frac{p-4}{4p}} \|f\|_2.
\tag{1.1.8}$$

Next, write

$$|\mathfrak{S}_\ell| \leq \int \left[\sum_{|m|\leq N} e^{iQ(m)t}\right] e^{-i\ell t} \varphi(t)\,dt,
\tag{1.1.8'}$$

where φ is compactly supported and $\hat\varphi \geq 0$, $\hat\varphi \geq 1$ on $[-1,1]$.
   Assume $p \leq 8$, so that $\frac{p}{p-4} \geq 2$ and from the Hausdorff-Young inequality again

$$\left(\sum |\mathfrak{S}_\ell|^{\frac{p}{p-4}}\right)^{\frac{p-4}{p}} \lesssim \left[\int_{\text{loc}} \prod_{j=1}^{3} \left|\sum_{0\leq m\leq N} e^{i\theta_j m^2 t}\right|^{\frac{p}{4}} dt\right]^{\frac{4}{p}}$$

$$\lesssim \left[\int_{\text{loc}} \left|\sum_{0\leq m\leq N} e^{im^2 t}\right|^{\frac{3p}{4}} dt\right]^{\frac{4}{p}}.
\tag{1.1.9}$$

Since $p > \frac{16}{3}$, $q = \frac{3p}{4} > 4$ and

$$\int_{\text{loc}} \left|\sum_{0\leq m\leq N} e^{im^2 t}\right|^q dt \sim N^{q-2}
\tag{1.1.10}$$

(immediate from Hardy-Littlewood).
   Therefore,

$$(1.1.9) \lesssim N^{3-\frac{8}{p}},$$

and substituting in (1.1.8), we obtain (1.1.3)

$$\|e^{it\Delta}f\|_{L_t^p L_x^4} \leq C N^{\frac{3}{4}-\frac{2}{p}} \|f\|_2$$

for $p \leq 8$. For $p > 8$, the result simply follows from

$$\|e^{it\Delta}f\|_{L_t^p L_x^4} \leq N^{2(\frac{1}{8}-\frac{1}{p})} \|e^{it\Delta}f\|_{L_t^8 L_x^4}.
\tag{1.1.11}$$

This proves Proposition 1.

Remarks.

   1. For $p = \frac{16}{3}$, we have the inequality

$$\|e^{it\Delta}f\|_{L_t^{16/3} L_x^4} \leq N^{\frac{3}{8}+} \|f\|_2
\tag{1.1.12}$$

   assuming $\operatorname{supp}\hat f \subset B(0,N)$.
   2. Inequalities (1.1.3) and (1.1.12) remain valid if $\operatorname{supp}\hat f \subset B(a,N)$ with $a \in \mathbb{Z}^3$ arbitrary.

Indeed,

$$|e^{it\Delta}f| = \left| \sum_{|m|\leq N} \hat{f}(a+m)e^{i[(x+2(\theta_1 a_1+\theta_2 a_2+\theta_3 a_3)t).m+Q(m)t]} \right|$$

so that

$$\|e^{it\Delta}f\|_{L_t^p L_x^4} = \left\| \sum_{|m|\leq N} \hat{f}(a+m)e^{i(x.m+Q(m)t)} \right\|_{L_t^p L_x^4}.$$

1.2 APPLICATION TO THE 3D NLS

Consider the defocusing 3D NLS

$$iu_t + \Delta u - u|u|^{p-2} = 0 \tag{1.2.1}$$

on $\mathbb{T}^3$ and with Δ as in (1.1.2).

Assume $4 \leq p < 6$.

PROPOSITION 1.2 (1.2.1) *is locally and globally wellposed in H^1 for $p < 6$.*

Sketch of Proof. Using $X_{s,b}$-spaces (see [B1]), the issue of bounding the nonlinearity reduces to an estimate on an expression

$$\| |e^{it\Delta}\phi_1| |e^{it\Delta}\phi_2| |e^{it\Delta}\psi|^{p-2}\|_1,$$

with $\|\phi_1\|_2, \|\phi_2\|_2 \leq 1$ and $\|\psi\|_{H^1} \leq 1$.

Thus we need to estimate

$$\| |e^{it\Delta}\phi_1| |e^{it\Delta}\psi|^{\frac{p-2}{2}} \|_2. \tag{1.2.2}$$

By dyadic restriction of the Fourier transform, we assume further

$$\text{supp}\,\hat{\phi}_1 \subset B(0, 2M)\backslash B(0, M) \tag{1.2.3}$$
$$\text{supp}\,\hat{\psi} \subset B(0, 2N)\backslash B(0, N) \tag{1.2.4}$$

for some dyadic $M, N > 1$.

Write

$$(1.2.2) \leq \|[e^{it\Delta}\phi_1][e^{it\Delta}\psi](1 + |e^{it\Delta}\psi|^2)^{\frac{p}{4}-1}\|_2, \tag{1.2.5}$$

where $(1 + |z|^2)^{\frac{p}{4}-1}$ is a smooth function of z.

If in (1.2.3), (1.2.4), $M > N$, partition $\mathbb{Z}^3$ in boxes I of size N and write

$$\phi_1 = \sum_I P_I\phi_1,$$

and by almost orthogonality

$$(1.2.5) \lesssim \left[\sum_I \| |e^{it\Delta}P_I\phi_1| |e^{it\Delta}\psi|(1 + |e^{it\Delta}\psi|^2)^{\frac{p}{4}-1}\|_2^2 \right]^{1/2}. \tag{1.2.6}$$

For fixed I, estimate

$$\| \, |e^{it\Delta} P_I \phi_1| \, |e^{it\Delta} \psi| (1 + |e^{it\Delta} \psi|^2)^{\frac{p}{4} - 1} \|_2$$

$$\leq \|e^{it\Delta} P_I \phi_1\|_{L_t^{16/3} L_x^4} \|e^{it\Delta} \psi\|_{L_t^{16/3} L_x^4} \left(1 + \|e^{it\Delta} \psi\|_{L_t^{8(\frac{p}{2}-2)} L_x^\infty}\right)^{\frac{p}{2}-2}, \quad (1.2.7)$$

and in view of (1.1.12) and Remarks (1), (2) above and (1.2.4),

$$\|e^{it\Delta} P_I \phi_1\|_{L_t^{16/3} L_x^4} \leq N^{\frac{3}{8}+} \|P_I \phi_1\|_2 \qquad (1.2.8)$$

$$\|e^{it\Delta} \psi\|_{L_t^{16/3} L_x^4} \leq N^{\frac{3}{8}+} N^{-1} \|\psi\|_{H^1} < N^{-\frac{5}{8}+}. \qquad (1.2.9)$$

To bound the last factor in (1.2.7), distinguish the cases

(A) $4 \leq p \leq \frac{16}{3}$

Then $8(\frac{p}{2} - 2) \leq \frac{16}{3}$ and by (1.2.9)

$$\|e^{it\Delta} \psi\|_{L_t^{8(\frac{p}{2}-2)} L_x^\infty} \leq \|e^{it\Delta} \psi\|_{L_t^{16/3} L_x^\infty} \leq N^{3/4} \|e^{it\Delta} \psi\|_{L_t^{16/3} L_x^4} < N^{1/8+}. \qquad (1.2.10)$$

Substitution of (1.2.8)–(1.2.10) in (1.2.7) gives

$$N^{-\frac{1}{4}+} N^{\frac{1}{8}(\frac{p}{2}-2)+} \|P_I \phi_1\|_2 \leq N^{-\frac{1}{6}+} \|P_I \phi_1\|, \qquad (1.2.11)$$

hence

$$(1.2.6) < N^{-\frac{1}{6}+}. $$

(B) $\frac{16}{3} < p < 6$

$$\|e^{it\Delta} \psi\|_{L_t^{8(\frac{p}{2}-2)} L_x^\infty} \leq N^{\frac{3}{8} - \frac{1}{2p-8} + \frac{3}{4}} \|e^{it\Delta} \psi\|_{L_t^{16/3} L_x^4} < N^{\frac{1}{2} - \frac{1}{2(p-4)} +} \qquad (1.2.12)$$

and

$$(1.2.7) \leq N^{\frac{p}{4} - \frac{3}{2} +} \|P_I \phi_1\|_2 \qquad (1.2.13)$$

$$(1.2.6) < N^{\frac{p}{4} - \frac{3}{2} +}. $$

This proves Proposition 1.2.

1.3 IMPROVED L^4-BOUND

It follows from (1.1.12) that

$$\|e^{it\Delta} f\|_{L_{t,x}^4} \leq N^{\frac{3}{8}+} \|f\|_2 \text{ if supp } \hat{f} \subset B(0, N). \qquad (1.3.1)$$

In this section, we will obtain the following first improvement:

$$\|e^{it\Delta} f\|_{L^4_{t,x}} \leq N^{\frac{7}{20}} \|f\|_2 \text{ for supp } \hat{f} \subset B(0, N).$$ (1.3.2)

Restrict $\hat{f}$ to a one level set, thus

$$\hat{f} = \hat{f} \chi_{\Omega_\mu}$$ (1.3.3)

with

$$\Omega_\mu = \{n \in [-N, N]^3 \mid |\hat{f}(n)| \sim \mu\}$$

$$|\Omega_\mu| \leq \mu^{-2}.$$ (1.3.4)

In what follows, we assume f of the form (1.3.3).

LEMMA 1.1

$$\|e^{it\Delta} f\|_{L^4_{x,t}} < \mu^{1/6} N^{\frac{1}{2}+}$$ (1.3.5)

Proof. From estimates (1.1.4) and (1.1.5′) with $p = 4$ and letting

$$c_n = \begin{cases} \mu & \text{if } n \in \Omega\mu \\ 0 & \text{otherwise} \end{cases}$$

we get the following bound on $\|e^{it\Delta} f\|_4^2$:

$$\mu^2 \left[\sum_{a \in \mathbb{Z}^3} \sum_{\ell \in \mathbb{Z}} |(a + \mathfrak{S}_\ell) \cap (2\Omega_\mu) \cap (2a - 2\Omega_\mu)|^2 \right]^{1/2}.$$ (1.3.6)

Recall also estimate (1.1.9) for $p = \frac{16}{3}$,

$$\left(\sum |\mathfrak{S}_\ell|^4 \right)^{1/4} < N^{\frac{3}{2}+}.$$ (1.3.7)

Hence, if we denote for $L \geq 1$ (a dyadic integer)

$$\mathcal{L}_L = \{\ell \in \mathbb{Z} \mid |\mathfrak{S}_\ell| \sim N^{\frac{3}{2}+} L^{-1/4}\},$$ (1.3.8)

it follows that

$$|\mathcal{L}_L| < L.$$ (1.3.9)

Estimate (1.3.6) by

$$\mu^2 \left[\sum_{\ell \in \mathbb{Z}} |\mathfrak{S}_\ell| \sum_a |(a + \mathfrak{S}_\ell) \cap (2\Omega_\mu) \cap (2a - 2\Omega_\mu)| \right]^{1/2}$$ (1.3.10)

and restrict in (1.3.10) the ℓ-summation to $\mathcal{L}_L$.

There are the following two bounds:

$$\mu^2 \left[\sum_{\ell \in \mathcal{L}_L} |\mathfrak{S}_\ell| \sum_a |(a + \mathfrak{S}_\ell) \cap (2\Omega_\mu)| \right]^{1/2} \leq \mu^2 \left[\sum_{\ell \in \mathcal{L}_L} |\mathfrak{S}_\ell|^2 |\Omega_\mu| \right]^{1/2}$$

$$< \mu N^{\frac{3}{2}+} L^{1/4} \qquad (1.3.11)$$

and also

$$\mu^2 N^{\frac{3}{4}+} L^{-1/8} \left[\sum_{\ell,a} |(a + \mathfrak{S}_\ell) \cap (2\Omega_\mu) \cap (2a - 2\Omega_\mu)| \right]^{1/2}$$

$$< \mu^2 N^{\frac{3}{4}+} L^{-1/8} |\Omega_\mu| < N^{\frac{3}{4}+} L^{-1/8}. \qquad (1.3.12)$$

Taking the minimum of (1.3.11), (1.3.12), we obtain $\mu^{1/3} N^{1+}$. Summing over dyadic values of $L \lesssim N^2$, the estimate follows.

Next, we need a discrete maximal inequality of independent interest.

LEMMA 1.2 *Consider the following maximal function on $\mathbb{Z}^3$*

$$F^*(x) = \max_{1 < \ell < N^2} \sum_{|Q(y) - \ell| \leq 1} F(x + y). \qquad (1.3.13)$$

For

$$\lambda > N^{\frac{1}{2}} \|F\|_2 \qquad (1.3.14)$$

we have

$$|[F^* > \lambda]| < N^{\frac{3}{2}+} \|F\|_2^2 \lambda^{-2}. \qquad (1.3.15)$$

$(\|F\|_2$ *denotes* $(\sum_{x \in \mathbb{Z}^3} |F(x)|^2)^{1/2}).$

Proof. Let $A = [F^* > \lambda] \subset \mathbb{Z}^3$. Thus for $x \in A$, there is ℓ_x s.t.

$$\langle F, \chi_{x + \mathfrak{S}_{\ell_x}} \rangle > \lambda.$$

Estimate as usual

$$\lambda.|A| \leq \left\langle F, \sum_{x \in A} \chi_{x + \mathfrak{S}_{\ell_x}} \right\rangle$$

$$\leq \|F\|_2 \left\| \sum_{x \in A} \chi_{x + \mathfrak{S}_{\ell_x}} \right\|_2$$

$$= \|F\|_2 [|A| \max_\ell |\mathfrak{S}_\ell| + |A|^2 \max_{x \neq y} |(x + \mathfrak{S}_x) \cap (y + \mathfrak{S}_y)|]^{1/2}. \quad (1.3.16)$$

Use the crude bound $|\mathfrak{S}_\ell| < N^{\frac{3}{2}+}$ from (1.3.7) and denote

$$K = \max_{x,y \in \mathbb{Z}^3, x \neq y} |(x + \mathfrak{S}_x) \cap (y + \mathfrak{S}_y)|. \qquad (1.3.17)$$

From (1.3.16), we conclude that

$$|A| < N^{\frac{3}{2}+}\|F\|_2^2\lambda^{-2} \tag{1.3.18}$$

if

$$\lambda > \|F\|_2\, K^{1/2}. \tag{1.3.19}$$

It remains to evaluate K.

If $n \in \mathbb{Z}^3$ lies in $(x + \mathfrak{S}_{\ell_x}) \cap (y + \mathfrak{S}_{\ell_y})$, then

$$|Q(x - n) - \ell_x| \le 1$$
$$|Q(y - n) - \ell_y| \le 1,$$

and subtracting

$$|2\theta_1(x_1 - y_1)n_1 + 2\theta_2(x_2 - y_2)n_2 + 2\theta_3(x_3 - y_3)n_3$$
$$-Q(x) + Q(y) + \ell_x - \ell_y| \le 2. \tag{1.3.20}$$

Since $x \ne y$ in $\mathbb{Z}^3$, $|x - y| \ge 1$ and (1.3.20) restricts n to a 1-neighborhood $\prod_{(1)}$ of some plane $\prod$. Therefore (fig. 1.1.),

$$\left|(x + \mathfrak{S}_x) \cap (y + \mathfrak{S}_y)\right| < \max_{\ell,\prod} \left|\mathfrak{S}_\ell \cap \prod_{(1)}\right| \tag{1.3.21}$$

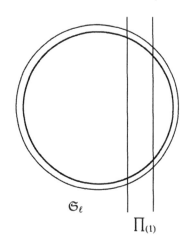

Fig. 1.1.

Recall that $\mathfrak{S}_\ell$ is a $\frac{1}{\sqrt{\ell}}$-neighborhood of a "regular" ellipsoid $\mathcal{E}$ of size $\sqrt{\ell}$. Estimate the number of lattice points $|\mathfrak{S}_\ell \cap \prod_{(1)}|$ in $\mathfrak{S}_\ell \cap \prod_{(1)}$ by the area of $\mathcal{E}$ inside $\prod_{(1)}$. By affine transformation, we may assume $\mathcal{E}$ a sphere of radius at most N. A simple calculation shows that this area is at most $\sim N$. Hence $K \lesssim N$ and (1.3.18) holds if (1.3.14).

This proves Lemma 1.2.

Remark. The number K in (1.3.17) allows more refined estimates that will be pointed out later.

Taking in Lemma 1.2 $F = \chi_{2\Omega_\mu}$, we get

COROLLARY 1.3 *If* $\lambda > N^{\frac{1}{2}}\mu^{-1}$, *then*

$$|\{a \in \mathbb{Z}^3 | \max_{\ell \lesssim N^2} |(a + \mathfrak{S}_\ell) \cap 2\Omega_\mu| > \lambda\}| < N^{\frac{3}{2}+}(\mu\lambda)^{-2}. \tag{1.3.22}$$

Now we establish

LEMMA 1.4

$$\|e^{it\Delta} f\|_4 < N^{\frac{3}{16}+} + N^{\frac{1}{8}}\mu^{-\frac{1}{4}}. \tag{1.3.23}$$

Proof. We return to (1.3.6).

Denote for dyadic λ

$$A_\lambda = \{a \in \mathbb{Z}^3 \cap [-N, N]^3 | \max_\ell |(a + \mathfrak{S}_\ell) \cap (2\Omega_\mu)| \sim \lambda\}.$$

For $a \in A_\lambda$, there are at most $\mu^{-2}\lambda^{-1}$ values of $\ell \in \mathbb{Z}$ s.t.

$$|(a + \mathfrak{S}_\ell) \cap (2\Omega_\mu)| > \lambda \tag{1.3.24}$$

(since the $\mathfrak{S}_\ell$ are disjoint).

We estimate

$$\sum_{a \in A_\lambda} \sum_{\substack{\ell \in \mathbb{Z} \\ (1.3.24)}} |(a + \mathfrak{S}_\ell) \cap (2\Omega_\mu) \cap (2a - 2\Omega_\mu)|^2 \tag{1.3.25}$$

distinguishing the following two cases:

CASE 1.1 $\lambda \leq N^{\frac{1}{2}}\mu^{-1}$.

Write

$$(1.3.25) \leq \lambda \sum_a \sum_{\ell \in \mathbb{Z}} |(a + \mathfrak{S}_\ell) \cap (2\Omega_\mu) \cap (2a - 2\Omega_\mu)|$$

$$< \lambda |\Omega_\mu|^2 < N^{\frac{1}{2}}\mu^{-5}. \tag{1.3.26}$$

CASE 1.2 $\lambda > N^{\frac{1}{2}}\mu^{-1}$.

Then (1.3.22) *applies and* $|A_\lambda| < N^{\frac{3}{2}+}(\mu\lambda)^{-2}$. *Hence*

$$(1.3.25) < |A_\lambda|\mu^{-2}\lambda < N^{\frac{3}{2}+}\mu^{-4}\lambda^{-1}. \tag{1.3.27}$$

Since there is also the obvious bound given by (1.3.26)

$$(1.3.25) < \lambda\mu^{-4}, \tag{1.3.28}$$

we obtain

$$(1.3.25) < N^{\frac{3}{4}+}\mu^{-4}. \tag{1.3.29}$$

Substitution of (1.3.26), (1.3.29) *implies*

$$\|e^{it\Delta}f\|_4^2 \le (1.3.6) < N^{\frac{1}{4}}\mu^{-\frac{1}{2}} + N^{\frac{3}{8}+}.$$

PROPOSITION 1.3 $\|e^{it\Delta}f\|_{L^4_{x,t}} \le N^{\frac{7}{20}+}\|f\|_2$ *if* $\operatorname{supp}\hat{f} \subset B(0,N)$.

Proof. With f as above, it follows from Lemma 1.1 and 1.4 that

$$\|e^{it\Delta}f\|_4 < N^{\frac{3}{16}+} + \min(\mu^{\frac{1}{6}}N^{\frac{1}{2}+}, N^{\frac{1}{8}}\mu^{-\frac{1}{4}}) < N^{\frac{7}{20}+}.$$

As a corollary of Proposition 1.3, we get the following wellposedness result for cubic NLS in 3D.

PROPOSITION 1.4 *Consider* $iu_t + \Delta u \pm u|u|^2 = 0$ *on* $\mathbb{T}^3$ *and with* Δ *as above. There is local wellposedness for* $u(0) \in H^s(\mathbb{T}^3)$, $s > \frac{7}{10}$.

1.4 A REFINEMENT OF PROPOSITION 3

Our purpose is to improve upon Lemma 1.2 by a better estimate on the quantity K in (1.3.17), thus

$$|(x + \mathcal{E}_\varepsilon) \cap (x' + \mathcal{E}'_\varepsilon) \cap \mathbb{Z}^3|, \tag{1.4.1}$$

where $\mathcal{E}, \mathcal{E}'$ are nondegenerated ellipsoids centered at 0 of size $\sim R < N$ and $\varepsilon = \frac{1}{R}$ refers to an ε'-neighborhood, $x \ne x'$ in $\mathbb{Z}^3$.

The main ingredients are versions of the the standard Jarnick argument to estimate the number of lattice points on a curve (cf. [Ja]). Here we will have to deal with neighborhoods.

We start with a 2-dimensional result.

LEMMA 1.5 *Let* $\mathcal{E}$ *be a "regular" oval in* $\mathbb{R}^2$ *of size* R. *Then*

$$\max_a |B(a, R^{1/3}) \cap \mathcal{E}_{\frac{1}{R}} \cap \mathbb{Z}^2| < C. \tag{1.4.2}$$

In particular

$$|\mathcal{E}_{\frac{1}{R}} \cap \mathbb{Z}^2| < CR^{2/3} \tag{1.4.3}$$

and for all $\rho > 1$

$$|B(a, \rho) \cap \mathcal{E}_{\frac{1}{R}} \cap \mathbb{Z}^2| < C\rho^{2/3} \tag{1.4.4}$$

($\mathcal{E}_{\frac{1}{R}}$ *denotes a* $\frac{1}{R}$-*neighborhood of* $\mathcal{E}$).

Proof. Let P_1, P_2, P_3 be noncolinear points in $B(a, cR^{1/3}) \cap \mathcal{E}_{\frac{1}{R}} \cap \mathbb{Z}^2$, letting c be a sufficiently small constant. Following Jarnick's argument,

$$0 \neq \text{ area triangle } (P_1, P_2, P_3) = \frac{1}{2} \left| \begin{array}{ccc} 1 & 1 & 1 \\ P_1 & P_2 & P_3 \end{array} \right| \in \frac{1}{2} \mathbb{Z}_+$$

and hence

$$\text{area } (P_1, P_2, P_3) \geq \frac{1}{2}. \tag{1.4.5}$$

Take P_1', P_2', $P_3' \in \mathcal{E}$ so that $|P_j - P_j'| < \frac{1}{R}$. Clearly,

$$\left| \left| \begin{array}{ccc} 1 & 1 & 1 \\ P_1 & P_2 & P_3 \end{array} \right| - \left| \begin{array}{ccc} 1 & 1 & 1 \\ P_1' & P_2' & P_3' \end{array} \right| \right| < R^{\frac{1}{3}} R^{-1} \ll 1$$

so that

$$\text{area } (P_1', P_2', P_3') > \frac{1}{4}. \tag{1.4.6}$$

On the other hand, obviously

$$\text{area } (P_1', P_2', P_3') \leq cR^{1/3} \frac{R^{2/3}}{R} \ll 1 \tag{1.4.7}$$

a contradiction. This proves (1.4.2), observing that if Λ is a line, then clearly $\Lambda \cap \mathcal{E}_{\frac{1}{R}}$ is at most of bounded length. Hence $|\mathcal{E}_{\frac{1}{R}} \cap \Lambda \cap \mathbb{Z}^2| < C$.

Partitioning $\mathcal{E}_{\frac{1}{R}}$ in sets of size $cR^{1/3}$ (1.4.3) follows.

Finally, estimate (1.4.4) by $\min(1 + \rho R^{-1/3}, R^{2/3}) \lesssim \rho^{2/3}$.

Remark. Projecting on one of the coordinate planes, Lemma 1.5 applies equally well to a regular oval $\mathcal{E}$ in a 2-plane $\prod$ in $\mathbb{R}^3$ and

$$\max_a |B(a, R^{1/3}) \cap \mathcal{E}_{\frac{1}{R}} \cap \mathbb{Z}^3| < C \tag{1.4.8}$$

and

$$\max_a |B(a, \rho) \cap \mathcal{E}_{\frac{1}{R}} \cap \mathbb{Z}^3| < C\rho^{2/3}, \tag{1.4.9}$$

where $\mathcal{E}$ is of size R and $\mathcal{E}_{\frac{1}{R}}$ denotes an $\frac{1}{R}$-neighborhood of $\mathcal{E}$.

There is an obvious extension of (1.4.2) in dimension 3. One has

LEMMA 1.6 *Let $\mathcal{E}$ be a 2-dim regular oval in $\mathbb{R}^3$ of size R. Then, for all $a \in \mathbb{R}^3$ and appropriate c, $B(a, cR^{1/4}) \cap \mathcal{E}_{\frac{1}{R}} \cap \mathbb{Z}^3$ does not contain 4 noncoplanar points.*

Proof. If P_1, P_2, P_3, P_4 are noncoplanar points in $B(a, cR^{1/4}) \cap \mathcal{E}_{\frac{1}{R}} \cap \mathbb{Z}^3$ and $|P_j - P_j'| < \frac{1}{R}$, $P_j' \in \mathcal{E}$, write

$$\frac{1}{6}\mathbb{Z}_+ \ni \text{Vol} (P_1, P_2, P_3, P_4) = \text{Vol} (P_1', P_2', P_3', P_4') + 0(R^{1/2}R^{-1}),$$

and hence

$$\text{Vol}\,(P_1', P_2', P_3', P_4') > \frac{1}{7}. \tag{1.4.10}$$

On the other hand, this volume may be estimated by the volume of the cap obtained as convex hull $\text{conv}(\mathcal{E} \cap B(a, cR^{1/4})$ bounded by $(cR^{1/4})^2 \frac{R^{\frac{1}{2}}}{R} \ll 1$. This proves Lemma 1.6.

We now return to (1.3.21) and estimate $|\mathcal{E}_{\frac{1}{R}} \cap \prod_{(1)} \cap \mathbb{Z}^3|$, where $\prod_{(1)}$ is a 1-neighborhood of a plane $\prod$ in $\mathbb{R}^3$. Our purpose is to show

LEMMA 1.7

$$\left| \mathcal{E}_{\frac{1}{R}} \cap \prod_{(1)} \cap \mathbb{Z}^3 \right| < R^{2/3+}. \tag{1.4.11}$$

Proof. (see fig. 1.2.).

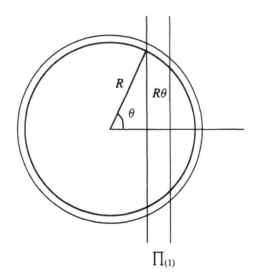

$$\prod_{(1)}$$

Fig. 1.2.

Thus $\mathcal{E} \cap \prod_{(1)}$ is a truncated conical region of base-size $R\theta$, slope θ and height 1, for some $\theta > \frac{1}{R}$.

Consider first the case $\theta < R^{-1/4}$. Partition $\mathcal{E}_{\frac{1}{R}} \cap \prod_{(1)}$ in $\sim R^{-\frac{1}{2}}(R\theta)\frac{1}{\theta}$ regions $\mathcal{D}$ of size $cR^{1/4}$. According to Lemma 1.6, $\mathcal{D} \cap \mathbb{Z}^3$ consists of coplanar points, therefore lying in some plane $\mathcal{P} \subset \mathbb{R}^3$ and

$$\mathcal{D} \cap \mathbb{Z} = \mathcal{D} \cap \mathcal{P} \cap \mathcal{E}_{\frac{1}{R}} \cap \mathbb{Z}^3. \tag{1.4.12}$$

$\mathcal{P} \cap \mathcal{E}$ is an ellipse $\mathcal{E}'$ of size r (we may assume $r \gg 1$) and we claim that $\mathcal{P} \cap \mathcal{E}_{\frac{1}{R}} \subset$
$\mathcal{E}'_{\frac{1}{r}} = \frac{1}{r}$-neighborhood of $\mathcal{E}'$. To see this, we may by affine transformation assume
$\mathcal{E}$ to be a sphere of radius R, in which case it is a straightforward calculation.

From (1.4.12) and the preceding,

$$|\mathcal{D} \cap \mathbb{Z}^3| \le |\mathcal{D} \cap \mathcal{E}'_{\frac{1}{r}} \cap \mathbb{Z}^3|$$
$$< C(\text{diam } \mathcal{D})^{2/3}$$
$$< CR^{1/6}, \tag{1.4.13}$$

applying (1.4.9) to $\mathcal{E}'_{\frac{1}{r}}$ in the plane $\mathcal{P}$.

We conclude that for $\theta < R^{-1/4}$,

$$\left| \mathcal{E}_{\frac{1}{R}} \cap \prod\nolimits_{(1)} \cap \mathbb{Z}^3 \right| < CR^{\frac{1}{2}} R^{\frac{1}{6}} < CR^{\frac{2}{3}} \tag{1.4.14}$$

and hence (1.4.12).

Assume next that $\theta > R^{-1/4}$.

Let $D > 1$ be such that $B(a, D) \cap \mathcal{E}_{\frac{1}{R}} \cap \prod_{(1)} \cap \mathbb{Z}^3$ (for some $a \in \mathbb{R}^3$) contains 4
noncoplanar points P_1, P_2, P_3, P_4. Assume

$$D < (\theta R)^{1/2}. \tag{1.4.15}$$

Repeating the argument in Lemma 1.6, let $|P_j - P'_j| \le \frac{1}{R}$, $P'_j \in B(a, D+1) \cap \mathcal{E} \cap$
$\prod_{(2)}$.

By (1.4.15),

$$\text{Vol}\,(P'_1, P'_2, P'_3, P'_4) > \frac{1}{6} - 0(D^2 R^{-1}) > \frac{1}{7}. \tag{1.4.16}$$

Considering sections parallel to $\prod$, write an upper bound on the left side of (1.4.16)
by

$$\text{Vol}\left(\text{conv}(B(a, 2D) \cap \mathcal{E} \cap \prod\nolimits_{(2)}\right) \le D\frac{D^2}{R\theta}. \tag{1.4.17}$$

Together with (1.4.16), (1.4.17) implies

$$D \gtrsim (R\theta)^{1/3} \gtrsim \frac{1}{\theta}, \tag{1.4.18}$$

which therefore holds independently from assumption (1.4.15).

Next, we consider a cover of $\mathcal{E}_{\frac{1}{R}} \cap \prod_{(1)}$ by essentially disjoint balls $B(a_\alpha, D_\alpha)$
chosen in such a way that the following properties hold:

1. (1.4.19) All elements of $B(a_\alpha, D_\alpha) \cap \mathcal{E}_{\frac{1}{R}} \cap \prod_{(1)} \cap \mathbb{Z}^3$ are coplanar.
2. (1.4.20) $B(a_\alpha, 2D_\alpha) \cap \mathcal{E}_{\frac{1}{R}} \cap \prod_{(1)} \cap \mathbb{Z}^3$ contains 4 noncoplanar points.

By (1.4.18), $D_\alpha > \frac{1}{\theta}$. Fixing a dyadic size $\theta R > D > \frac{1}{\theta}$ and considering α's such
that

$$D_\alpha \sim D, \tag{1.4.21}$$

their number is at most

$$\frac{R\theta}{D}. \tag{1.4.22}$$

Proceeding as earlier, let $\mathcal{P}$ be a plane containing the elements of

$$B(a_\alpha, D_\alpha) \cap \mathcal{E}_{\frac{1}{R}} \cap \prod_{(1)} \cap \mathbb{Z}^3 \tag{1.4.23}$$

and $\mathcal{E}'$ an ellipse of size r in $\mathcal{P}$ such that $\mathcal{E}' = \mathcal{P} \cap \mathcal{E}, \mathcal{E}'_{\frac{1}{r}} \supset \mathcal{E}_{\frac{1}{R}} \cap \mathcal{P}$.

Let P_1 be any point in (1.4.23) and denote τ the tangent plane to $\mathcal{E}$ at P_1, ψ the angle of τ and P. Thus (fig. 1.3.),

$$r \sim R\psi$$

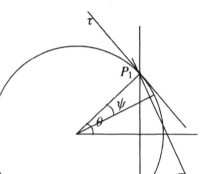

Fig. 1.3.

If $\psi \gtrsim \theta$, then $r \sim R\psi \gtrsim R\theta$ and we estimate

$$\left| B(P_1, 2D) \cap \prod_{(1)} \cap \mathcal{E}'_{\frac{1}{r}} \cap \mathbb{Z}^3 \right| \lesssim D.r^{-1/3}. \tag{1.4.24}$$

The corresponding contribution to $\mathcal{E}_{\frac{1}{R}} \cap \prod_{(1)} \cap \mathbb{Z}^3$ is at most

$$\frac{R\theta}{D} D(R\theta)^{-1/3} \lesssim R^{2/3}. \tag{1.4.25}$$

Assume thus $\psi \ll \theta$, in which case $\theta \approx$ angle $(\prod, \mathcal{P})$ and

$$\operatorname{diam}\left(\mathcal{E}' \cap \prod_{(1)} \right) \sim \sqrt{\frac{r}{\theta}}. \tag{1.4.26}$$

Estimate

$$\left| B(P_1, 2D) \cap \prod_{(1)} \cap \mathcal{E}'_{\frac{1}{r}} \cap \mathbb{Z}^3 \right| \lesssim \min\left(D, \sqrt{\frac{r}{\theta}}\right) r^{-1/3} < \theta^{-1/3} D^{1/3}, \quad (1.4.27)$$

which collected contribution to $\mathcal{E} \cap \prod_{(1)} \cap \mathbb{Z}$ is bounded by

$$\frac{R\theta}{D} \theta^{-1/3} D^{1/3} = \frac{R\theta^{2/3}}{D^{2/3}} \le \frac{R}{D^{2/3}}. \quad (1.4.28)$$

We may thus assume $D < R^{1/2}$.

Assume (1.4.23) contains K points and denote $d < D$ the diameter of (1.4.23). Hence, from (1.4.26),

$$\theta d^2 \lesssim r. \quad (1.4.29)$$

Partitioning $\mathcal{E}'$ in arcs of size $\frac{d}{K}$, we get thus a set $\mathcal{E}'_{\frac{1}{r}} \cap B(P_1, \frac{d}{K}) \cap \mathbb{Z}^3$ containing 3 noncollinear points P_1, P_2, P_3 from (1.4.23). Recalling assumption (1.4.20) there is $P \in B(P_1, 2D) \cap \mathcal{E}_{\frac{1}{R}} \cap \prod_{(1)} \cap \mathbb{Z}^3$ such that P_1, P_2, P_3, P are noncoplanar and therefore

$$\text{Vol}(P_1, P_2, P_3, P) \ge \frac{1}{6}. \quad (1.4.30)$$

Estimate from above (since $P_1, P_2, P_3 \in \mathcal{P}$)

$$\text{Vol}(P_1, P_2, P_3, P) \le \text{area}(P_1, P_2, P_3) \, \text{dist}(P, \mathcal{P})$$
$$\lesssim \frac{1}{r}\left(\frac{d}{K}\right)^3 \text{dist}(P, \mathcal{P}). \quad (1.4.31)$$

(We use here the fact that $P_1, P_2, P_3 \in \mathcal{E}'_{1/r}$ and $\text{diam}\{P_1, P_2, P_3\} < \frac{d}{K}$.)

It remains to estimate $\text{dist}(P, \mathcal{P})$.

Letting τ be again the tangent plane at P_1, write

$$\text{dist}(P, \mathcal{P}) \le |P - \overline{P}| + \text{dist}(\overline{P}, \mathcal{P}), \quad (1.4.32)$$

where $\overline{P} \in \tau$

$$|P - \overline{P}| = \text{dist}(P, \tau) \lesssim \frac{D^2}{R} < 1, \quad (1.4.33)$$

and hence

$$\overline{P} \in \tau \cap \prod_{(2)} \cap B(P_1, 2D + 1).$$

We may assume $P_1 \in \prod$. Denote ℓ_0 the line

$$\ell_0 = \prod \cap \tau \quad (1.4.34)$$

and ℓ_1 the line

$$\ell_1 = \mathcal{P} \cap \tau. \quad (1.4.35)$$

Thus $P_1 \in \ell_0 \cap \ell_1$

$$\text{dist}(\overline{P}, \mathcal{P}) = \text{dist}(\overline{P}, \ell_1) \text{ angle}(\tau, \mathcal{P}) = \text{dist}(\overline{P}, \ell_1)\psi$$
$$\sim \frac{r}{R} \text{dist}(\overline{P}, \ell_1)$$
$$\sim \frac{r|\overline{P} - P_1|}{R} \text{angle}([P_1, \overline{P}], \ell_1). \quad (1.4.36)$$

By assumption, there is a point $P_4 \in$ (1.4.23) s.t.

$$\frac{d}{2} < |P_1 - P_4| \le d. \quad (1.4.37)$$

Thus $P_4 \in \mathcal{P} \cap \prod_{(1)}$. Estimate

$$\text{angle}([P_1, \overline{P}], \ell_1) \le \text{angle}([P_1, P_4], \ell_1) + \text{angle}([P_1, \overline{P}], [P_1, P_4])$$
$$= (1.4.38) + (1.4.39).$$

Since $\text{dist}(P_4, \tau) \sim \frac{d^2}{R}$, we have

$$\frac{d^2}{R} \sim \text{dist}(P_4, \ell_1).\psi$$

and

$$(1.4.38) \sim \frac{1}{d}\text{dist}(P_4, \ell_1) \sim \frac{d}{R\psi} \sim \frac{d}{r}. \quad (1.4.40)$$

Estimate

$$(1.4.39) \le \text{angle}([P_1, \overline{P}], \ell_0) + \text{angle}([P_1, P_4], \ell_0). \quad (1.4.41)$$

Since $\text{dist}(P_4, \prod) \le 1$ and $\text{dist}(P_4, \tau) \sim \frac{d^2}{R} < 1$, we get

$$2 \ge \text{dist}(P_4, \ell_0) \text{ angle}\left(\tau, \prod\right) = \theta . \text{ dist}(P_4, \ell_0)$$

$$\text{angle}([P_1, P_4], \ell_0) \lesssim \frac{1}{\theta d}. \quad (1.4.42)$$

Similarly,

$$\text{angle}([P_1, \overline{P}], \ell_0) \lesssim \frac{1}{\theta|P_1 - \tilde{P}|}, \quad (1.4.43)$$

and hence

$$(1.4.39) \lesssim \frac{1}{\theta d} + \frac{1}{\theta|P_1 - \tilde{P}|}. \quad (1.4.44)$$

It follows that

$$\text{angle}([P_1, \overline{P}], \ell_1) \lesssim \frac{d}{r} + \frac{1}{\theta d} + \frac{1}{\theta|P_1 - \tilde{P}|} \overset{(1.4.29)}{<} \frac{1}{\theta d} + \frac{1}{\theta|P_1 - \tilde{P}|}. \quad (1.4.45)$$

Recalling (1.4.36),

$$\text{dist}\,(\overline{P}, \mathcal{P}) \lesssim \frac{rD}{R\theta d} + \frac{r}{R\theta} \sim \frac{rD}{R\theta d} \tag{1.4.46}$$

and, by (1.4.32),

$$\text{dist}\,(P, \mathcal{P}) \lesssim \frac{D^2}{R} + \frac{rD}{Rd\theta}. \tag{1.4.47}$$

Substituting in (1.4.31) gives by (1.4.30)

$$1 \lesssim \frac{d^3 D^2}{r R K^3} + \frac{d^2 D}{\theta R K^3} \stackrel{(1.4.29)}{\lesssim} \frac{dD^2}{\theta R K^3}$$

$$K \lesssim \frac{D}{(\theta R)^{1/3}}. \tag{1.4.48}$$

Multiplying with (1.4.22), we obtain again

$$\frac{R\theta}{D} \frac{D}{(\theta R)^{1/3}} \le R^{2/3} \tag{1.4.49}$$

as a bound on $|\mathcal{E}_{\frac{1}{R}} \cap \prod_{(1)} \cap \mathbb{Z}^3|$.

   This proves Lemma 1.7.

   Lemma 1.7 allows for the following improvement of Lemma 1.2.

LEMMA 1.2′.   *Let F^* be the discrete maximal function* (1.3.13). *Then*

$$|[F^* > \lambda]| < N^{\frac{3}{2}+} \|F\|_2^2 \lambda^{-2} \tag{1.4.50}$$

provided

$$\lambda > N^{\frac{1}{3}+} \|F\|_2. \tag{1.4.51}$$

Proof. Returning to the proof of Lemma 1.2, Lemma 1.7 implies the bound on K introduced in (1.3.17)

$$K < N^{\frac{2}{3}+}, \tag{1.4.52}$$

and (1.3.19) becomes (1.4.51) instead of (1.3.14).

   Hence (1.3.32) in Corollary 1.1 holds under the assumption

$$\lambda > N^{\frac{1}{3}+} \mu^{-1}, \tag{1.4.53}$$

which leads to the following improved Lemma 1.3 and Propositions 1.5, 1.6.

LEMMA 1.4′.

$$\|e^{it\Delta} f\|_4 < N^{\frac{3}{16}+} + N^{\frac{1}{12}+} \mu^{-\frac{1}{4}}. \tag{1.4.54}$$

PROPOSITION 1.3′.

$$\|e^{it\Delta} f\|_4 \le N^{\frac{1}{3}+} \|f\|_2 \text{ if supp } \hat{f} \subset B(0, N). \tag{1.4.55}$$

PROPOSITION 1.4′.   *The 3D cubic NLS $iu_t + \Delta u \pm u|u|^2 = 0$ on $\mathbb{T}^3$ with Δ as in (1.1.2) is locally well-posed for $u(0) \in H^s(\mathbb{T}^3)$, $s > \frac{2}{3}$.*

REFERENCES

[B1]   J. Bourgain, *Fourier transform restriction phenomena for certain lattice subsets and applications to nonlinear evolution equations*, GAFA 3(1993), no. 2, 107–156.

[B2]   J. Bourgain, *Exponential sums and nonlinear Schrödinger equations*, GAFA 3 (1993), no. 2, 157–178.

[B-G-T]   N. Burq, P. Gerard, N. Tzvetkov, *Strichartz' inequalities and the nonlinear Schrödinger equation on compact manifolds*, Amer. J. Math. 126 (2004), no. 3, 569–605.

[B-P]   E. Bombieri, J. Pila, *The number of integral points on arcs and ovals*, Duke Math. J. 59 (1989), no. 2, 337–357.

[Ja]   V. Jarnick, *Über die Gitterpunkte auf konvexen Curven*, Math. Z. 24 (1926), 500–518.

Chapter Two

Diffusion Bound for a Nonlinear Schrödinger Equation

J. Bourgain and W.-M. Wang

2.1 INTRODUCTION AND STATEMENT OF THE THEOREM

We study diffusion for a nonlinear lattice Schrödinger equation. This problem falls within the same general category of bounds on the higher Sobolev norms (H^1 and beyond) for the continuum nonlinear Hamiltonian PDE in a compact domain, e.g., a circle or a finite interval with Dirichlet boundary conditions; see, e.g., [B]. (Recall that typically L^2 norm is conserved. So H^1 is the first nontrivial norm to consider.)

As in previous papers [BW1,2], the nonlinear random Schrödinger equation is the medium where we make the construction, the random variables (potential) being the needed parameters. Here we work with a slightly tempered equation in $1 - d$:

$$i\dot{q}_j = v_j q_j + \epsilon(q_{j-1} + q_{j+1}) + \lambda_j q_j |q_j|^2 = 0, \quad j \in \mathbb{Z}, \tag{2.1.1}$$

where we take, for instance, $V = \{v_j\}$ to be independent, randomly chosen variables in $[0, 1]$ (uniform distribution). The multiplier $\{\lambda_j\}_{j \in \mathbb{Z}}$ satisfies the condition

$$|\lambda_j| < \epsilon(|j| + 1)^{-\tau}, \tag{2.1.2}$$

with $\tau > 0$ fixed and arbitrarily small. Note that $\tau = 0$, $\lambda_j = 1$ for all $j \in \mathbb{Z}$, is the standard lattice random Schrödinger equation.

As in the case $\lambda_j = 1$ for all $j \in \mathbb{Z}$, (2.1.1) is a Hamiltonian equation of motion, with the Hamiltonian

$$H(q, \bar{q}) = \frac{1}{2}\left(\sum v_j |q_j|^2 + \epsilon \sum (\bar{q}_j q_{j+1} + q_j \bar{q}_{j+1}) + \frac{1}{2} \sum \lambda_j |q_j|^4 \right) \tag{2.1.3}$$

(see (2.2.3, 2.2.4) for more details). We want to study its time evolution. Specifically, we want to bound

$$\sum_{j \in \mathbb{Z}} j^2 |q_j(t)|^2 \tag{2.1.4}$$

in terms of t as $t \to \infty$ for initial conditions satisfying

$$\sum_{j \in \mathbb{Z}} j^2 |q_j(0)|^2 < \infty. \tag{2.1.5}$$

Note that typically $q_j(0) \ll \lambda_j$ as $j \to \infty$.

The expression in (2.1.4) is sometimes called the "diffusion norm." The ℓ^2 norm $\sum_{j \in \mathbb{Z}} |q_j|^2$ is a conserved quantity for (2.1.1) (see (2.2.5)). The initial condition (2.1.5) shows at $t = 0$ the concentration on the lower modes q_j, with $|j|$ not too large. The diffusion norm (2.1.4) measures the propagation into higher ones, q_j, $|j| \gg 1$. If one interprets j as an index of Fourier series, then (2.1.4) is the equivalent of H^1 norm, for example for nonlinear Schrödinger (NLS) on a circle. So in fact one could also pursue higher moments:

$$\sum_{j \in \mathbb{Z}} j^{2s} |q_j(t)|^2$$

for $s > 1$, which correspond to H^s norms.

We have the following bound on the diffusion norm (2.1.4):

THEOREM. *Given* $\tau > 0$, $\kappa > 0$ *and taking* $0 < \epsilon < \epsilon(\tau, \kappa)$, *the following is true almost surely in* V. *If at* $t = 0$, *the initial datum* $\{q_j(0)\}_{j \in \mathbb{Z}}$ *satisfies*

$$\sum_{j \in \mathbb{Z}} j^2 |q_j(0)|^2 < \infty, \tag{2.1.6}$$

then

$$\sum_{j \in \mathbb{Z}} j^2 |q_j(t)|^2 < t^\kappa \quad as \; t \to \infty. \tag{2.1.7}$$

The bound in (2.1.7) shows that if there is propagation, it is very slow $\sim t^{\kappa/2}$. If the initial datum is in H^1, then the growth of H^1 norm in time cannot be faster than $t^{\kappa/2}$. (Recall also that $\sim t^{1/2}$ is diffusive, $\sim t$ is ballistic.) In [BW2], we constructed time quasi-periodic solutions to the standard random Schrödinger equation on $\mathbb{Z}^d$, i.e., (2.1.1), when $\lambda_j = 1$ for all j and in any dimension on a set of V of positive measure and for a corresponding appropriate set of small initial data with compact support. Clearly such initial data are a subset of q_j satisfying (2.1.6). The present theorem is an attempt to address the growth of Sobolev norm for more generic initial data.

The proof of the above theorem is, however, unlike [BW2]. We rely on the underlying Hamiltonian structure and make symplectic transforms to render (2.1.3) into a normal form amenable to the proof. The main feature of this normal form, to be spelled out completely in (2.2.10–2.2.13) is that it contains energy barriers centered at some $\pm j_0 \in \mathbb{Z}$, $j_0 > 1$ of width $\sim \log j_0$, where the terms responsible for mode propagation are small $\sim 1/j_0^C$ ($C \sim 1/\kappa$); see (2.2.19, 2.2.20). In (2.1.3), the mode propagation terms are $\epsilon(\bar{q}_j q_{j+1} + q_j \bar{q}_{j+1})$, where ϵ, even though small, *does not* decay with j.

We traverse the usual path in order to reach the desired normal form in (2.2.10–2.2.13). The symplectic transformations are generated by auxiliary polynomial Hamiltonians (cf. (2.3.13)) related to the H in (2.1.3). This is the content of sections 2.2 and 2.3.

Here we only want to point out that condition (2.1.2) plays an essential role in the construction. It enables us to work only with monomials of *bounded* degrees. The small divisors arising in the process (cf. (2.3.13)) are then controlled by shifting in V. We estimate the probability measures of the lower bounds on these small divisors in section 2.4, where we also probabilistically determine j_0. Here the $1 - d$ setting plays a crucial role.

The specific form of the nonlinearity, on the other hand, is inessential. It can be more general (see section 2.2). Due to the polynomial nature, these symplectic transforms preserve the ℓ^2 norm (cf. (2.5.7)). So it remains a conserved quantity.

In the subject of random Schrödinger equations, whether in $d = 1$ or $d > 1$, even in the linear case, where $\lambda_j = 0$, no argument for any diffusive behavior, i.e.,

$$\sum j^2 |q_j(t)|^2 \gtrsim t^\alpha, \quad (\alpha > 0), \tag{2.1.8}$$

seems to be known. What is better documented in the linear case, $\lambda_j = 0$ is the "complement" of (2.1.8), i.e., Anderson localization (A. L.) (see, e.g., [FS, vDK]),

$$\sum j^2 |q_j(t)|^2 < \infty \quad \text{as } t \to \infty \tag{2.1.9}$$

for the initial conditions satisfying (2.1.5), $\epsilon \ll 1$ and appropriate probability distribution for $V = \{v_j\}$. So it is tempting to ask whether, in fact, for (2.1.1) the bound (2.1.7) could be bettered to

$$\sup_t \sum j^2 |q_j(t)|^2 < \infty, \tag{2.1.10}$$

which would be a nonlinear version of "dynamical localization," well known to hold in the linear case (i.e., $\lambda_j = 0$); see, e.g., [DG].

The problem of retaining (2.1.7) when we only assume $|\lambda_j| < \delta$, which corresponds to the usual random Schrödinger equation, remains largely open. In [BW2], time quasi-periodic solutions were constructed for $\delta \ll 1$ and carefully chosen initial conditions, which are special cases of (2.1.6). For this special set of initial conditions, manifestly (2.1.10) holds. Does this tempt us to put forth (2.1.10) for initial conditions satisfying (2.1.6) as a conjecture?

2.2 STRUCTURE OF TRANSFORMED HAMILTONIANS

Recall from section 2.1 the tempered nonlinear random Schrödinger equation:

$$i\dot{q}_j = v_j q_j + \epsilon(q_{j-1} + q_{j+1}) + \lambda_j q_j |q_j|^2 = 0, \quad j \in \mathbb{Z}, \tag{2.2.1}$$

where $V = \{v_j\}$ is a family of i.i.d. random variables in $[0, 1]$ with uniform distribution, $0 < \epsilon \ll 1$. The multiplier λ_j satisfies the condition

$$|\lambda_j| < \epsilon(|j| + 1)^{-\tau}, \tag{2.2.2}$$

with $\tau > 0$ fixed and arbitrarily small. As mentioned in section 2.1, (2.2.2) (previously (2.1.2)) is crucial for the construction below.

We recast (2.2.1) as a Hamiltonian equation of motion, with the Hamiltonian

$$H(q, \bar{q}) = \frac{1}{2}\left(\sum v_j |q_j|^2 + \epsilon \sum (\bar{q}_j q_{j+1} + q_j \bar{q}_{j+1}) + \frac{1}{2} \sum \lambda_j |q_j|^4 \right). \tag{2.2.3}$$

(2.2.1) can be rewritten as

$$i\dot{q}_j = 2\frac{\partial H}{\partial \bar{q}_j}.$$ (2.2.4)

Remark. The connection with the usual canonical variables (ξ, x) is $q = \xi + ix$, $\bar{q} = \xi - ix$. The equation of motion in the (ξ, x) coordinates is

$$\dot{\xi} = \frac{\partial H}{\partial x}, \quad \dot{x} = -\frac{\partial H}{\partial \xi},$$

which can be rewritten as a single equation, namely (2.2.4).

As in the linear case $(\lambda_j = 0)$, the ℓ^2 norm is conserved, since

$$
\begin{aligned}
\frac{\partial}{\partial t}\sum_{j\in\mathbb{Z}}|q_j|^2 &= \sum q_j\dot{\bar{q}}_j + \dot{q}_j\bar{q}_j \\
&= 2i\left(\sum_{j\in\mathbb{Z}}q_j\frac{\partial H}{\partial q_j} - \sum_{j\in\mathbb{Z}}\frac{\partial H}{\partial \bar{q}_j}\bar{q}_j\right) \\
&= 2i\left[\sum_{j\in\mathbb{Z}}(q_j\bar{q}_{j-1} + q_j\bar{q}_{j+1}) - \sum_{j\in\mathbb{Z}}(q_{j+1}\bar{q}_j + q_{j-1}\bar{q}_j)\right] \\
&= 0.
\end{aligned}
$$ (2.2.5)

It is worth noting that the cancellation of the two sums is due to the fact that the sum is over $\mathbb{Z}$.

The nonlinear term $\sum \lambda_j|q_j|^4$ by itself conserves the individual action variables $|q_j|^2$. It is the combination with the angular variable ϵ term in (2.2.3), which is the culprit for diffusion into higher modes q_j, $j \gg 1$.

In the linear case, diffusion is obstructed by the random potential $V = \{v_j\}$ by proving A. L., i.e., the existence of a complete set of ℓ^2 eigenfunctions which are well localized with respect to the canonical basis of $\mathbb{Z}$ or more generally $\mathbb{Z}^d$. In the nonlinear case, we again use the random potential $V = \{v_j\}$ to obstruct energy transfer from low to high modes by creating "zones" in $\mathbb{Z}$, where the only mode coupling term is of order $\mathcal{O}(\sum |j|^{-C}|q_j|^2) \ll \epsilon$. This obstruction is achieved by invoking the usual process of symplectic transformations, to be described shortly.

Here it suffices to remark that due to the finite range (or more generally sufficiently short range) nature of the Hamiltonian in (2.2.3), we only need one such zone, say centered at $\pm j_0$ (depending on t and determined probabilistically) of width $W(j_0) \ll j_0$, in order to control the time derivative of the truncated sum of higher modes:

$$\frac{\partial}{\partial t}\sum_{|j|>j_0}|q_j(t)|^2$$ (2.2.6)

(cf. (2.2.5) and the comment afterwards), which in turn enables us to control the sum

$$\sum_{j\in\mathbb{Z}}j^2|q_j(t)|^2.$$ (2.2.7)

As mentioned earlier, the decay assumption on the nonlinearity in (2.2.2) is crucial. It permits us to deal with monomials of bounded degrees. This is, in fact, what prevents us from being able to make a statement for $\lambda_j = 1$, which corresponds to the standard nonlinear Schrödinger equation. On the other hand, the specific form of the nonlinearity in (2.2.1) is inessential, as long as it is of finite range or sufficiently short range and of bounded degree, for example, $|q_j|^4$ may be replaced by $\Re|q_j|^2 \bar{q}_j |q_{j+1}|^2 q_{j+1}$ in the finite-range case and

$$|q_j|^2 \sum_k e^{-\frac{C}{\epsilon}|j-k|} |q_k|^2$$

in the short-range case. Finally, in the measure estimates, we make crucial use of the $1-d$ setting, i.e., the barrier zones are of width and hence *volume* $W(j_0) \ll j_0$.

In what follows, we will deal extensively with monomials (polynomials) in q_j. So we first introduce some notions. Rewrite any monomial in the form

$$\prod_{j\in\mathbb{Z}} q_j^{n_j} \bar{q}_j^{n'_j}. \tag{2.2.8}$$

Let $n = \{n_j, n'_j\}_{j\in\mathbb{Z}} \in \mathbb{N}^{\mathbb{Z}} \times \mathbb{N}^{\mathbb{Z}}$. We use three notions: support, diameter, and degree:

$$\begin{aligned} \operatorname{supp} n &= \{j \,|\, n_j \neq 0 \text{ or } n'_j \neq 0\} \\ \Delta(n) &= \operatorname{diam}\{\operatorname{supp} n\} \\ |n| &= \sum_{j\in\operatorname{supp} n} n_j + n'_j. \end{aligned} \tag{2.2.9}$$

For example, the monomial $q_{k_0}\bar{q}_{k_0+1}$ in H in (2.2.3) has $\operatorname{supp} n = \{k_0, k_0 + 1\}$, $\Delta(n) = 1$, $|n| = 2$; while the monomial $|q_{k_0}|^4$ has $\operatorname{supp} n = \{k_0\}$, $\Delta(n) = 0$, $|n| = 4$. Note that in both cases: $\sum_{\operatorname{supp} n} n_j = \sum_{\operatorname{supp} n} n'_j$, which is a general feature of polynomial Hamiltonians.

If, furthermore, $n_j = n'_j$ for all $j \in \operatorname{supp} n$, then the monomial is called *resonant*. Otherwise it is *nonresonant*. $|q_{k_0}|^4$ is resonant, $q_{k_0}\bar{q}_{k_0+1}$ is nonresonant. It is important to observe at this point that nonresonant monomials contribute to the truncated sum in (2.2.6), while resonant ones do not, in view of (2.2.5).

2.2.1 Structure of Transformed Hamiltonians

To control the sum in (2.2.6), (which leads to control of the sum in (2.2.7)), we transform H in (2.2.1) to H' of the form

$$H' = \frac{1}{2} \sum_{j\in\mathbb{Z}} (v_j + w_j)|q_j|^2 \tag{2.2.10}$$

$$+ \sum_{n\in\mathbb{N}^{\mathbb{Z}}\times\mathbb{N}^{\mathbb{Z}}} c(n) \prod_{\operatorname{supp} n} q_j^{n_j} \bar{q}_j^{n'_j} \tag{2.2.11}$$

$$+ \sum_{n\in\mathbb{N}^{\mathbb{Z}}\times\mathbb{N}^{\mathbb{Z}}} d(n) \prod_{\operatorname{supp} n} |q_j|^{2n_j} \tag{2.2.12}$$

$$+ \mathcal{O}\left(\sum_{j\in\mathbb{Z}} |j|^{-C} |q_j|^2 \right), \tag{2.2.13}$$

where (2.2.11) consists of nonresonant monomials, $\sum n_j = \sum n'_j$ ($n_j \neq n'_j$ for some j), (2.2.12) consists of resonant monomials of degree at least 4. In (2.2.13) (and in the sequel), C stands for a fixed large constant, in fact $C \sim 1/\kappa$, κ as in the theorem. The coefficients $c(n)$, $d(n)$ satisfy the bounds

$$|c(n)| + |d(n)| < \exp(-\rho\{\Delta(n) + 1)\log\frac{1}{\epsilon} + \tau(|n| - 2)\log(|j| + 1)\}), \quad (2.2.14)$$

where $j = \min(|\min\{\operatorname{supp} n\}|, |\max\{\operatorname{supp} n\}|)$ and $\rho > 1/10$.

The transformation from H to H' is symplectic, $H' = H \circ \Gamma$, with Γ symplectic. It is a finite step iterative process, (2.2.13) is the remainder. Let H_s, Γ_s be the Hamiltonian and the transformation at step s, $H_{s+1} = H_s \circ \Gamma_s$. At each step s, Γ_s is the symplectic transformation generated by an appropriate polynomial Hamiltonian F. H_{s+1} is the time-1 map, computed by using a convergent Taylor series of successive Poisson brackets of H_s and F (see section 2.3).

At the initial step, $H_1 \overset{\text{def}}{=} H$ given by (2.2.1) satisfies

$$(2.2.10): \quad w_j = 0; \quad (2.2.15)$$
$$(2.2.11): \quad |n| = 2, \ \Delta(n) = 1, \ |c(n)| \leq \epsilon; \quad (2.2.16)$$
$$(2.2.12): \quad |n| = 4, \ \Delta(n) = 0, \ |d(n)| < \epsilon(|j| + 1)^{-\tau}; \quad (2.2.17)$$
$$(2.2.13) = 0.$$

So that (2.2.14) holds with $\rho = 1/2$. Along the iteration process $\rho = \rho_s$ will slightly vary, but (2.2.14) will be shown to hold for $c = c_s$, $d = d_s$ with $\rho_s > 1/10$. Note also that (2.2.14) implies in particular that

$$|d(n)| < \epsilon^{1/10}(|j| + 1)^{-\tau/5} \quad (2.2.18)$$

(compare with (2.2.2, 2.2.17)).

The purpose of H' given by (2.2.10–2.2.13) is to manifest an "energy barrier." Assume we have fixed a large $j_0 \in \mathbb{Z}^+$. We require further that

$$|c(n)| < \delta \quad \text{if } \operatorname{supp} n \cap \{[-b, -a] \cup [a, b]\} \neq \emptyset, \quad (2.2.19)$$

where $\delta < j_0^{-C}$, C as in (2.2.13) and

$$\left[j_0 - \frac{1}{2}\log j_0, \ j_0 + \frac{1}{2}\log j_0\right] \subset [a, b] \subset [j_0 - \log j_0, \ j_0 + \log j_0]. \quad (2.2.20)$$

Remark. The construction of H' depends on j_0, which is determined probabilistically and dependent on t: $j_0 = j_0(t)$, $j_0 \to \infty$ as $t \to \infty$.

At each step s, in (2.2.19, 2.2.20) we make the replacement: $c = c_s$, $\delta = \delta_s$, $a = a_s$, $b = b_s$. δ_s will be shown in section 2.3 to satisfy

$$\delta_{s+1} = \delta_s^{19/10} + j_0^{-\tau/20}\delta_s. \quad (2.2.21)$$

From (2.2.16), $\delta_1 = \epsilon$. We terminate the construction at step $s_* \sim \log\log j_0$ such that

$$\delta_{s_*} < j_0^{-C}, \tag{2.2.22}$$

same C as in (2.2.13).

Remark. Looking ahead, we comment here that the first term in (2.2.21) comes from Poisson brackets of polynomials with coefficients c, while the second one comes from polynomials with coefficients c and d (cf. (2.2.18, 2.2.19)).

From (2.2.14), we may restrict ourselves to monomials $\prod q_j^{n_j} \bar{q}_j^{n_j'}$ satisfying

$$\Delta(n) \lesssim C \frac{\log(|j|+1)}{\log \frac{1}{\epsilon}}, \tag{2.2.23}$$

$$|n| \lesssim \frac{C}{\tau} \quad \text{(hence bounded degree).} \tag{2.2.24}$$

The others are captured by (2.2.13) in H'. (2.2.19, 2.2.20, 2.2.23) then show that the construction of H' from H only involves modes $j \in \mathbb{Z}$ for which $\big||j| - j_0\big| \lesssim \log j_0$. So in particular $w_j = 0$ unless $\big||j| - j_0\big| \lesssim \log j_0$. This point will be important in the measure estimates.

H' can now be used to control the sum in (2.2.6) as follows. We check in section 2.5 that the symplectic transform Γ preserves $\sum j^2 |q_j|^2$ up to a factor of 2. So q_j in (2.2.6) could be taken as the new q_j. (2.2.23) and (2.2.5) then give that

$$\frac{\partial}{\partial t} \sum_{|j| \geq j_0} |q_j(t)|^2 = 2i \sum_{\big||j|-j_0\big| \lesssim C \frac{\log |j_0|}{\log \frac{1}{\epsilon}}} \sum_{\substack{n \\ \text{supp } n \cap j \neq \emptyset}}$$

$$\times (n_j - n_j') c(n) \prod_{k \in \text{supp } n} q_k^{n_k} \bar{q}_k^{n_k'} + \mathcal{O}(j_0^{-C}) \sim j_0^{-C} \tag{2.2.25}$$

by using (2.2.19). This is because the sum

$$\sum_{\big||j|-j_0\big| > C \frac{\log |j_0|}{\log \frac{1}{\epsilon}}} \sum_{\substack{n \\ \text{supp } n \cap j \neq \emptyset}} (n_j - n_j') c(n) \prod_{k \in \text{supp } n} q_k^{n_k} \bar{q}_k^{n_k'} = 0 \tag{2.2.26}$$

by shifting the index j and using (2.2.23). Note that (2.2.10, 2.2.12) do not contribute.

2.2.2 Some Preliminary Comments on the Measure Estimates

In (2.2.10), $w_j = w_j(V)$ and all coefficients, in particular $c(n)$, $d(n)$ depend on V. Let $W = \{w_j\}_{j \in \mathbb{Z}^d}$. We assume

$$|\nabla_V c(n)| + |\nabla_V d(n)| \lesssim 1, \tag{2.2.27}$$

and

$$\left\| \frac{\partial W}{\partial V} \right\|_{\ell^2 \to \ell^2} < \sqrt{\epsilon}. \tag{2.2.28}$$

Note that the initial Hamiltonian H in (2.2.1) trivially satisfies (2.2.27, 2.2.28).

As mentioned previously, $w_j = 0$ unless $\big||j| - j_0\big| \lesssim \log j_0$, (2.2.28) implies that the frequency modulation map $V \to \tilde{V} = V + W$ satisfies

$$j_0^{-\sqrt{\epsilon}} < (1 - \sqrt{\epsilon})^{\log j_0} < \left| \det \frac{\partial \tilde{V}}{\partial V} \right| < (1 + \sqrt{\epsilon})^{\log j_0} < j_0^{\sqrt{\epsilon}}. \tag{2.2.29}$$

The nonresonance estimates in section 2.3 on symplectic transforms are expressed in terms of $\tilde{V}$. So (2.2.29) will be important in section 2.4, where these nonresonance conditions are translated into probabilistic estimates in V.

It is essential at this point to point out that, if j_0 is fixed, the desired normal form H' as in (2.2.10–2.2.13), satisfying (2.2.14, 2.2.19, 2.2.20) can only be achieved with small probability in V. However, by varying j_0 in a dyadic interval, H' may be achieved with large probability.

2.3 ANALYSIS AND ESTIMATES OF THE SYMPLECTIC TRANSFORMATIONS

We now make explicit the symplectic transformations that were often hinted at in the previous sections. The analysis is straightforward. It is a finite-step induction. The main objective is to check that $c(n)$, $d(n)$ satisfy (2.2.14, 2.2.19, 2.2.27) and W satisfy (2.2.28) at each step s.

At the first step: $s = 1$,

$$H_1 = H = \frac{1}{2}\left(\sum v_j |q_j|^2 + \epsilon \sum (\bar{q}_j q_{j+1} + q_j \bar{q}_{j+1}) + \frac{1}{2} \sum \lambda_j |q_j|^4 \right) \tag{2.3.1}$$

from (2.2.3). Let η_j denote the canonical basis of $\mathbb{Z}$, (2.2.14, 2.2.19) are satisfied with

$$c(n) = c(\eta_j \times \eta_{j+1}) = \frac{\epsilon}{2}, \quad |n| = 2, \ \Delta(n) = 1, \ \mathrm{supp}\, n = \{j, \ j+1, \ j \in \mathbb{Z}\},$$
$$= 0 \quad \text{otherwise}. \tag{2.3.2}$$

$$d(n) = d(\eta_j \times \eta_j) = \frac{\lambda_j}{4}, \ |\lambda_j| < \epsilon(|j| + 1)^{-\tau}, \quad |n| = 4, \ \Delta(n) = 0,$$
$$\mathrm{supp}\, n = j, \ j \in \mathbb{Z},$$
$$= 0 \quad \text{otherwise}. \tag{2.3.3}$$

(2.2.27) is trivially satisfied with

$$|\nabla_V(c)| = |\nabla_V(d)| = 0; \tag{2.3.4}$$

and so is (2.2.28):

$$W = 0; \quad \frac{\partial W}{\partial V} = 0. \tag{2.3.5}$$

Assume that we have obtained at step s, the Hamiltonian H_s in the form (2.2.10–2.2.13), satisfying (2.2.14, 2.2.19, 2.2.20, 2.2.27, 2.2.28)$_s$, i.e., (2.2.14, 2.2.19, 2.2.20, 2.2.27, 2.2.28) at step s. Our aim is to produce H_{s+1} possessing the corresponding properties at step $s + 1$. We first take care of the propagation of the first three properties.

(2.2.14, 2.2.19, 2.2.20)$_s$ state that

$$|c(n)| + |d(n)| < \exp\left(-\rho_s\left\{(\Delta(n)+1)\log\frac{1}{\epsilon} + \tau(|n|-2)\log(|j|+1)\right\}\right), \tag{2.3.6}$$

with $\rho_s > 1/10$, moreover,

$$|c(n)| < \delta_s \quad \text{if supp } n \cap \{[-b_s, -a_s] \cup [a_s, b_s]\} \neq \emptyset, \tag{2.3.7}$$

with

$$\left[j_0 - \frac{1}{2}\log j_0, j_0 + \frac{1}{2}\log j_0\right] \subset [a_s, b_s] \subset [j_0 - \log j_0, j_0 + \log j_0], \tag{2.3.8}$$

δ_s is defined inductively as in (2.2.21):

$$\delta_s = \delta_{s-1}^{19/10} + j_0^{-\tau/20}\delta_{s-1} \quad (s \geq 2), \tag{2.3.9}$$

$\delta_1 = \epsilon/2$.

We satisfy (2.3.7) at step $s+1$ constructively by removing those $c(n)$ with $\delta_{s+1} < |c(n)| < \delta_s$, supp $n \cap \{[-b_s, -a_s] \cup [a_s, b_s]\} \neq \emptyset$ and a corresponding reduction of $[-b_s, -a_s] \cup [a_s, b_s]$ to $[-b_{s+1}, -a_{s+1}] \cup [a_{s+1}, b_{s+1}]$ with $a_{s+1} > a_s, b_{s+1} < b_s$, so that

$$|c(n)| < \delta_{s+1} \text{ if supp } n \cap \{[-b_{s+1}, -a_{s+1}] \cup [a_{s+1}, b_{s+1}]\} \neq \emptyset. \tag{2.3.10}$$

We proceed as follows. Denoting in H_s (2.2.10–2.2.13),

$$\tilde{v}_j = v_j + w_j^{(s)} \text{ and } H_0 = \sum_{j\in\mathbb{Z}} \tilde{v}_j|q_j|^2, \tag{2.3.11}$$

we define, following the standard approach,

$$H_{s+1} = H_s \circ \Gamma_F, \tag{2.3.12}$$

where Γ_F is the symplectic transformation obtained from the Hamiltonian function

$$F \sim \sum_{\substack{\text{supp } n \subset [a_s,b_s] \cup [-b_s,-a_s] \\ |c(n)| > \delta_{s+1}}} \frac{c(n)}{\sum(n_j - n'_j)\tilde{v}_j} \prod q_j^{n_j} \bar{q}_j^{n'_j}, \tag{2.3.13}$$

$(\sum(n_j - n'_j)\tilde{v}_j \neq 0.)$

Recall that H_{s+1} is the time-1 map, and by Taylor series

$$H_{s+1} = H_s \circ \Gamma_F = H_0$$

$$+ (2.2.11) + \{H_0, F\} \qquad\qquad (2.2.11')$$

$$+ (2.2.12)$$

$$+ \{(2.2.11), F\} + \frac{1}{2!}\{\{(2.2.11), F\}, F\} + \cdots \qquad (2.3.14)$$

$$+ \{(2.2.12), F\} + \frac{1}{2!}\{\{(2.2.12), F\}, F\} + \cdots \qquad (2.3.15)$$

$$+ (2.2.13) \circ \Gamma_F. \qquad\qquad (2.2.13')$$

Note that $\{H_0, F\} \sim F$, terminating the series corresponding to H_0. Using (2.3.13), (2.2.13') has the same property as (2.2.13). We now define a_{s+1}, b_{s+1}, in order that (2.2.11') satisfy (2.3.10). Observe that if $|c(n)| > \delta_{s+1}$ as in (2.3.13), then by (2.2.14)

$$\Delta(n) < 10\frac{\log \frac{1}{\delta_{s+1}}}{\log \frac{1}{\epsilon}}. \qquad\qquad (2.3.16)$$

Thus defining

$$a_{s+1} = a_s + 20\frac{\log \frac{1}{\delta_{s+1}}}{\log \frac{1}{\epsilon}}, \quad b_{s+1} = b_s - 20\frac{\log \frac{1}{\delta_{s+1}}}{\log \frac{1}{\epsilon}}, \qquad (2.3.17)$$

$\{H_0, F\}$ removes in (2.2.11) all monomials for which $|c(n)| > \delta_{s+1}$ and supp $n \cap \{[-b_{s+1}, -a_{s+1}] \cup [a_{s+1}, b_{s+1}]\} \neq \emptyset$. So (2.2.11') satisfies (2.3.10) by construction.

Returning to (2.3.13), we impose the small divisor condition

$$\left| \sum (n_j - n'_j)\tilde{v}_j \right| > \delta_s^{\frac{1}{100s^2}}, \qquad\qquad (2.3.18)$$

which will lead to measure estimates of this construction in section 2.4. Using (2.3.18) in (2.3.13), the main task of the rest of the section is to estimate (2.3.14, 2.3.15) and show that they satisfy $(2.2.14)_{s+1}$ with $\rho_{s+1} > 1/10$. These estimates also determine δ_{s+1}.

2.3.1 Monomials in (2.3.14)

We start with the lowest-order Poisson bracket $\{(2.2.11), F\}$, which produces monomials of the form

$$\left\{ \prod q_j^{m_j} \bar{q}_j^{m'_j}, \prod q_j^{n_j} \bar{q}_j^{n'_j} \right\}$$

$$\sim \sum_k (m_k n'_k - m'_k n_k) q_k^{m_k + n_k - 1} \bar{q}_k^{m'_k + n'_k - 1} \prod_{j \neq k} q_j^{m_j + n_j} \bar{q}_j^{m'_j + n'_j}, \qquad (2.3.19)$$

with coefficient

$$\frac{c(m)c(n)}{\sum(n_j - n'_j)\tilde{v}_j},$$

(2.3.20)

where

$$\text{supp } n \subset [a_s, b_s] \cup [-b_s, -a_s], \quad \text{supp } m \cap \text{supp } n \neq \emptyset.$$

(2.3.21)

Hence supp $m \cap ([a_s, b_s] \cup [-b_s, -a_s]) \neq \emptyset$ and from (2.3.7), both

$$|c(m)|, \ |c(n)| < \delta_s.$$

(2.3.22)

Remark. This is, in fact, the other reason to restrict n to have the support in (2.3.13).

The monomials in (2.3.19) correspond to multi-indices μ, where

$$\Delta(\mu) \leq \Delta(m) + \Delta(n),$$

(2.3.23)

$$|\mu| = |m| + |n| - 2.$$

(2.3.24)

Recalling (2.2.24), the prefactors in (2.3.19) may be bounded as

$$|m_k n'_k - m'_k n_k| \leq (|m| + |n|)^2 \leq (|\mu| + 2)^2 \leq 2\left(\frac{C}{\tau}\right)^2.$$

(2.3.25)

The number of realizations of a *fixed* monomial $\prod q_j^{\mu_j} \bar{q}_j^{\mu'_j}$ in $\{(2.2.11), F\}$ is easily seen to be bounded by

$$2^{|\mu|}(\Delta(m) \wedge \Delta(n)) < \exp\left(\frac{\mathcal{O}(1)C}{\tau}\right)(\Delta(m) + \Delta(n)).$$

(2.3.26)

To bound the coefficient in (2.3.20), we define

$$\rho_{s+1} = \rho_s\left(1 - \frac{1}{10s^2}\right).$$

(2.3.27)

Using the small divisor bound (2.3.18) and (2.2.14)$_s$, we have

$$(3.20) \leq \delta_s^{-\frac{1}{100s^2}}(|c(m)||c(n)|)^{\frac{1}{10s^2}}(|c(m)||c(n)|)^{(1-\frac{1}{10s^2})}$$

$$\leq \delta_s^{-\frac{1}{100s^2}}(|c(m)||c(n)|)^{\frac{1}{10s^2}}$$

(2.3.28)

$$\exp\left(-\rho_{s+1}\left\{(\Delta(\mu) + 2)\log\frac{1}{\epsilon} + \tau(|\mu| - 2)\log j_0\right\}\right),$$

which is starting to have the flavor of (2.2.14)$_{s+1}$.

Taking into account the bound in (2.3.25) on the prefactor, the bound in (2.3.26) on entropy and using (2.3.28), (2.2.14)$_s$ to bound $\Delta(m)$, $\Delta(n)$ in terms of $c(m)$,

$c(n)$, we obtain the following bound for $g_1(\mu)$, the coefficient of the $\prod q_j^{\mu_j} \bar{q}_j^{\mu'_j}$ factor in $\{(2.2.11), F\}$:

$$g_1(\mu) \le \delta_s^{-\frac{1}{100s^2}} 2^{|\mu|} \left(\log \frac{1}{|c(m)||c(n)|} \right) (|c(m)||c(n)|)^{\frac{1}{10s^2}}$$

$$\exp\left(-\rho_{s+1} \left\{ (\Delta(\mu) + 1) \log \frac{1}{\epsilon} + \tau(|\mu| - 2) \log j_0 \right\} \right)$$

$$\lesssim \delta_s^{-\frac{1}{100s^2}} 2^{|\mu|} (|\log \delta_s^2|) \delta_s^{\frac{2}{10s^2}} \exp\left(-\rho_{s+1} \left\{ (\Delta(\mu) + 1) \log \frac{1}{\epsilon} \right. \right.$$

$$\left. \left. + \tau(|\mu| - 2) \log j_0 \right\} \right). \tag{2.3.29}$$

From (2.2.21), $\log \log \delta_s^{-1} \sim s$, (2.3.22) then permits us to estimate

$$g_1(\mu) < (2.3.29) < \delta_s^{\frac{1}{20s^2}} \exp\left(-\rho_{s+1} \left\{ (\Delta(\mu) + 1) \log \frac{1}{\epsilon} + \tau(|\mu| - 2) \log j_0 \right\} \right). \tag{2.3.30}$$

To bound $g(\mu)$, the coefficient in front of $\prod q_j^{\mu_j} \bar{q}_j^{\mu'_j}$ factor in (2.3.14), we need to take into account the higher-order poisson brackets in the Taylor series. We illustrate it on

$$\frac{1}{2!} \{\{(2.11, F\}, F\},$$

through which the general structure of the estimates will hopefully become clear.

A fixed monomial $\prod_{\text{supp}\mu} q_j^{\mu_j} \bar{q}_j^{\mu'_j}$ in $\{\{(2.11, F\}, F\}$ is now the confluence of three sources, denoted by m, n, p with

$$|\mu| = |m| + |n| + |p| - 4. \tag{2.3.31}$$

Let

$$|w| = |m| + |n| - 2. \tag{2.3.32}$$

Then

$$|\mu| = |w| + |p| - 2, \tag{2.3.33}$$

$$\Delta(w) \le \Delta(m) + \Delta(n), \tag{2.3.34}$$

$$\Delta(\mu) \le \Delta(w) + \Delta(p). \tag{2.3.35}$$

Continuing the previous terminology, the coefficient is

$$c(m)c(n)c(p), \quad |c(m)|, \ |c(n)|, \ |c(p)| < \delta_s; \tag{2.3.36}$$

the prefactor is a sum of terms of the form

$$(m_k n'_k - m'_k n_k)[(m_j + n_j)p'_\ell - (m'_j + n'_j)p_\ell] \quad \text{if } j \ne k$$

$$\text{or } (m_k n'_k - m'_k n_k)[(m_k + n_k - 1)p'_\ell - (m'_k + n'_k - 1)p_\ell], \tag{2.3.37}$$

(cf. (2.3.19)). Using (2.3.31), the prefactor can be bounded by

$$C_{|\mu|}^3 |\mu|^3 \lesssim |\mu|^6; \tag{2.3.38}$$

the entropy can be bounded by

$$\sum_{|w|=2}^{|\mu|} 2^{|w|}(\Delta(m) \wedge \Delta(n)) \, 2^{|\mu|}(\Delta(w) \wedge \Delta(p)) \tag{2.3.39}$$
$$\lesssim [2^{|\mu|}(\Delta(m) + \Delta(n) + \Delta(p))]^2.$$

(2.3.36, 2.3.38, 2.3.39) then give that $g_2(\mu)$, the $\prod q_j^{\mu_j} \bar{q}_j^{\mu'_j}$ factor from

$$\frac{1}{2!}\{\{(2.11, F\}, F\}$$

can be bounded by

$$g_2(\mu) \le \frac{1}{2!}\left(\delta_s^{-\frac{1}{100s^2}}\right)^2 [2^{|\mu|}(\Delta(m) + \Delta(n) + \Delta(p))]^2 |\mu|^6$$
$$[c(m)c(n)c(p)]^{\frac{1}{10s^2}} \exp\left(-\rho_{s+1}\left\{(\Delta(\mu)+1)\log\frac{1}{\epsilon} + \tau(|\mu|-2)\log j_0\right\}\right)$$
$$\lesssim \left(\delta_s^{-\frac{1}{100s^2}}\right)^2 \left(\frac{2^{|\mu|}\log \delta_s^3}{2!}\right)^2 \delta_s^{\frac{3}{10s^2}}$$
$$\times \exp\left(-\rho_{s+1}\left\{(\Delta(\mu)+1)\log\frac{1}{\epsilon} + \tau(|\mu|-2)\log j_0\right\}\right)$$
$$< \left(\delta_s^{\frac{1}{20s^2}}\right)^2 \exp\left(-\rho_{s+1}\left\{(\Delta(\mu)+1)\log\frac{1}{\epsilon} + \tau(|\mu|-2)\log j_0\right\}\right). \tag{2.3.40}$$

From (2.3.30, 2.3.40), the structure of the estimates on the Poisson brackets in (2.3.14) is clear and we obtain that the $\prod q_j^{\mu_j} \bar{q}_j^{\mu'_j}$ factor in (2.3.14) is bounded by

$$g(\mu) \le \delta_s^{\frac{1}{20s^2}} \exp\left(-\rho_{s+1}\left\{(\Delta(\mu)+1)\log\frac{1}{\epsilon} + \tau(|\mu|-2)\log j_0\right\}\right). \tag{2.3.41}$$

In particular, (2.2.14) remains valid with ρ_{s+1} replacing ρ_s. From (2.3.20, 2.3.22, 2.3.25, 2.3.26, 2.3.36, 2.3.37, 2.3.38, 2.3.39), there is also the bound

$$g(\mu) < \delta_s^{2-\frac{1}{50s^2}} < \delta_s^{\frac{19}{10}} \tag{2.3.42}$$

for the $\prod q_j^{\mu_j} \bar{q}_j^{\mu'_j}$ factor in (2.3.14), which will be part of δ_{s+1} (cf. (2.3.9)).

2.3.2 Monomials in (2.3.15)

We proceed as in the estimate of (2.3.14). For the lowest-order bracket $\{(2.2.12), F\}$, we get the coefficient

$$\frac{d(m)c(n)}{\sum(n_j - n'_j)\tilde{v}_j} \tag{2.3.43}$$

in lieu of (2.3.20), (2.3.21) remains satisfied, and hence, from (2.3.6, 2.3.7),

$$|c(n)| < \delta_s$$
$$|d(m)| \leq j_0^{-\tau/5}$$

(2.3.44)

(2.3.23, 2.3.24) remain valid, the prefactors are bounded as

$$|m_k(n_k' - n_k)| \leq (|\mu| + 2)^2 \leq 2\left(\frac{C}{\tau}\right)^2$$

(2.3.45)

as in (2.3.25). The entropy factor in (2.3.26) remains valid. So $\gamma_1(\mu)$, the coefficient of the $\prod q_j^{\mu_j} \bar{q}_j^{\mu_j'}$ monomial stemming from $\{(2.2.12), F\}$ is again bounded by the first line in (2.3.29) with $d(m)$ replacing $c(m)$. Since $s < s_\star \sim \log\log j_0$ from (2.2.22), the bound on the last line of (2.3.29) remains valid for $\gamma_1(\mu)$. Similarly, we prove that the bound in (2.3.41) carries through for $\gamma(\mu)$, the coefficient of the $\prod q_j^{\mu_j} \bar{q}_j^{\mu_j'}$ monomial from (2.3.15). Moreover, as in (2.3.14), $\gamma(\mu)$ is also bounded by

$$\gamma(\mu) < |d(m)| \left(\log\frac{1}{d(m)}\right) \delta_s^{1-\frac{1}{50s^2}} < j_0^{-\tau/6} \delta_s^{1-\frac{1}{50s^2}} < j_0^{-\tau/10} \delta_s.$$

(2.3.46)

Thus, in conclusion, (2.3.14, 2.3.15) satisfy $(2.2.14)_{s+1}$. Define

$$\delta_{s+1} = \delta_s^{19/10} + j_0^{-\tau/20} \delta_s$$

(2.3.47)

as in (2.2.21) and subdivide the (2.3.14, 2.3.15) terms according to (2.2.10–2.2.13). Since (2.2.11′) satisfies (2.3.10) by construction,

$$|\gamma(\mu)|, |g(\mu)| < \delta_{s+1},$$

(2.3.48)

(2.2.12) satisfies $(2.2.14)_s$, hence $(2.2.14)_{s+1}$, H_{s+1} satisfies (2.2.14, 2.2.19, $2.2.20)_{s+1}$.

We are now left to check the validity of properties $(2.2.27, 2.2.28)_{s+1}$ for H_{s+1}. In H_0 (2.3.11), we need to add resonant quadratic terms produced in (2.3.14, 2.3.15). Denoting these terms by w_j, $\tilde{v}_j$ is then perturbed to

$$\tilde{\tilde{v}}_j = \tilde{v}_j + w_j^{(s)},$$

(2.3.49)

where w_j satisfies, in particular,

$$|w_j^{(s)}| < \delta_{s+1}.$$

(2.3.50)

An important point is that therefore all nonresonance conditions imposed so far, and in particular (2.3.18), can be replaced by

$$\left|\sum (n_j - n_j') \tilde{\tilde{v}}_j\right| > \delta_t^{\frac{1}{100\tau^2}}$$

(2.3.51)

for all $t \le s$ and n satisfying

$$\text{supp } n \subset [a_t, \ b_t] \cup [-b_t \ - a_t],$$

$$\Delta(n) < 10 \frac{\log \frac{1}{\delta_{t+1}}}{\log \frac{1}{\epsilon}}.$$

We now check the V dependence for $g(\mu)$, $\gamma(\mu)$ in (2.3.14, 2.3.15). We illustrate the computation on first-order Poisson brackets. We have

$$|\nabla_V (2.3.20)| + |\nabla_V (2.3.43)|$$
$$\le [(|\nabla_V c(m)| + |\nabla_V d(m)|)|c(n)| + (|c(m)|$$
$$+ |d(m)|)|\nabla_V c(n)|] \sum (n_j - n'_j)\tilde{v}_j|^{-1}$$
$$+ \frac{c}{\tau}(|c(m)| + |d(m)|)|c(n)|$$
$$\times \left(\left| \sum (n_j - n'_j)\tilde{v}_j \right|^{-2} \|D\tilde{V}\|_{\ell^2 \to \ell^2} \right) < \delta_s^{-\frac{1}{100 s^2}} (\delta_s + j_0^{-\tau/5})$$
$$< \delta_s^{1/2} + j_0^{-\tau/10}, \qquad\qquad (2.3.52)$$

where we used (2.2.24, 2.2.27, 2.2.28, 2.3.22, 2.3.44).

Taking into account the factors (2.3.25, 2.3.26), (2.3.52) still remains valid for the first-order Poisson brackets. The higher-order brackets can be treated similarly and (2.3.52) remains essentially unchanged as the bound for $|\nabla_V g(\mu)| + |\nabla_V \gamma(\mu)|$. In particular, (2.2.27) remains valid and, moreover, from Schols' lemma,

$$\|DW^{(s)}\| \lesssim (\delta_s^{1/2} + j_0^{-\tau/10}) \frac{\log \frac{1}{\delta_{s+1}}}{\log \frac{1}{\epsilon}} \qquad\qquad (2.3.53)$$
$$< \delta_s^{1/3} + j_0^{-\tau/20}$$

as

$$\text{supp } n \simeq \frac{\log \frac{1}{\delta_{s+1}}}{\log \frac{1}{\epsilon}},$$

where δ_s satisfies (2.2.21) and $s \le s_\star \sim \log \log j_0$. Since $W = \sum_{s=1}^{s_\star} W^{(s)}$, (2.2.28) remains valid along the process.

Finally, we check (2.2.20) for intervals $[a_s, \ b_s]$. From (2.3.17, 2.2.21, 2.2.22)

$$|a_s - a_{s+1}| + |b_s - b_{s+1}| \lesssim \frac{\log \frac{1}{\delta_{s+1}}}{\log \frac{1}{\epsilon}}$$

$$|a - a_{s_\star}| + |b - b_{s_\star}| \lesssim \frac{1}{\log \frac{1}{\epsilon}} \left(\sum_{t \le s} \log \frac{1}{\delta_t} \right)$$

$$\lesssim \frac{\log \frac{1}{\delta_{s_\star}}}{\log \frac{1}{\epsilon}} \lesssim \frac{\log j_0}{\log \frac{1}{\epsilon}},$$

and (2.2.20) will hold from $a_0 = j_0 - \log j_0$, $b_0 = j_0 + \log j_0$. Hence, at step s_*, (2.2.19) is satisfied with $\delta = j_0^{-C}$. $\qquad\qquad\qquad\qquad\qquad\qquad\qquad\qquad\qquad\qquad\qquad\qquad\qquad\square$

2.4 ESTIMATES ON MEASURE

Recall that the estimates on the symplectic transformations in section 2.3, hence the Hamiltonian in its desired form H' in (2.2.10–2.2.13), depend crucially on the non-resonance condition

$$\left| \sum (n_j - n'_j) \tilde{v}_j \right| > \delta_s^{\frac{1}{100s^2}}, \qquad (2.4.1)$$

previously (2.3.18), where n satisfies

$$\text{supp } n \subset [a_s, b_s] \cup [-b_s, -a_s]$$
$$\subset [j_0 - \log j_0, \ j_0 + \log j_0] \cup [-j_0 - \log j_0, \ -j_0 + \log j_0] \quad (2.4.2)$$
$$\Delta(n) < 10 \frac{\log \frac{1}{\delta_{s+1}}}{\log \frac{1}{\epsilon}}, \quad |n| < \frac{C}{\tau} \qquad (2.4.3)$$

(cf. (2.2.24, 2.3.13, 2.3.16)). From the estimates on c, d, in section 2.3, in particular (2.3.49, 2.3.50), $\tilde{v}_j$ in (2.4.1), which is $\tilde{v}_j$ at step s, can be replaced by $\tilde{v}_j$ at step s_*, the last step, with n continue to satisfy (2.4.2, 2.4.3). This is convenient as we now only need to work with a fixed $\tilde{v}_j$, namely $\tilde{v}_j = \tilde{v}_j^{(s_*)}$.

The measure estimates are in terms of V, the original i.i.d. random variables. But we first make estimates in $\tilde{V}$ via (2.4.1) and then convert the estimates to estimates in V by incorporating the Jacobian in (2.2.29). Denote for a given n,

$$j_+(n) = \max\{j \in \mathbb{Z} \mid n_j - n'_j \neq 0\}. \qquad (2.4.4)$$

Note that $j_+(n)$ is the largest j with non zero contribution to (2.4.1). The set of acceptable $\tilde{V}$ contains

$$\mathcal{S} = \bigcap_{\|k\| - j_0| < \log j_0} \bigcap_{s=1,\dots,s_*} \bigcap_{\substack{n \text{ satisfies } (2.4.3) \\ j_+(n)=k}} \left[\tilde{V} \mid \left| \sum_{j \le k} (n_j - n'_j) \tilde{v}_j \right| > \delta_s^{\frac{1}{100s^2}} \right]. \quad (2.4.5)$$

Define

$$\mathcal{S}_k = \bigcap_{s=1,\dots,s_*} \bigcap_{\substack{n \text{ satisfies } (2.4.3) \\ j_+(n)=k}} \left[\tilde{V} \mid \left| \sum_{j \le k} (n_j - n'_j) \tilde{v}_j \right| > \delta_s^{\frac{1}{100s^2}} \right]. \qquad (2.4.6)$$

For each $\mathcal{S}_k$, the restriction on $\tilde{V}$ only relates to $(\tilde{v}_j)_{j \le k}$. Moreover, for fixed $(\tilde{v}_j)_{j < k}$ and n such that $j_+(n) = k$

$$\text{mes}_{\tilde{v}_k} \left[\left| \sum_{j \le k} (n_j - n'_j) \tilde{v}_j \right| < \delta_s^{\frac{1}{100s^2}} \right] < 2\delta_s^{\frac{1}{100s^2}}. \qquad (2.4.7)$$

Let $\mathcal{S}_k^c$ be the complement of the set $\mathcal{S}_k$ defined in (2.4.5). Its measure can then be estimated as

$$\text{mes } \mathcal{S}_k^c \leq \sum_{s=1}^{s_\star} \sum_{\substack{n \text{ satisfies } (2.4.3) \\ j_+(n)=k}} \text{mes}_{\tilde{v}_k}\left[\left|\sum_{j \leq k}(n_j - n_j')\tilde{v}_j\right| < \delta_s^{\frac{1}{100s^2}}\right]$$

$$< \sum_{s=1}^{s_\star}\left(10\frac{\log\frac{1}{\delta_{s+1}}}{\log\frac{1}{\epsilon}}\right)^{C/\tau} \delta_s^{\frac{1}{100s^2}}, \tag{2.4.8}$$

where we used (2.4.3) and the $\Delta(n)$-nomial formula $((\Delta(n))^{|n|})$ to estimate the entropy coming from n (the sum over n).

From (2.2.21, 2.2.22), the terms in (2.4.8) decay faster than a geometric sequence for $s_\star \sim \log\log j_0$ ($j_0 \gg (\frac{1}{2})^{1/C}$ as $\delta_{s_\star} \sim j_0^{-C} \ll \epsilon$). Since $\delta_1 = \epsilon$, we have

$$\text{mes } \mathcal{S}_k^c < \epsilon^{10^{-3}}, \tag{2.4.9}$$

assuming $\epsilon \ll 1$. Using (2.4.9) in (2.4.5), we obtain

$$\text{mes}_{\tilde{V}}\mathcal{S} > (1 - \epsilon^{10^{-3}})^{2\log j_0} \sim j_0^{-2\epsilon^{10^{-3}}}. \tag{2.4.10}$$

Using the Jacobian estimate in (2.2.29) to express the restrictions in the original random variables $V = \{v_j\}_{j \in \mathbb{Z}}$, we have

$$\text{mes}_V\mathcal{S} > j_0^{-\sqrt{\epsilon}}j_0^{-2\epsilon^{10^{-3}}} > j_0^{-3\epsilon^{10^{-3}}}. \tag{2.4.11}$$

$\mathcal{S}$ defined in (2.4.5), which is for a fixed j_0, thus corresponds to a rare event. To circumvent that, we allow j_0 to vary in a dyadic interval $[\bar{j}_0, 2\bar{j}_0]$, taking into account that the restriction in (2.4.11) only relates to $v_j|_{||j|-j_0|<\mathcal{O}(1)\log j_0}$ in view of (2.2.11, 2.4.3). Using independence, we then obtain that with probability, at least,

$$1 - \left(1 - j_0^{-3\epsilon^{10^{-3}}}\right)^{\frac{j_0}{2\mathcal{O}(1)\log j_0}} > 1 - e^{-\sqrt{j_0}}, \tag{2.4.12}$$

the condition in (2.4.5) holds for some $j_0 \in [\bar{j}_0, 2\bar{j}_0]$. For such j_0, the analysis in section 2.3 applies and H is transformed to H' in (2.2.10–2.2.13), satisfying (2.2.14, 2.2.19, 2.2.20). We are now poised to prove the theorem.

2.5 BOUND ON DIFFUSION

We first recapitulate the setting that we have achieved so far. For any given $\bar{j}_0 \in \mathbb{Z}_+$ (assumed large), there is $j_0 \sim \bar{j}_0$, in fact $j_0 \in [\bar{j}_0, 2\bar{j}_0]$, such that with probability

$$1 - e^{-\sqrt{j_0}}, \tag{2.5.1}$$

H in (2.2.3) is symplectically transformed into H':

$$H' = \frac{1}{2} \sum_{j \in \mathbb{Z}} \tilde{v}_j |q_j|^2 \tag{2.5.2}$$

$$+ \sum_{n \in \mathbb{N}^{\mathbb{Z}} \times \mathbb{N}^{\mathbb{Z}}} c(n) \prod_{\text{supp } n} q_j^{n_j} \bar{q}_j^{n'_j} \tag{2.5.3}$$

$$+ \sum_{n \in \mathbb{N}^{\mathbb{Z}} \times \mathbb{N}^{\mathbb{Z}}} d(n) \prod_{\text{supp } n} |q_j|^{2n_j} \tag{2.5.4}$$

$$+ \mathcal{O}\left(\sum_{j \in \mathbb{Z}} |j|^{-C} |q_j|^2 \right), \tag{2.5.5}$$

where $\tilde{v}_j$ are the modulated frequencies, $c(n)$ are the coefficients of the *nonresonant* monomials, i.e., in (2.5.3), for any n, there is $j \in \text{supp } n$, such that $n_j \neq n'_j$, $d(n)$ are the coefficients of the *resonant* monomials of degrees at least 4. (2.5.3) is responsible for diffusion into higher modes. The coefficients satisfy

$$|c(n)| < j_0^{-C} \quad \text{if}$$
$$\text{supp } n \cap \left[j_0 - \frac{1}{2} \log j_0, \; j_0 + \frac{1}{2} \log j_0 \right]$$
$$\cup \left[-j_0 - \frac{1}{2} \log j_0, \; -j_0 + \frac{1}{2} \log j_0 \right] \neq \emptyset, \tag{2.5.6}$$

where C, a fixed large constant, is the same as in (2.5.5).

The symplectic transformations from H to H' are generated by polynomial Hamiltonians F in (2.3.13), which give the vector fields:

$$\sum \frac{\partial F}{\partial \bar{q}_i} \frac{\partial}{\partial q_i} - \frac{\partial F}{\partial q_i} \frac{\partial}{\partial \bar{q}_i}, \tag{2.5.7}$$

where the sum is over

$$i \subset [a_s, b_s] \cup [-b_s, -a_s] \subset [j_0 - \log j_0, \; j_0 + \log j_0]$$
$$\cup [-j_0 - \log j_0, \; -j_0 + \log j_0]. \tag{2.5.8}$$

The vector fields in (2.5.7) preserve the ℓ^2 norm $\sum |q_j|^2$. From (2.2.5), H preserves the ℓ^2 norm. So ℓ^2 norm remains conserved. Moreover, since the symplectic transforms in (2.5.7) only concern neighborhoods about j_0 of size $\log j_0$, $\sum j^2 |q_j|^2$ is preserved up to a factor of 2 in the transformation process from H to H'.

We now return to the NLS in (2.2.1) and finish

Proof of the Theorem. Assume

$$\sum_{j \in \mathbb{Z}} j^2 |q_j(0)|^2 < C, \quad C \text{ as in (2.5.5).} \tag{2.5.9}$$

We want to bound the diffusion norm

$$\sum_{j \in \mathbb{Z}} j^2 |q_j(t)|^2 \tag{2.5.10}$$

in terms of t as $t \to \infty$. The coordinates $q(t) = \{q_j(t)\}_{j \in \mathbb{Z}}$ satisfy

$$i\dot{q} = \frac{\partial H}{\partial \bar{q}}. \qquad (2.5.11)$$

Fix $\bar{j}_0$, choose $j_0 \in [\bar{j}_0, 2\bar{j}_0] \sim \bar{j}_0$ as described above. Then on a set of V of probability

$$1 - e^{-\sqrt{j_0}},$$

H is symplectically transformed into H' of the form (2.5.2–2.5.5). To avoid confusion, denote the new coordinates in H' by q'. Equation (2.5.11) then becomes

$$i\dot{q}' = \frac{\partial H'}{\partial \bar{q}'}. \qquad (2.5.12)$$

We estimate (2.5.10) via estimating the truncated sum $\sum_{|k|>j_0} |q'_k(t)|^2$. Using (2.5.2–2.5.5) in the RHS of (2.5.12), we write

$$\frac{d}{dt}\left[\sum_{|k|>j_0} |q'_k(t)|^2\right] = 4\Im \sum_{|k|>j_0} \bar{q}'_k \frac{\partial H'}{\partial \bar{q}'_k}$$

$$\sim \sum_{n \in \mathbb{N}^{\mathbb{Z}} \times \mathbb{N}^{\mathbb{Z}}} c(n)\left(\sum_{|k|>j_0} (n_k - n'_k) \prod_{\text{supp } n} q'^{n_j}_j \bar{q}'^{n'_j}_j\right) \qquad (2.5.13)$$

$$+ \mathcal{O}(j_0^{-C}), \qquad (2.5.14)$$

where (2.5.13) is the contribution from (2.5.3), (2.5.14) from (2.5.5), (2.5.2, 2.5.3) do not contribute.

We analyze (2.5.13) further. Recall from (2.2.23) that the monomial in (2.5.3) satisfies

$$\Delta(n) \le C \frac{\log j_0}{\log \frac{1}{\epsilon}}, \qquad \text{same } C \text{ as in (2.5.5)}. \qquad (2.5.15)$$

So if

$$\text{supp } n \cap (-\infty, -j_0] \cup [j_0, \infty) \ne \emptyset, \qquad (2.5.16)$$

then

$$\text{supp } n \subset \left(-\infty, -j_0 + C\frac{\log j_0}{\log \frac{1}{\epsilon}}\right] \cup \left[j_0 - C\frac{\log j_0}{\log \frac{1}{\epsilon}}, \infty\right). \qquad (2.5.17)$$

On the other hand, if $|c(n)| \ge j_0^{-C}$, then (2.5.6) implies that

$$\text{supp } n \subset \left(-\infty, -j_0 - \frac{1}{2}\log j_0\right] \cup \left[j_0 + \frac{1}{2}\log j_0, \infty\right)$$

$$\subset (-\infty, -j_0) \cup (j_0, \infty). \qquad (2.5.18)$$

The last set in (2.5.18) is precisely the set that is summed over in (2.5.13). We have

$$\sum_{|k|>j_0} (n_k - n'_k) = \sum_{|k|>j_0} (n'_k - n_k) = 0, \tag{2.5.19}$$

(cf. (2.2.26) and the comments just below it). So only terms where $|c(n)| < j_0^{-C}$ contribute to (2.5.13), and we have

$$\left| \frac{d}{dt} \left[\sum_{|k|>j_0} |q'_k(t)|^2 \right] \right| \lesssim j_0^{-C}. \tag{2.5.20}$$

Integrating in t, we obtain

$$\sum_{|k|>j_0} |q'_k(t)|^2 < \sum_{|k|>j_0} |q'_k(0)|^2 + j_0^{-C}t. \tag{2.5.21}$$

Since the symplectic transformation only acts on a $\sim \log j_0$ neighborhood of $\pm j_0$ (cf. (2.3.13, 2.5.7, 2.5.8)), we have

$$\sum_{|k|>j_0+10\log j_0} |q_k|^2 < \sum_{|k|>j_0} |q'_k|^2 + j_0^{-C}. \tag{2.5.22}$$

Using (2.5.22) to translate (2.5.21) in terms of q_k, the original coordinates, and multiplying the resulting inequality by j_0^2 on both sides, we arrive at

$$j_0^2 \sum_{|k|>j_0+10\log j_0} |q_k(t)|^2 < j_0^2 \sum_{|k|>j_0-10\log j_0} |q_k(0)|^2 + j_0^{-C+2}t, \tag{2.5.23}$$

with probability at least $1 - e^{-\sqrt{j_0}}$ after taking into account (2.5.1). The inequality in (2.5.23) is a statement about the solution of (2.2.1), as can be seen as follows.

Recall that $j_0 \sim \bar{j}_0$ ($j_0 \in [\bar{j}_0, 2\bar{j}_0]$) and $\bar{j}_0$ is an arbitrary sufficiently large integer. To convert (2.5.23) into a bound on the diffusion norm (2.5.10), we take $\bar{j}_0 = 2^\ell$, $\ell \geq \ell_0$. Inequality (2.5.23) implies

$$4^\ell \sum_{|k|\geq 2^{\ell+1}} |q_k(t)|^2 < 4^\ell \sum_{|k|\geq 2^{\ell-1}} |q_k(0)|^2 + 2^{\ell(2-C)}t \tag{2.5.24}$$

with probability at least

$$1 - e^{-2^{\frac{\ell}{2}}}, \quad \ell \geq \ell_0. \tag{2.5.25}$$

We now sum over $\ell \geq \ell_0$. The LHS of (2.5.24) gives

$$\sum_{\ell\geq\ell_0} 4^\ell \sum_{|k|\geq 2^{\ell+1}} |q_k(t)|^2 > \frac{1}{4} \sum_{|k|\geq 2^{\ell_0+1}} k^2 |q_k(t)|^2, \tag{2.5.26}$$

where for each ℓ, we retain only the terms $|q_{\pm 2^{\ell+1}}(t)|^2$ in the sum over k.

To obtain an upper bound on

$$\sum_{\ell \geq \ell_0} 4^\ell \sum_{|k| \geq 2^{\ell-1}} |q_k(0)|^2, \tag{2.5.27}$$

we notice, for example, in the sum (2.5.27) the term containing $|q_{2^{\ell_0-1}}(0)|^2$ appears **once** as

$$4^{\ell_0} |q_{2^{\ell_0-1}}(0)|^2 = 4\{(2^{\ell_0-1})^2 |q_{2^{\ell_0-1}}(0)|^2\}, \tag{2.5.28}$$

here $k = 2^{\ell_0-1}$; when $k = 2^{\ell_0}$, the term containing $|q_{2^{\ell_0}}(0)|^2$ appears as

$$4\{(1 + 1/4)(2^{\ell_0})^2 |q_{2^{\ell_0}}(0)|^2\}; \tag{2.5.29}$$

when $k = 2^{\ell_0+m}$, it appears as

$$4\{(1 + 1/4 + 1/16 + \cdots + 1/4^{m+1})(2^{\ell_0+m})^2 |q_{2^{\ell_0+m}}(0)|^2\}. \tag{2.5.30}$$

Similar observations hold for k nondyadic, so we have

$$(2.5.27) < 6 \sum_{k \in \mathbb{Z}} k^2 |q_k(0)|^2\}. \tag{2.5.31}$$

Summing over $\ell \geq \ell_0$ in (2.5.24) and using (2.5.26, 2.5.27, 2.5.31), we obtain

$$\frac{1}{4} \sum_{|k| \geq 2^{\ell_0+1}} k^2 |q_k(t)|^2 < 6 \sum_{k \in \mathbb{Z}} k^2 |q_k(0)|^2 + 2 \cdot 2^{-\ell_0(C-2)} t. \tag{2.5.32}$$

Hence

$$\sum_{k \in \mathbb{Z}} k^2 |q_k(t)|^2 < 24C(1 + 4^{\ell_0+1}) + 2 \cdot 2^{-\ell_0(C-2)} t \tag{2.5.33}$$

with probability at least

$$1 - e^{-2^{(\frac{\ell_0}{2}-1)}}, \tag{2.5.34}$$

where we summed over (2.5.25).

Choosing

$$\ell_0 = \ell_0(t) \sim \log t^{(3/C)}, \tag{2.5.35}$$

(2.5.33) gives

$$\sum_{k \in \mathbb{Z}} k^2 |q_k(t)|^2 < t^{3/C}, \tag{2.5.36}$$

with probability at least

$$1 - e^{-t^{3/C}}. \tag{2.5.37}$$

An application of the Borel-Cantelli theorem for $t \in \mathbb{Z}^+$ and supplementing with (2.5.20) proves the theorem by choosing $C = 3/\kappa$. $\qquad\qquad\square$

REFERENCES

[B]   J. Bourgain, *Nonlinear Schrödinger equations*, Park City Lectures, 1999.

[BW1] J. Bourgain, W.-M. Wang, *Anderson localization for time quasi-periodic random Schrödinger and wave equations*, Commun. Math. Phys. (2004).

[BW2] J. Bourgain, W.-M. Wang, *Quasi-periodic solutions for nonlinear random Schrödinger equations*, submitted 2004.

[DG] S. De Bievre, F. Germinet, *Dynamical localization for discrete and continuous random Schrödinger operators*, Commun. Math. Phys. 194 (1998), 323–341.

[vDK] H. von Dreifus, A. Klein, *A new proof of localization in the Anderson tight binding model*, Commun. Math. Phys. 124 (1989), 285–299.

[FS] J. Fröhlich, T. Spencer, *Absence of diffusion in the Anderson tight binding model for large disorder or low energy*, Commun. Math. Phys. 88 (1983), 151–184.

Chapter Three

Instability of Finite Difference Schemes for Hyperbolic Conservation Laws

A. Bressan, P. Baiti, and H. K. Jenssen

3.1 INTRODUCTION

A *system of conservation laws* in one space dimension takes the form

$$u_t + f(u)_x = 0. \tag{3.1.1}$$

The components of the vector $u = (u_1, \ldots, u_n) \in \mathbb{R}^n$ are the *conserved quantities*, while the components of the function $f = (f_1, \ldots, f_n) : \mathbb{R}^n \mapsto \mathbb{R}^n$ are the corresponding *fluxes*. For smooth solutions, (3.1.1) is equivalent to the quasi-linear system

$$u_t + A(u)u_x = 0, \tag{3.1.2}$$

where $A(u) \doteq Df(u)$ is the $n \times n$ Jacobian matrix of the flux function f. We recall that the system is *strictly hyperbolic* if this Jacobian matrix $A(u)$ has n real distinct eigenvalues, $\lambda_1(u) < \cdots < \lambda_n(u)$ for every $u \in \mathbb{R}^n$. In this case, $A(u)$ admits a basis of eigenvectors $r_1(u), \ldots, r_n(u)$. In the strictly hyperbolic case, the Cauchy problem for (3.1.1) is well posed, within a class of functions having small total variation [7, 9]. We remark that, in general, even for smooth initial data the solution can develop shocks in finite time. The equation (3.1.1) must then be interpreted in distributional sense, namely,

$$\iint [u \, \phi_t + f(u) \, \phi_x] \, dx dt = 0,$$

for every test function $\phi \in \mathcal{C}_c^1$, continuously differentiable with compact support. In case of discontinuous solutions, uniqueness is obtained by imposing additional entropy admissibility conditions along shocks [16, 19]. As proved in [7], the Liu conditions characterize the unique solutions obtained as limits of vanishing viscosity approximations.

In addition to the celebrated Glimm scheme [13], various other approximation methods have been introduced and studied in more recent literature, namely: front tracking [1, 2, 8, 9, 11, 14], vanishing viscosity [7], relaxation approximations [15, 22, 6], and semidiscrete schemes [5]. Given an initial data

$$u(0, x) = \bar{u}(x) \tag{3.1.3}$$

with small BV norm, in all of the above cases one can prove that the approximate solutions retain small total variation for all times $t > 0$ and depend Lipschitz continuously on the initial data, in the $\mathbf{L}^1$ distance. Moreover, the approximations converge to the unique entropy weak solution of the hyperbolic system (3.1.1).

For computational purposes, the most important type of approximations are the fully discrete numerical schemes [17]. In this case, one starts by constructing a grid in the t-x plane with mesh Δt, Δx. An approximate solution $U_{k,j} \approx u(k\,\Delta t,\ j\,\Delta x)$ is then obtained by replacing partial derivatives in (3.1.1) with finite differences. For example, if for all u the eigenvalues of the Jacobian matrix $Df(u)$ satisfy

$$\left|\lambda_i(u)\right| < \Delta x/\Delta t \qquad i = 1, \ldots, n,$$

then one can use the Lax-Friedrichs scheme

$$U_{k+1,j} = U_{k,j} + \frac{\Delta t}{2\Delta x}\left[f(U_{k,j-1}) - f(U_{k,j+1})\right]. \qquad (3.1.4)$$

In the case where

$$0 < \lambda_i(u) < \Delta x/\Delta t \qquad i = 1, \ldots, n,$$

one can also use the upwind Godunov scheme

$$U_{k+1,j} = U_{k,j} + \frac{\Delta t}{\Delta x}\left[f(U_{k,j-1}) - f(U_{k,j+1})\right]. \qquad (3.1.5)$$

As for all previous methods, it is natural to expect that, if the initial data have small total variation, then the approximations constructed by finite difference schemes will have uniformly small variation for all positive times. Surprisingly, this is not true. Indeed, the analysis in [3, 4], has brought to light a subtle mechanism for the instability of fully discrete schemes, due to possible resonances between the speed of a shock and the ratio $\Delta x/\Delta t$ in the mesh of the grid. As shown by the counterexample in [4], these resonances can prevent the validity of a priori BV bounds and the $\mathbf{L}^1$ stability for these approximate solutions.

We remark that all previous results about BV stability for viscous, semidiscrete, and relaxation approximations relied on the local decomposition of a solution as a superposition of traveling wave profiles. To implement this approach, it is essential to work with a center manifold of traveling profiles smoothly depending on parameters. In case of the difference schemes (3.1.4) or (3.1.5), a traveling profile with speed σ is a continuous function $U = U(\xi)$ such that the assignment

$$U_{k,j} = U(j\Delta x - \sigma k\Delta t)$$

provides a solution to the equation (3.1.4) or (3.1.5), respectively. The existence of discrete traveling profiles was proved by Majda and Ralston [21] in the case of rational wave speeds, and by Liu and Yu [20] in the case of irrational, diofantine speeds. However, as remarked by Serre [23], these discrete profiles cannot depend continuously on the wave speed σ. In particular, for general $n \times n$ hyperbolic systems, no regular manifold of discrete traveling profiles exists. A detailed example, showing how continuous dependence fails for Lax-Friedrichs wave profiles, was constructed in [3]. Of course, this already implies that the techniques used in [7] for proving uniform BV bounds cannot be applied to fully discrete approximations. To further settle the issue, the recent counterexample in [4] shows that these a priori BV bounds simply cannot hold.

In the present chapter we review the main ideas in [3] and [4]. Section 3.2 contains a formal analysis, explaining how the discrete traveling wave profile for the Godunov scheme can fail to depend continuously on parameters, as the wave speed approaches a given rational number.

In Section 3.3 we outline the construction of a Godunov solution for a strictly hyperbolic 2×2 system, where the total variation is amplified by an arbitrarily large factor.

3.2 INSTABILITY OF DISCRETE TRAVELING WAVE PROFILES

Here and in the next section, we study a 2×2 system of the form

$$\begin{cases} u_t + f(u)_x = 0 \\ v_t + \frac{1}{2}v_x + g(u)_x = 0. \end{cases} \tag{3.2.1}$$

We assume that $f'(u) \in]1/2, 1[$, so that the system is strictly hyperbolic, with both characteristic speeds contained inside the interval $[0, 1]$. To fix the ideas, let $\Delta t = \Delta x = 1$. The Godunov (upwind) approximations then take the form

$$u_{k+1,j} = u_{k,j} + \left[f(u_{k,j-1}) - f(u_{k,j}) \right], \tag{3.2.2}$$

$$v_{k+1,j} = \frac{v_{k,j-1} + v_{k,j}}{2} + g_{k,j}, \tag{3.2.3}$$

where

$$g_{k,j} = g(u_{k,j-1}) - g(u_{k,j}).$$

Note that the u-component of the solution satisfies a scalar difference equation. Moreover, the v-component satisfies a linear difference equation with source terms $g_{k,j}$ derived from the first equation. The solution of (3.2.3) can be explicitly computed in terms of binomial coefficients:

$$v_{m,i} = \sum_{0 \le k < m, \ i-(m-k) \le j \le i} B(m-k, \ i-j) \, g_{k,j}, \tag{3.2.4}$$

where

$$B(m, \ell) = \frac{m!}{\ell! \, (m - \ell)!} \cdot 2^{-m}. \tag{3.2.5}$$

Assume that the u-component is a traveling shock profile connecting the states u^-, u^+, with speed $\sigma \in]0, 1[$. In other words, $u_{k,j} = U(k - \sigma j)$, with

$$\lim_{\xi \to \pm\infty} U(\xi) = u^{\pm}, \qquad \sigma = \frac{f(u^+) - f(u^-)}{u^+ - u^-} > 0.$$

We are interested in the oscillations of the v-component of the discretized solution, at (3.2.3). In the following, we will choose a function g such that $g'(u) = 0$ for all u outside a small neighborhood of $(u^+ + u^-)/2$. More precisely:

$$g(u) = \begin{cases} 1 & \text{if } u < (u^+ + u^- - \varepsilon)/2, \\ 0 & \text{if } u > (u^+ + u^- + \varepsilon)/2. \end{cases} \tag{3.2.6}$$

As a consequence, all the source terms $g_{k,j}$ vanish, except within a thin strip around the line $\{U(x - \sigma t) = (u^+ + u^-)/2\}$.

To achieve a better understanding of the solution of (3.2.4), two approximations can be performed:

(i) The binomial coefficients in (3.2.5) can be replaced by a Gauss kernel.

(ii) Choosing $\varepsilon > 0$ small, we can assume that the source term $g_{k,j}$ is nonzero only at the integer points (k, j) immediately to the left of the line $x = \sigma t$. More precisely,

$$g_{n,j} = \begin{cases} 1 & \text{if } j = [[\sigma n]], \\ 0 & \text{otherwise.} \end{cases} \tag{3.2.7}$$

Here $[[s]]$ denotes the largest integer $\leq s$. After a linear rescaling of variables, in place of (3.2.3) we are led to study the heat equation with point sources

$$v_t - v_{xx} = \delta_{n,[[\sigma n]]}. \tag{3.2.8}$$

In turn, the above equation can be compared with

$$v_t - v_{xx} = \delta_{n,\sigma n}, \tag{3.2.9}$$

$$v_t - v_{xx} = \delta_{t,\sigma t}, \tag{3.2.10}$$

see fig. 3.1. Note the difference between these three equations: In (3.2.10) the source term acts continuously in time, along the straight line $x = \sigma t$. In (3.2.9), at every integer time $t = n$ the source consists of a unit mass at the point $(n, \sigma n)$ (the white dots in fig. 3.1). On the other hand, in (3.2.8) these sources are located at the points with integer coordinates $(n, [[\sigma n]])$, immediately to the left of the line $x = \sigma t$ (the black dots in fig. 3.1).

The traveling wave solution of (3.2.10) contains no downstream oscillations. Indeed, $v(t, x) = \phi(x - \sigma t)$, where

$$\phi(y) = \int_0^\infty G(t, y + \sigma t)\, dt = \begin{cases} e^{-\sigma y}/\sigma & \text{if } y \geq 0. \\ 1/\sigma & \text{if } y < 0. \end{cases}$$

Here $G(t, x) = e^{-x^2/4t}/\sqrt{4\pi t}$ is the standard Gauss kernel.

Concerning the solution of (3.2.9), by repeated integration by parts one can show that downstream oscillations are rapidly decreasing. Namely, for every $k \geq 1$ one has

$$|v_x(t, \sigma t - y)| = \mathcal{O}(1) \cdot y^{-k} \qquad \text{as } y \to \infty. \tag{3.2.11}$$

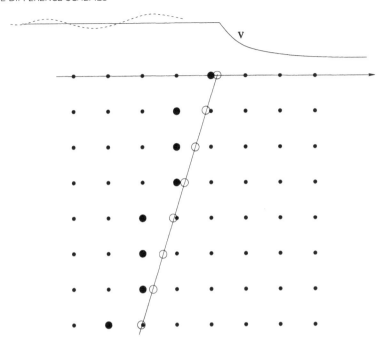

Fig. 3.1. The heat equations with various singular sources.

Thanks to (3.2.11), the analysis of downstream oscillations in the solution of (3.2.8) can be based on a comparison with (3.2.9). This will show that, if the speed of the source is close to a rational but not exactly rational, then resonances will occur. For example, assume that $\sigma = 1 + \varepsilon$, with $0 < \varepsilon \ll 1$. Consider a traveling wave solution, so that $v(t, x) = \Psi^\sigma(x - \sigma t)$ for every integer time t and every $x \in \mathbb{R}$. In this case, we have the explicit representation

$$v(t, x) = \sum_{-\infty < n < t} G\big(t - n, x - [[\sigma n]]\big).$$

For $y \ll 0$, calling $\Psi^\sigma(y)$ the value of the solution profile at a distance $|y|$ downstream from the shock, we find

$$
\begin{aligned}
\Psi^\sigma(y) &\doteq \sum_{-\infty < n < 0} G\big(-n, y - [[\sigma n]]\big) \\
&= \sum_{-\infty < n < 0} G\big(-n, y - \sigma n\big) \\
&\quad - \sum_{-\infty < n < 0} \Big[G\big(-n, y - \sigma n\big) - G\big(-n, y - [[\sigma n]]\big) \Big] \\
&\approx \frac{1}{\sigma} - \sum_{-\infty < n < 0} G_x(-n, y - \sigma n)\big(\sigma n - [[\sigma n]]\big) \\
&\approx \frac{1}{\sigma} + \int_0^\infty G_x(t, y + \sigma t)\big(\varepsilon t - [[\varepsilon t]]\big)\, dt.
\end{aligned}
$$

(3.2.12)

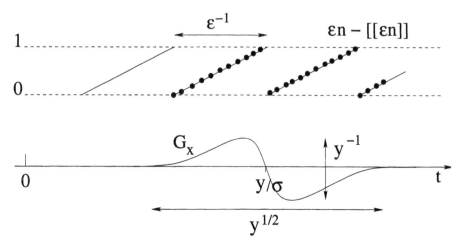

Fig. 3.2. Amplitude of oscillations generated by almost periodic sources.

The amplitude of the two factors in the integral on the right hand side of (3.2.12) is illustrated in fig. 3.2. Clearly, the function $t \mapsto \varepsilon t - [[\varepsilon t]]$ takes values in $[0, 1]$ and has period ε^{-1}. We observe that

$$\int_0^\infty G_x(t, y + \sigma t)\, dt = 0, \qquad \int_0^\infty |G_x(t, y + \sigma t)|\, dt = \mathcal{O}(1) \cdot |y|^{-1/2}.$$

A direct analysis of the last integral in (3.2.12) shows that nontrivial oscillations occur when $|y|$ has the same order of magnitude as ε^{-2}. More precisely, call $y_\varepsilon \doteq -\varepsilon^{-2}$, $t_\varepsilon = \varepsilon^{-2}/\sigma$, so that $G_x(t_\varepsilon, y_\varepsilon + \sigma t_\varepsilon) = 0$. As y ranges over the interval $I_\varepsilon \doteq [y_\varepsilon/2, 3y_\varepsilon/2]$, the function $y \mapsto \Psi^{1+\varepsilon}(y)$ oscillates several times. The amplitude of each oscillation is $\geq c_0\, \varepsilon$, and the distance between a peak and the next one is $\approx \varepsilon^{-1}$. On the whole interval I_ε we thus have $\approx \varepsilon^{-1}$ oscillations. The total variation of $\Psi^{1+\varepsilon}$ on the interval I_ε remains uniformly positive, as $\varepsilon \to 0$. Since $y_\varepsilon \to -\infty$ as $\varepsilon \to 0+$, the above analysis shows that the family of profiles $\Psi^{1+\varepsilon}$ cannot converge in the BV norm, as $\varepsilon \to 0$. In particular, discrete traveling profiles cannot depend smoothly on the wave speed.

3.3 LACK OF BV BOUNDS FOR GODUNOV SOLUTIONS

In this section, we consider the 2×2 system

$$u_t + \big[\ln(1 + e^u)\big]_x = 0, \tag{3.3.1}$$

$$v_t + \frac{1}{2}v_x + g(u)_x = 0. \tag{3.3.2}$$

We choose a right state u^+ and a left state u^-, with $u^- > u^+ > 0$, in such a way that the corresponding shock for the scalar conservation law (3.3.1) travels with rational speed. For sake of definiteness, we choose

$$\sigma_0 = \frac{\ln(1 + e^{u^-}) - \ln(1 + e^{u^+})}{u^- - u^+} = \frac{5}{6}. \tag{3.3.3}$$

In equation (3.3.2), we consider a smooth function g, with $g'(u) = 0$ everywhere except on a small subinterval of $[u^+, u^-]$. More precisely, for some $0 < \delta_0 \ll u^- - u^+$, we assume that

$$g(u) = \begin{cases} 1 & \text{if} \quad u \le (u^+ + u^- - \delta_0)/2, \\ 0 & \text{if} \quad u \ge (u^+ + u^- + \delta_0)/2. \end{cases} \tag{3.3.4}$$

On a large time interval $[0, T]$, we now consider a solution of (3.3.1) having a shock located along a curve γ such that

$$\gamma(t) = \frac{5}{6}t - 2\sqrt{T - t} \qquad \text{for all } t \in [0, T - T^{2/3}]. \tag{3.3.5}$$

Note that this solution can be obtained by adding a weak compression wave to the left of the shock. This compression wave impinges on the shock and slightly increases its speed as time goes by. The solution corresponds to an initial condition of the form

$$u(0, x) = \begin{cases} u^- + \phi(x) & \text{if} \quad x < 0, \\ u^+ & \text{if} \quad x > 0. \end{cases} \tag{3.3.6}$$

Observe that

$$\text{Tot.Var.}\{\dot{\gamma}\} = \frac{1}{\sqrt{T - T^{2/3}}} - \frac{1}{\sqrt{T}} \tag{3.3.7}$$

becomes arbitrarily small as $T \to \infty$. Therefore, choosing T large, we can assume that the C^1 norm of the perturbation ϕ is as small as we like.

Next, consider the corresponding approximate solution of (3.3.1)–(3.3.2) computed by the Godunov scheme, with step size $\Delta t = \Delta x = 1$, i.e.,

$$u_{k+1,j} = u_{k,j} + \ln\left(1 + e^{u_{k,j-1}}\right) - \ln\left(1 + e^{u_{k,j}}\right), \tag{3.3.8}$$

$$v_{k+1,j} = \frac{v_{k,j-1} + v_{k,j}}{2} + g_{k,j}, \tag{3.3.9}$$

where

$$g_{k,j} = g(u_{k,j-1}) - g(u_{k,j}).$$

We claim that, when the initial data provide a discrete approximation to (3.3.6), at the later integer time $m = T$ the corresponding Godunov solutions satisfy

$$V(m) \doteq \sum_i |v_{m,i} - v_{m,i-1}| \ge c_1 \cdot \ln T. \tag{3.3.10}$$

Letting $T \to \infty$, we thus obtain a sequence of Godunov solutions where:

(i) The initial data are a vanishingly small perturbation of a discrete shock profile joining u^- with u^+.

(ii) At suitably large times, by (3.3.10) the total variation becomes arbitrarily large.

We outline below the two main steps in the construction.

Step 1. We first need to construct an explicit solution of the nonlinear difference equation (3.3.8), which approximates an exact solution of (3.3.1) having a shock along γ. Thanks to the special choice of the flux function in (3.3.1), we can use a nonlinear transformation due to P. Lax [16], analogous to the Hopf-Cole transformation. Namely, if the positive numbers $z_{n,j}$ provide a solution to the linear difference equation

$$z_{n+1,j} = \frac{z_{n,j} + z_{n,j-1}}{2}, \tag{3.3.11}$$

then a solution of (3.3.8) is provided by

$$u_{n,j} = \ln \left(\frac{z_{n,j-1}}{z_{n,j}} \right). \tag{3.3.12}$$

Explicit solutions of (3.3.11) in the form of discrete traveling profiles are easy to obtain. In particular, if

$$z(t, x) = e^{-b[x - \sigma(b),t]}, \qquad \sigma(b) \doteq \frac{\ln(1 + e^b) - \ln 2}{b},$$

then $z_{m,j} \doteq z(m, j)$ provide a solution to (3.3.11). Since (3.3.11) is linear homogeneous, any integral combination of these traveling profiles will provide yet another solution, say

$$z(t, x) = 1 + \int_{-\infty}^{T} a(\xi) e^{-\xi [x - \sigma(\xi)t]} d\xi. \tag{3.3.13}$$

By a judicious choice of the positive function $a(\cdot)$ in (3.3.13), we obtain a smooth function

$$u(t, x) \doteq \ln \left(\frac{z(t, x)}{z(t, x-1)} \right), \tag{3.3.14}$$

which closely approximates the desired solution of (3.3.1). In particular, using the Laplace asymptotic method, one can prove that the level curve

$$\{(t, x) ; \ u(t, x) = (u^+ + u^-)/2\}$$

is very close to the shock curve γ at (3.3.5). By the Lax formula, the numbers $u_{m,i} \doteq u(m, i)$ provide a solution to the finite difference equation (3.3.8).

Step 2. The analysis of the oscillations produced by the source $g_{m,j}$ in the v-component of the solution relies on the same ideas outlined in Section 3.2. In the explicit formula

$$v_{m,i} = \sum_{0 \le k < m,\ i-(m-k) \le j \le i} B(m-k, i-j)\, g_{k,j}, \qquad (3.3.15)$$

two approximations can be performed:

   (i) The binomial coefficients $B(m, \ell)$ can be replaced with a Gaussian kernel.
   (ii) By our special choice of the function g, we can assume that $g_{k,j}$ is nonzero only at the integer points (k, j) immediately to the left of the curve $x = \gamma(t)$. More precisely,

$$g_{n,j} = \begin{cases} 1 & \text{if } \quad j = [[\gamma(n)]], \\ 0 & \text{otherwise.} \end{cases} \qquad (3.3.16)$$

By the previous analysis, if the sources $g_{k,j}$ are concentrated on the points with integer coordinates to the left of a line $x = [(5/6) + \varepsilon](t - T)$, for $t \in [T - 2\varepsilon^{-2},\ T-\varepsilon^{-2}]$, then at time $m = T$ solution v will contain a "packet" of downstream oscillations. More precisely, on the interval $I_\varepsilon \doteq [-2\varepsilon^{-2}/3,\ -\varepsilon^{-2}/3]$ the total variation satisfies the lower bound

$$\sum_{i \in I_\varepsilon} |v_{m,i} - v_{m,i-1}| \ge c_0, \qquad (3.3.17)$$

for some constant $c_0 > 0$ independent of ε.

We now observe that, if the sources are located along the curve γ at (3.3.5) whose speed is not constant, arbitrarily many of these oscillation packets can be obtained. Indeed, call

$$\varepsilon_0 \doteq T^{1/3}, \qquad\qquad \varepsilon_k \doteq 2^{-k}\varepsilon_0.$$

Note that the sources occurring within the time interval $J_k \doteq [T-2\varepsilon_k^{-2},\ T-\varepsilon_k^{-2}]$ are located at integer points next to the curve γ, and this curve has speed $\dot\gamma(t) \approx 5/6+\varepsilon_k$ when $t \in J_k$. Therefore, at the terminal time $m = T$, these sources are responsible for a "packet of oscillations", located on the interval $I_{\varepsilon_k} \doteq [-2\varepsilon_k^{-2}/3,\ \varepsilon_k^{-2}/3]$. Note that, for T large, we have

$$T - 2\varepsilon_k^{-2} = T - 2^{2k+1}T^{2/3} > 0$$

provided that $\ln T > 3(2k + 1)\ln 2$. We thus conclude that (3.3.10) holds with $c_1 = c_0/6(k + 1)$. For all details of these estimates we refer to [4]. (Fig. 3.3.)

3.4 CONCLUDING REMARKS

The above analysis was concerned with Godunov solutions to a very specific 2×2 system of conservation laws. Dealing with the special flux function $f(u) = \ln(1 + e^u)$ enabled us to use the explicit Lax formula. However, this is only an expedient to simplify the analysis. We believe that similar instability results hold for all finite difference schemes, in connection with generic hyperbolic systems.

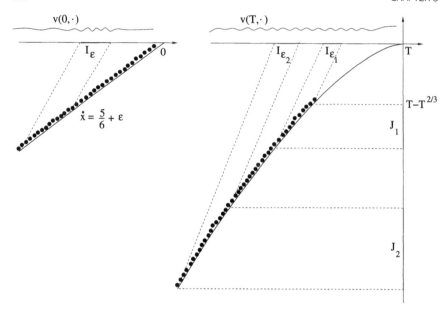

Fig. 3.3. Packets of oscillations generated by sources at integer points with speed close to rational.

The counterexample points out a basic limitation of analytic theory: there is no hope to achieve a rigorous proof of convergence of finite difference approximations by means of a priori BV bounds.

On the other hand, the present analysis should not have worrisome consequences for the practical performance of numerical schemes. Indeed, it appears that the initial data yielding large total amounts of oscillations are very rare: they have to be carefully constructed, so the the shocks present in the solution have exactly the appropriate speed that resonates with the grid. For "generic" initial data, we do not expect such resonances. Moreover, even in our example, the total variation is large but spread out over many grid points. If a sequence of approximate solutions were constructed, letting the mesh Δt, $\Delta x \to 0$, one would still recover the correct solution in the limit.

REFERENCES

[1] F. Ancona and A. Marson, A front tracking algorithm for general nonlinear hyperbolic systems, Preprint, 2005.

[2] P. Baiti and H. K. Jenssen, On the front tracking algorithm, *J. Math. Anal. Appl.* 217 (1998), 395–404.

[3] P. Baiti, A. Bressan, and H. K. Jenssen, Instability of travelling wave profiles for the Lax-Friedrichs scheme, *Discr. Cont. Dynam. Syst.* 13 (2005), 877–899.

[4] P. Baiti, A. Bressan, and H. K. Jenssen, An instability of the Godunov scheme, *Comm. Pure Appl. Math.* 59 (2006), 1604–1638.

[5] S. Bianchini, BV solutions of the semidiscrete upwind scheme, *Arch. rational Mech. Anal.* 167 (2003), 1–81.

[6] S. Bianchini, Hyperbolic limit of the Jin-Xin relaxation model, *Comm. Pure Appl. Math.* 59 (2006), 688–753.

[7] S. Bianchini and A. Bressan, Vanishing viscosity solutions of nonlinear hyperbolic systems, *Annals of Mathematics* (2) 161 (2005), no. 1, 223–342.

[8] A. Bressan, Global solutions to systems of conservation laws by wave-front tracking, *J. Math. Anal. Appl.* 170 (1992), 414–432.

[9] A. Bressan, *Hyperbolic Systems of Conservation Laws: The One Dimensional Cauchy Problem*, Oxford University Press, 2000.

[10] A. Bressan and H. K. Jenssen, On the convergence of Godunov scheme for nonlinear hyperbolic systems, *Chinese Ann. Math.* B - 21 (2000), 1–16.

[11] C. Dafermos, Polygonal approximations of solutions of the initial value problem for a conservation law, *J. Math. Anal. Appl.* 38 (1972), 33–41.

[12] X. Ding, G. Q. Chen, and P. Luo, Convergence of the fractional step Lax-Friedrichs scheme for the isentropic system of gas dynamics, *Comm. Math. Phys.* 121 (1989), 63–84.

[13] J. Glimm, Solutions in the large for nonlinear hyperbolic systems of equations, *Comm. Pure Appl. Math.* 18 (1965), 697–715.

[14] H. Holden and N. H. Risebro, *Front Tracking for Hyperbolic Conservation Laws*, Springer-Verlag, New York 2002.

[15] S. Jin and Z. Xin, The relaxation schemes for systems of conservation laws in arbitrary space dimensions, *Comm. Pure Appl. Math.* 48 (1955), 235–277.

[16] P. Lax, Hyperbolic systems of conservation laws II, *Comm. Pure Appl. Math.* 10 (1957), 537–566.

[17] R. LeVeque, *Numerical Methods for Conservation Laws*, Birkhäuser, 1990.

[18] R. LeVeque and B. Temple, Stability of Godunov's method for a class of 2×2 systems of conservation laws. *Trans. Amer. Math. Soc.* 288 (1985), 115–123.

[19] T. P. Liu, The entropy condition and the admissibility of shocks, *J. Math. Anal. Appl.* 53 (1976), 78–88.

[20] T. P. Liu and S. H. Yu, Continuum shock profiles for discrete conservation laws I. Construction. *Comm. Pure Appl. Math.* 52 (1999), 85–127.

[21] A. Majda and J. Ralston, Discrete shock profiles for systems of conservation laws. *Comm. Pure Appl. Math.* 32 (1979), 445–482.

[22] R. Natalini, Recent results on hyperbolic relaxation problems, in *Analysis of Systems of Conservation Laws*, ed. H. Freistühler, Chapman and Hall/CRC, 1998, pp. 128–198.

[23] D. Serre, Remarks about the discrete profiles of shock waves. *Mat. Contemp.* 11 (1996), 153–170.

Chapter Four

Nonlinear Elliptic Equations with Measures Revisited

H. Brezis, M. Marcus, and A. C. Ponce

4.0 INTRODUCTION

Let $\Omega \subset \mathbb{R}^N$ be a bounded domain with smooth boundary. Let $g : \mathbb{R} \to \mathbb{R}$ be a continuous, nondecreasing function such that $g(0) = 0$. In this paper we are concerned with the problem

$$\begin{cases} -\Delta u + g(u) = \mu & \text{in } \Omega, \\ u = 0 & \text{on } \partial\Omega, \end{cases} \qquad (4.0.1)$$

where μ is a measure. The study of (4.0.1) when $\mu \in L^1(\Omega)$ was initiated by Brezis-Strauss [BS]; their main result asserts that for *every* $\mu \in L^1$ and *every* g as above, problem (4.0.1) admits a unique weak solution (see Theorem 4.B.2 in Appendix 4B below). The right concept of weak solution is the following:

$$\begin{cases} u \in L^1(\Omega), g(u) \in L^1(\Omega) \text{ and} \\ -\int_\Omega u\Delta\zeta + \int_\Omega g(u)\zeta = \int_\Omega \zeta \, d\mu \quad \forall \zeta \in C^2(\bar{\Omega}), \zeta = 0 \text{ on } \partial\Omega. \end{cases} \qquad (4.0.2)$$

It will be convenient to write

$$C_0(\bar{\Omega}) = \{\zeta \in C(\bar{\Omega}); \zeta = 0 \text{ on } \partial\Omega\}$$

and

$$C_0^2(\bar{\Omega}) = \{\zeta \in C^2(\bar{\Omega}); \zeta = 0 \text{ on } \partial\Omega\},$$

and to say that (4.0.1) holds in the sense of $(C_0^2)^*$. We will often omit the word "weak" and simply say that u is a solution of (4.0.1), meaning (4.0.2). It follows from standard (linear) regularity theory that a weak solution u belongs to $W_0^{1,q}(\Omega)$ for every $q < \frac{N}{N-1}$ (see, e.g., [S] and Theorem 4.B.1 below).

The case where μ is a measure turns out to be much more subtle than one might expect. It was observed in 1975 by Ph. Bénilan and H. Brezis (see [B1], [B2], [B3], [B4], [BB], and Theorem 4.B.6 below) that if $N \geq 3$ and $g(t) = |t|^{p-1}t$ with $p \geq \frac{N}{N-2}$, then (4.0.1) has *no solution* when $\mu = \delta_a$, a Dirac mass at a point $a \in \Omega$. On the other hand, it was also proved (see Theorem 4.B.5 below) that if $g(t) = |t|^{p-1}t$ with $p < \frac{N}{N-2}$ (and $N \geq 2$), then (4.0.1) has a solution for any measure μ. Later, Baras-Pierre [BP] (see also [GM]) *characterized* all measures μ

for which (4.0.1) admits a solution. Their necessary and sufficient condition for the existence of a solution when $p \geq \frac{N}{N-2}$ can be expressed in two equivalent ways:

$$\begin{cases} \mu \text{ admits a decomposition } \mu = f_0 - \Delta v_0 \text{ in the } (C_0^2)^*\text{-sense,} \\ \text{with } f_0 \in L^1 \text{ and } v_0 \in L^p, \end{cases} \tag{4.0.3}$$

or

$$|\mu|(A) = 0 \quad \text{for every Borel set } A \subset \Omega \text{ with } \operatorname{cap}_{2,p'}(A) = 0, \tag{4.0.4}$$

where $\operatorname{cap}_{2,p'}$ denotes the capacity associated to $W^{2,p'}$.

Our goal in this paper is to analyze the nonexistence mechanism and to describe what happens if one "forces" (4.0.1) to have a solution in cases where the equation "refuses" to possess one. The natural approach is to introduce an approximation scheme. For example, μ is kept fixed and g is truncated. Alternatively, g is kept fixed and μ is approximated, e.g., via convolution. It was originally observed by one of us (see [B4]) that if $N \geq 3$, $g(t) = |t|^{p-1}t$, with $p \geq \frac{N}{N-2}$, and $\mu = \delta_a$, with $a \in \Omega$, then all "natural" approximations (u_n) of (4.0.1) converge to $u \equiv 0$. And, of course, $u \equiv 0$ is *not* a solution of (4.0.1) corresponding to $\mu = \delta_a$! It is this kind of phenomenon that we propose to explore in full generality. We are led to study the convergence of the approximate solutions (u_n) under various assumptions on the sequence of data.

Concerning the function g we will assume *throughout the rest of the paper* (except in Section 4.7) that $g : \mathbb{R} \to \mathbb{R}$ is continuous, nondecreasing, and that

$$g(t) = 0 \quad \forall t \leq 0. \tag{4.0.5}$$

Remark 4.1. Assumption (4.0.5) is harmless when the data μ is nonnegative, since the corresponding solution u is nonnegative by the maximum principle and it is only the restriction of g to $[0, \infty)$ that is relevant. However, when μ is a signed measure it is worthwhile to remove assumption (4.0.5), and this is done in Section 4.7 below.

By a *measure* μ we mean a continuous linear functional on $C_0(\bar{\Omega})$, or equivalently a finite measure on $\bar{\Omega}$ such that $|\mu|(\partial\Omega) = 0$ (see Appendix 4C below). The space of measures is denoted by $\mathcal{M}(\Omega)$ and is equipped with the standard norm

$$\|\mu\|_{\mathcal{M}} = \sup \left\{ \int_\Omega \varphi \, d\mu; \varphi \in C_0(\bar{\Omega}) \text{ and } \|\varphi\|_{L^\infty} \leq 1 \right\}.$$

By a (weak) *solution* u of (4.0.1) we mean that (4.0.2) holds. A (weak) *subsolution* u of (4.0.1) is a function u satisfying

$$\begin{cases} u \in L^1(\Omega), g(u) \in L^1(\Omega) \text{ and} \\ -\int_\Omega u \Delta \zeta + \int_\Omega g(u)\zeta \leq \int_\Omega \zeta \, d\mu \quad \forall \zeta \in C_0^2(\bar{\Omega}), \zeta \geq 0 \text{ in } \Omega. \end{cases} \tag{4.0.6}$$

We will say that $\mu \in \mathcal{M}(\Omega)$ is a *good measure* if (4.0.1) admits a solution. If μ is a good measure, then equation (4.0.1) has exactly one solution u (see Corollary 4.B.1 in Appendix 4B). We denote by $\mathcal{G}$ the set of good measures (relative to g).

Remark 4.2. In many places throughout this paper, the quantity $\int_\Omega \zeta \, d\mu$, with $\zeta \in C_0^2(\bar\Omega)$, plays an important role. Such an expression makes sense even for measures μ that are not bounded but merely locally bounded in Ω, and such that $\int_\Omega \rho_0 \, d|\mu| < \infty$, where $\rho_0(x) = d(x, \partial\Omega)$. Many of our results remain valid for such measures provided some of the statements (and the proofs) are slightly modified. In this case, the condition $g(u) \in L^1(\Omega)$ in (4.0.2) (and also in (4.0.6)) must be replaced by $g(u)\rho_0 \in L^1(\Omega)$. Since we have not pursued this direction, we shall leave the details to the reader.

In Section 4.1 we will introduce the first approximation method, namely μ is fixed and g is "truncated." In the sequel we denote by (g_n) a sequence of functions $g_n :$ $\mathbb{R} \to \mathbb{R}$ which are continuous, nondecreasing, and satisfy the following conditions:

$$0 \le g_1(t) \le g_2(t) \le \cdots \le g(t) \quad \forall t \in \mathbb{R}, \tag{4.0.7}$$
$$g_n(t) \to g(t) \quad \forall t \in \mathbb{R}. \tag{4.0.8}$$

(Recall that, by Dini's lemma, conditions (4.0.7) and (4.0.8) imply that $g_n \to g$ uniformly on compact subsets of $\mathbb{R}$).

If $N \ge 2$, we assume in addition that each g_n has subcritical growth, i.e., that there exist $C > 0$ and $p < \frac{N}{N-2}$ (possibly depending on n) such that

$$g_n(t) \le C(|t|^p + 1) \quad \forall t \in \mathbb{R}. \tag{4.0.9}$$

A good example to keep in mind is $g_n(t) = \min\{g(t), n\}, \forall t \in \mathbb{R}$.

Our first result is

PROPOSITION 4.1 *Given any measure $\mu \in \mathcal{M}(\Omega)$, let u_n be the unique solution of*

$$\begin{cases} -\Delta u_n + g_n(u_n) = \mu & in \ \Omega, \\ u_n = 0 & on \ \partial\Omega. \end{cases} \tag{4.0.10}$$

Then $u_n \downarrow u^*$ in Ω as $n \uparrow \infty$, where u^* is the largest subsolution of (4.0.1). Moreover, we have

$$\left| \int_\Omega u^* \Delta\zeta \right| \le 2\|\mu\|_{\mathcal{M}} \|\zeta\|_{L^\infty} \quad \forall \zeta \in C_0^2(\bar\Omega) \tag{4.0.11}$$

and

$$\int_\Omega g(u^*) \le \|\mu\|_{\mathcal{M}}. \tag{4.0.12}$$

An important consequence of Proposition 4.1 is that u^* *does not depend on the choice of the truncating sequence* (g_n). *It is an intrinsic object that will play an important role in the sequel.* In some sense, u^* is the "best one can do" (!) in the absence of a solution.

Remark 4.3. If μ is a good measure, then u^* coincides with the unique solution u of (4.0.1); this is an easy consequence of standard comparison arguments (see Corollary 4.B.2 in Appendix 4B).

We now introduce the basic concept of *reduced measure*. From (4.0.11), (4.0.12), and the density of $C_0^2(\bar\Omega)$ in $C_0(\bar\Omega)$ (easy to check), we see that there exists a unique measure $\mu^* \in \mathcal{M}(\Omega)$ such that

$$-\int_\Omega u^* \Delta\zeta + \int_\Omega g(u^*)\zeta = \int_\Omega \zeta \, d\mu^* \quad \forall \zeta \in C_0^2(\bar\Omega). \quad (4.0.13)$$

We call μ^* the reduced measure associated to μ. Clearly, μ^* is always a good measure. Since u^* is a subsolution of (4.0.1), we have

$$\mu^* \le \mu. \quad (4.0.14)$$

Even though we have not indicated the dependence on g we emphasize that μ^* *does depend* on g (see Section 4.8 below).

One of our main results is

THEOREM 4.1 *The reduced measure μ^* is the largest good measure $\le \mu$.*

Here is an easy consequence:

COROLLARY 4.1 *We have*

$$0 \le \mu - \mu^* \le \mu^+ = \sup\{\mu, 0\}. \quad (4.0.15)$$

In particular,

$$|\mu^*| \le |\mu| \quad (4.0.16)$$

and

$$[\mu \ge 0] \Longrightarrow [\mu^* \ge 0]. \quad (4.0.17)$$

Indeed, every measure $\nu \le 0$ is a good measure since the solution v of

$$\begin{cases} -\Delta v = \nu & \text{in } \Omega, \\ v = 0 & \text{on } \partial\Omega, \end{cases}$$

satisfies $v \le 0$ in Ω, and therefore by (4.0.5)

$$-\Delta v + g(v) = \nu \quad \text{in } (C_0^2)^*.$$

In particular, $-\mu^-$ is a good measure (recall that $\mu^- = \sup\{-\mu, 0\}$). Since $-\mu^- \le \mu$, we deduce from Theorem 4.1 that

$$-\mu^- \le \mu^*,$$

and consequently

$$\mu - \mu^* \le \mu + \mu^- = \mu^+.$$

Our next result asserts that the measure $\mu - \mu^*$ is concentrated on a small set:

THEOREM 4.2 *There exists a Borel set $\Sigma \subset \Omega$ with cap $(\Sigma) = 0$ such that*

$$(\mu - \mu^*)(\Omega \setminus \Sigma) = 0. \quad (4.0.18)$$

Here and throughout the rest of the paper "cap" denotes the Newtonian (H^1) capacity with respect to Ω.

Remark 4.4. Theorem 4.2 is optimal in the following sense. Given any measure $\mu \geq 0$ concentrated on a set of zero capacity, there exists some g such that $\mu^* = 0$ (see Theorem 4.14 below). In particular, $\mu - \mu^*$ can be *any* nonnegative measure concentrated on a set of zero capacity.

Here is a useful

DEFINITION. *A measure* $\mu \in \mathcal{M}(\Omega)$ *is called diffuse if* $|\mu|(A) = 0$ *for every Borel set* $A \subset \Omega$ *such that* $\operatorname{cap}(A) = 0$.

An immediate consequence of Corollary 4.1 and Theorem 4.2 is

COROLLARY 4.2 *Every diffuse measure* $\mu \in \mathcal{M}(\Omega)$ *is a good measure.*

Indeed, let Σ be as in Theorem 4.2, so that $\operatorname{cap}(\Sigma) = 0$ and

$$(\mu - \mu^*)(\Omega \setminus \Sigma) = 0.$$

On the other hand, (4.0.15) implies

$$(\mu - \mu^*)(\Sigma) \leq \mu^+(\Sigma) = 0,$$

since μ is diffuse. Therefore,

$$(\mu - \mu^*)(\Omega) = 0,$$

so that $\mu = \mu^*$ and thus μ is a good measure.

Remark 4.5. The converse of Corollary 4.2 is *not* true. In Example 4.5 (see Section 4.8 below) the measure $\mu = c\delta_a$, with $0 < c \leq 4\pi$ and $a \in \Omega$, is a good measure, but it is not diffuse—$\operatorname{cap}(\{a\}) = 0$, while $\mu(\{a\}) = c > 0$. See, however, Theorem 4.5.

Remark 4.6. Recall that a measure μ is diffuse if and only if $\mu \in L^1 + H^{-1}$; more precisely, there exist $f_0 \in L^1(\Omega)$ and $v_0 \in H_0^1(\Omega)$ such that

$$\int_\Omega \zeta \, d\mu = \int_\Omega f_0 \zeta - \int_\Omega \nabla v_0 \cdot \nabla \zeta \quad \forall \zeta \in C_0(\bar{\Omega}) \cap H_0^1. \tag{4.0.19}$$

The implication $[\mu \in L^1 + H^{-1}] \Rightarrow [\mu \text{ diffuse}]$ is due to Grun-Rehomme [GRe]. (In fact, he proved only that $[v \in H^{-1}] \Rightarrow [v \text{ diffuse}]$, but L^1-functions are diffuse measures—since $[\operatorname{cap}(A) = 0] \Rightarrow [|A| = 0]$—and the sum of two diffuse measures is diffuse). The converse $[\mu \text{ diffuse}] \Rightarrow [\mu \in L^1 + H^{-1}]$ is due to Boccardo-Gallouët-Orsina [BGO1] (and was suggested by earlier results of Baras-Pierre [BP] and Gallouët-Morel [GM]). As a consequence of Corollary 4.2 we obtain that, for every measure μ of the form (4.0.19), the problem

$$\begin{cases} -\Delta u + g(u) = \mu & \text{in } \Omega, \\ \qquad\qquad u = 0 & \text{on } \partial\Omega, \end{cases} \tag{4.0.20}$$

admits a unique solution. In fact, the same conclusion was already known for any *distribution* in $L^1 + H^{-1}$, not necessarily in $\mathcal{M}(\Omega)$. (The proof, which combines techniques from Brezis-Browder [BBr] and Brezis-Strauss [BS], is sketched in Appendix 4B below; see Theorem 4.B.4). A very useful sharper version of the [BGO1] decomposition is the following:

THEOREM 4.3 *Assume $\mu \in \mathcal{M}(\Omega)$ is a diffuse measure. Then, there exist $f \in L^1(\Omega)$ and $v \in C_0(\bar{\Omega}) \cap H_0^1$ such that*

$$\int_\Omega \zeta \, d\mu = \int_\Omega f\zeta - \int_\Omega \nabla v \cdot \nabla \zeta \quad \forall \zeta \in C_0(\bar{\Omega}) \cap H_0^1. \qquad (4.0.21)$$

In addition, given any $\delta > 0$, then f and v can be chosen so that

$$\|f\|_{L^1} \le \|\mu\|_{\mathcal{M}}, \quad \|v\|_{L^\infty} \le \delta\|\mu\|_{\mathcal{M}} \quad and \quad \|v\|_{H^1} \le \delta^{1/2}\|\mu\|_{\mathcal{M}}. \qquad (4.0.22)$$

The proof of Theorem 4.3 is presented in Appendix 4D below.

In Section 4.2 we present some basic properties of the good measures. Here is a first one:

THEOREM 4.4 *Suppose μ_1 is a good measure. Then any measure $\mu_2 \le \mu_1$ is also a good measure.*

We now deduce a number of consequences:

COROLLARY 4.3 *Let $\mu \in \mathcal{M}(\Omega)$. If μ^+ is diffuse, then μ is a good measure.*

In fact, by Corollary 4.2, μ^+ diffuse implies that μ^+ is a good measure. Since $\mu \le \mu^+$, it follows from Theorem 4.4 that μ is a good measure.

COROLLARY 4.4 *If μ_1 and μ_2 are good measures, then so is $v = \sup\{\mu_1, \mu_2\}$.*

Indeed, by Theorem 4.1 we have $\mu_1 \le v^*$ and $\mu_2 \le v^*$. Thus $v \le v^* \le v$, and hence $v = v^*$ is good measure.

COROLLARY 4.5 *The set $\mathcal{G}$ of good measures is convex.*

Indeed, let $\mu_1, \mu_2 \in \mathcal{G}$. For any $t \in [0, 1]$, we have

$$t\mu_1 + (1 - t)\mu_2 \le \sup\{\mu_1, \mu_2\}.$$

Applying Corollary 4.4 and Theorem 4.4, we deduce that $t\mu_1 + (1 - t)\mu_2 \in \mathcal{G}$.

COROLLARY 4.6 *For every measure $\mu \in \mathcal{M}(\Omega)$ we have*

$$\|\mu - \mu^*\|_{\mathcal{M}} = \min_{v \in \mathcal{G}} \|\mu - v\|_{\mathcal{M}}. \qquad (4.0.23)$$

Moreover, μ^* is the unique good measure that achieves the minimum.

Proof. Let $v \in \mathcal{G}$ and write

$$|\mu - v| = (\mu - v)^+ + (\mu - v)^- \ge (\mu - v)^+ = \mu - \inf\{\mu, v\}.$$

But $\tilde{v} = \inf\{\mu, v\} \in \mathcal{G}$ by Theorem 4.4. Applying Theorem 4.1 we find $\tilde{v} \leq \mu^*$. Hence

$$|\mu - v| \geq \mu - \tilde{v} \geq \mu - \mu^* \geq 0,$$

and therefore

$$\|\mu - v\|_{\mathcal{M}} \geq \|\mu - \mu^*\|_{\mathcal{M}},$$

which gives (4.0.23). In order to establish uniqueness, assume $v \in \mathcal{G}$ attains the minimum in (4.0.23). Note that $\inf\{\mu, v\}$ is a good measure $\leq \mu$ and

$$\|\mu - \inf\{\mu, v\}\|_{\mathcal{M}} \leq \|\mu - v\|_{\mathcal{M}}.$$

Thus, $v = \inf\{\mu, v\} \leq \mu$. By Theorem 4.1, we deduce that $v \leq \mu^* \leq \mu$. Since v achieves the minimum in (4.0.23), we must have $v = \mu^*$.

As we have already pointed out, the set $\mathcal{G}$ of good measures associated to (4.0.1) depends on the nonlinearity g. Sometimes, in order to emphasize this dependence, we shall denote $\mathcal{G}$ by $\mathcal{G}(g)$. By Corollary 4.3, if $\mu \in \mathcal{M}(\Omega)$ and μ^+ is diffuse, then $\mu \in \mathcal{G}(g)$ for every g satisfying (4.0.5). The converse is also true. More precisely,

THEOREM 4.5 *Let $\mu \in \mathcal{M}(\Omega)$. Then $\mu \in \mathcal{G}(g)$ for every g if and only if μ^+ is diffuse.*

We also have a characterization of good measures in the spirit of the Baras-Pierre result (4.0.3):

THEOREM 4.6 *A measure $\mu \in \mathcal{M}(\Omega)$ is a good measure if and only if μ admits a decomposition*

$$\mu = f_0 - \Delta v_0 \quad \text{in } \mathcal{D}'(\Omega),$$

with $f_0 \in L^1(\Omega)$, $v_0 \in L^1(\Omega)$ and $g(v_0) \in L^1(\Omega)$.

COROLLARY 4.7 *We have*

$$\mathcal{G} + L^1(\Omega) \subset \mathcal{G}.$$

In Section 4.3 we discuss some properties of the mapping $\mu \mapsto \mu^*$. For example, we show that for every $\mu, v \in \mathcal{M}(\Omega)$, we have

$$(\mu^* - v^*)^+ \leq (\mu - v)^+. \tag{4.0.24}$$

Inequality (4.0.24) implies, in particular, that

$$[\mu \leq v] \Longrightarrow [\mu^* \leq v^*] \tag{4.0.25}$$

and

$$|\mu^* - v^*| \leq |\mu - v|. \tag{4.0.26}$$

In Section 4.4 we examine another approximation scheme. We now keep g fixed but we smooth μ via convolution. Let $\mu_n = \rho_n * \mu$ and let u_n be the solution of

$$\begin{cases} -\Delta u_n + g(u_n) = \mu_n & \text{in } \Omega, \\ \qquad\qquad u_n = 0 & \text{on } \partial\Omega. \end{cases} \qquad (4.0.27)$$

We prove (assuming in addition g is convex) that $u_n \to u^*$ in $L^1(\Omega)$, where u^* is given by Proposition 4.1. In Section 4.5 we discuss other convergence results.

Theorem 4.5 is established in Section 4.6. In Section 4.7 we extend Proposition 4.1 to deal with the case where $\mu \in \mathcal{M}(\Omega)$ is a signed measure, but assumption (4.0.5) is no longer satisfied. Finally, in Section 4.8 we present several examples where the measure μ^* can be explicitly identified, and in Section 4.9 we propose various directions of research.

Part of the results in this paper were announced in [BMP].

4.1 CONSTRUCTION OF u^* AND μ^*. PROOFS OF PROPOSITION 4.1 AND THEOREMS 4.1, 4.2

We start with the

Proof of Proposition 4.1. Using Corollary 4.B.2 in Appendix 4B we see that the sequence (u_n) is nonincreasing. Also (see Corollary 4.B.1)

$$\|g_n(u_n)\|_{L^1} \le \|\mu\|_{\mathcal{M}}$$

and thus

$$\|\Delta u_n\|_{\mathcal{M}} \le 2\|\mu\|_{\mathcal{M}}.$$

Consequently,

$$\|u_n\|_{L^1} \le C\|\mu\|_{\mathcal{M}}.$$

Therefore, (u_n) tends in L^1 to a limit denoted u^*. By Dini's lemma, $g_n \uparrow g$ uniformly on compact sets; thus,

$$g_n(u_n) \to g(u^*) \quad \text{a.e.}$$

Hence, $g(u^*) \in L^1(\Omega)$, (4.0.11)–(4.0.12) hold and, by Fatou's lemma,

$$-\int_\Omega u^* \Delta\zeta + \int_\Omega g(u^*)\zeta \le \int_\Omega \zeta\, d\mu \quad \forall \zeta \in C_0^2(\bar\Omega),\, \zeta \ge 0 \text{ in } \Omega.$$

Therefore, u^* is a subsolution of (4.0.1). We claim that u^* is the *largest* subsolution. Indeed, let v be any subsolution of (4.0.1). Then

$$-\Delta v + g_n(v) \le -\Delta v + g(v) \le \mu \quad \text{in } (C_0^2)^*.$$

By comparison (see Corollary 4.B.2),

$$v \le u_n \quad \text{a.e.}$$

and, as $n \to \infty$,

$$v \le u^* \quad \text{a.e.}$$

Hence, u^* is the largest subsolution.

Recall (see [FTS], or Appendix 4A below) that any measure μ on Ω can be uniquely decomposed as a sum of two measures, $\mu = \mu_d + \mu_c$ ("d" stands for diffuse and "c" for concentrated), satisfying $|\mu_d|(A) = 0$ for every Borel set $A \subset \Omega$ such that cap $(A) = 0$, and $|\mu_c|(\Omega \setminus F) = 0$ for some Borel set $F \subset \Omega$ such that cap $(F) = 0$. Note that a measure μ is diffuse if and only if $\mu_c = 0$, i.e., $\mu = \mu_d$.

A key ingredient in the proof of Theorems 4.1 and 4.2 is the following version of Kato's inequality (see [K]) due to Brezis-Ponce [BP2].

THEOREM 4.7 (Kato's inequality when Δv is a measure) *Let $v \in L^1(\Omega)$ be such that Δv is a measure on Ω. Then, for every open set $\omega \subset\subset \Omega$, Δv^+ is a measure on ω and the following holds:*

$$(\Delta v^+)_d \ge \chi_{[v \ge 0]}(\Delta v)_d \quad \text{in } \omega, \tag{4.1.1}$$

$$(-\Delta v^+)_c = (-\Delta v)_c^+ \quad \text{in } \omega. \tag{4.1.2}$$

Note that the right-hand side of (4.1.1) is well defined because the function v is quasi-continuous. More precisely, if $v \in L^1(\Omega)$ and Δv is a measure, then there exists $\tilde{v} : \Omega \to \mathbb{R}$ quasi-continuous such that $v = \tilde{v}$ a.e. in Ω (see [A1] and also [BP1, Lemma 4.1]). Recall that $\tilde{v}$ is quasi-continuous if and only if, given any $\varepsilon > 0$, one can find an open set $\omega_\varepsilon \subset \Omega$ such that cap $(\omega_\varepsilon) < \varepsilon$ and $\tilde{v}|_{\Omega \setminus \omega_\varepsilon}$ is continuous. In particular, $\tilde{v}$ is finite q.e. (= quasi-everywhere = outside a set of zero capacity). It is easy to see that $\chi_{[\tilde{v} \ge 0]}$ is integrable with respect to the measure $|(\Delta v)_d|$. When $v \in L^1$ and Δv is a measure, we will systematically replace v by its quasi-continuous representative.

Here are two consequences of Theorem 4.7 that will be used in the sequel. The first one was originally established by Dupaigne-Ponce [DP] and it is equivalent to (4.1.2):

COROLLARY 4.8 ("inverse" maximum principle) *Let $v \in L^1(\Omega)$ be such that Δv is a measure. If $v \ge 0$ a.e. in Ω, then*

$$(-\Delta v)_c \ge 0 \quad \text{in } \Omega.$$

Another corollary is the following

COROLLARY 4.9 *Let $u \in L^1(\Omega)$ be such that Δu is a measure. Then,*

$$\Delta T_k(u) \le \chi_{[u \le k]}(\Delta u)_d + (\Delta u)_c^+ \quad \text{in } \mathcal{D}'(\Omega).$$

Here, $T_k(s) = k - (k - s)^+$ for every $s \in \mathbb{R}$.

Proof. Let $\omega \subset\subset \Omega$. Applying (4.1.1) and (4.1.2) to $v = k - u$, yields

$$(\Delta T_k(u))_d = -(\Delta v^+)_d \le -\chi_{[v \ge 0]}(\Delta v)_d = \chi_{[u \le k]}(\Delta u)_d \quad \text{in } \omega$$

and

$$(\Delta T_k(u))_{\mathrm{c}} = (\Delta u)_{\mathrm{c}}^+ \quad \text{in } \omega.$$

Combining these two facts, we conclude that

$$\Delta T_k(u) \le \chi_{[u \le k]}(\Delta u)_{\mathrm{d}} + (\Delta u)_{\mathrm{c}}^+ \quad \text{in } \mathcal{D}'(\omega).$$

Since $\omega \subset\subset \Omega$ was arbitrary, the result follows.

Let u^* be the largest subsolution of (4.0.1), and define $\mu^* \in \mathcal{M}(\Omega)$ by (4.0.13). We have the following

LEMMA 4.1 *The reduced measure μ^* satisfies*

$$\mu^* \ge \mu_{\mathrm{d}} - \mu_{\mathrm{c}}^-.$$

Proof. Let (u_n) be the sequence constructed in Proposition 4.1. By Corollary 4.9, we have

$$\Delta T_k(u_n) \le \chi_{[u_n \le k]}(\Delta u_n)_{\mathrm{d}} + (\Delta u_n)_{\mathrm{c}}^+ \quad \text{in } \mathcal{D}'(\Omega). \tag{4.1.3}$$

Since u_n satisfies (4.0.10),

$$(\Delta u_n)_{\mathrm{d}} = g_n(u_n) - \mu_{\mathrm{d}} \quad \text{and} \quad (\Delta u_n)_{\mathrm{c}} = -\mu_{\mathrm{c}}.$$

Inserting into (4.1.3) gives

$$-\Delta T_k(u_n) \ge \chi_{[u_n \le k]}\{\mu_{\mathrm{d}} - g_n(u_n)\} - \mu_{\mathrm{c}}^-$$
$$\ge \chi_{[u_n \le k]}\mu_{\mathrm{d}} - g_n(T_k(u_n)) - \mu_{\mathrm{c}}^- \quad \text{in } \mathcal{D}'(\Omega).$$

For every $n \ge 1$ we have $u^* \le u_n \le u_1$, so that

$$[u^* \le k] \supset [u_n \le k] \supset [u_1 \le k]$$

and

$$\chi_{[u_n \le k]}\mu_{\mathrm{d}} \ge \chi_{[u_1 \le k]}\mu_{\mathrm{d}}^+ - \chi_{[u^* \le k]}\mu_{\mathrm{d}}^-.$$

Thus,

$$-\Delta T_k(u_n) + g_n(T_k(u_n)) \ge \chi_{[u_1 \le k]}\mu_{\mathrm{d}}^+ - \chi_{[u^* \le k]}\mu_{\mathrm{d}}^- - \mu_{\mathrm{c}}^- \quad \text{in } \mathcal{D}'(\Omega). \tag{4.1.4}$$

By dominated convergence,

$$g_n(T_k(u_n)) \to g(T_k(u^*)) \quad \text{in } L^1(\Omega), \quad \text{as } n \to \infty.$$

As $n \to \infty$ in (4.1.4), we get

$$-\Delta T_k(u^*) + g(T_k(u^*)) \ge \chi_{[u_1 \le k]}\mu_{\mathrm{d}}^+ - \chi_{[u^* \le k]}\mu_{\mathrm{d}}^- - \mu_{\mathrm{c}}^- \quad \text{in } \mathcal{D}'(\Omega).$$

Let $k \to \infty$. Since both sets $[u_1 = +\infty]$ and $[u^* = +\infty]$ have zero capacity (recall that u_1 and u^* are quasi-continuous and, in particular, both functions are finite q.e.), we conclude that

$$\mu^* = -\Delta u^* + g(u^*) \ge \mu_{\mathrm{d}}^+ - \mu_{\mathrm{d}}^- - \mu_{\mathrm{c}}^- = \mu_{\mathrm{d}} - \mu_{\mathrm{c}}^-.$$

This establishes the lemma.

Proof of Theorems 4.1 and 4.2. It follows from (4.0.14) and Lemma 4.1 that

$$\mu_d - \mu_c^- \le \mu^* \le \mu.$$

By taking the diffuse parts, we have

$$(\mu^*)_d = \mu_d. \tag{4.1.5}$$

Thus $\mu - \mu^* = (\mu - \mu^*)_c$, which proves Theorem 4.2.

We now turn to the proof of Theorem 4.1. Let λ be a good measure $\le \mu$. We must prove that $\lambda \le \mu^*$. Denote by v the solution of (4.0.1) corresponding to λ,

$$\begin{cases} -\Delta v + g(v) = \lambda & \text{in } \Omega, \\ v = 0 & \text{on } \partial\Omega. \end{cases}$$

By (4.1.5),

$$\lambda_d \le \mu_d = (\mu^*)_d.$$

Since u^* is the largest subsolution of (4.0.1), we also have

$$v \le u^* \quad \text{a.e.}$$

By the "inverse" maximum principle,

$$\lambda_c = (-\Delta v)_c \le (-\Delta u^*)_c = (\mu^*)_c.$$

Therefore $\lambda \le \mu^*$. This establishes Theorem 4.1.

The following lemma will be used later on:

LEMMA 4.2 *Given a measure $\mu \in \mathcal{M}(\Omega)$, let (u_n) be the sequence defined in Proposition 4.1. Then,*

$$g_n(u_n) \overset{*}{\rightharpoonup} g(u^*) + (\mu - \mu^*) = g(u^*) + (\mu - \mu^*)_c \quad \text{weak* in } \mathcal{M}(\Omega).$$

Proof. Let $\zeta \in C_0^2(\bar{\Omega})$. For every $n \ge 1$, we have

$$\int_\Omega g_n(u_n)\zeta = \int_\Omega u_n \Delta\zeta + \int_\Omega \zeta \, d\mu.$$

By Proposition 4.1, $u_n \to u^*$ in $L^1(\Omega)$. Thus,

$$\lim_{n\to\infty} \int_\Omega g_n(u_n)\zeta = \int_\Omega u^* \Delta\zeta + \int_\Omega \zeta \, d\mu = \int_\Omega g(u^*)\zeta + \int_\Omega \zeta \, d(\mu - \mu^*).$$

In other words,

$$g_n(u_n) \overset{*}{\rightharpoonup} g(u^*) + (\mu - \mu^*) \quad \text{weak* in } \mathcal{M}(\Omega).$$

Since $(\mu^*)_d = \mu_d$, the result follows.

4.2 GOOD MEASURES: PROOFS OF THEOREMS 4.4, 4.6

We start with

LEMMA 4.3 *If μ is a good measure with solution u, and u_n is given by (4.0.10), then*

$$u_n \to u \quad \text{in } W_0^{1,1}(\Omega) \quad \text{and} \quad g_n(u_n) \to g(u) \quad \text{in } L^1(\Omega).$$

Proof. We have

$$-\Delta u_n + g_n(u_n) = \mu \quad \text{and} \quad -\Delta u + g(u) = \mu \quad \text{in } (C_0^2)^*,$$

so that

$$-\Delta(u_n - u) + g_n(u_n) - g(u) = 0 \quad \text{in } (C_0^2)^*.$$

Thus

$$-\Delta(u_n - u) + g_n(u_n) - g_n(u) = g(u) - g_n(u) \quad \text{in } (C_0^2)^*.$$

Hence, by standard estimates (see Proposition 4.B.3),

$$\int_\Omega |g_n(u_n) - g_n(u)| \le \int_\Omega |g(u) - g_n(u)| \to 0.$$

Thus,

$$\int_\Omega |g_n(u_n) - g(u)| \le 2 \int_\Omega |g(u) - g_n(u)| \to 0.$$

In other words, $g_n(u_n) \to g(u)$ in $L^1(\Omega)$. This clearly implies that $\Delta(u_n - u) \to 0$ in $L^1(\Omega)$ and thus $u_n \to u$ in $W_0^{1,1}(\Omega)$.

We now turn to the

Proof of Theorem 4.4. Let $u_{1,n}, u_{2,n} \in L^1(\Omega)$ be such that

$$\begin{cases} -\Delta u_{i,n} + g_n(u_{i,n}) = \mu_i & \text{in } \Omega, \\ u_{i,n} = 0 & \text{on } \partial\Omega, \end{cases}$$

for $i = 1, 2$. Since $\mu_2 \le \mu_1$, we have

$$u_{2,n} \le u_{1,n} \quad \text{a.e.}$$

Thus $g_n(u_{2,n}) \le g_n(u_{1,n}) \to g(u_1^*)$ strongly in L^1 by Lemma 4.3. Hence $g_n(u_{2,n}) \to g(u_2^*)$ strongly in L^1 and we have

$$-\Delta u_2^* + g(u_2^*) = \mu_2 \quad \text{in } (C_0^2)^*,$$

i.e., μ_2 is a good measure.

A simple property of $\mathcal{G}$ is

PROPOSITION 4.2 *The set $\mathcal{G}$ of good measures is closed with respect to strong convergence in $\mathcal{M}(\Omega)$.*

Proof. Let (μ_k) be a sequence of good measures such that $\mu_k \to \mu$ strongly in $\mathcal{M}(\Omega)$. For each $k \geq 1$, let u_k be such that

$$\begin{cases} -\Delta u_k + g(u_k) = \mu_k & \text{in } \Omega, \\ \qquad\qquad u_k = 0 & \text{on } \partial\Omega. \end{cases}$$

By standard estimates (see Corollary 4.B.1),

$$\int_\Omega |g(u_{k_1}) - g(u_{k_2})| \leq \|\mu_{k_1} - \mu_{k_2}\|_{\mathcal{M}} \tag{4.2.1}$$

and

$$\int_\Omega |u_{k_1} - u_{k_2}| \leq C\|\Delta(u_{k_1} - u_{k_2})\|_{\mathcal{M}} \leq 2C\|\mu_{k_1} - \mu_{k_2}\|_{\mathcal{M}}. \tag{4.2.2}$$

By (4.2.1) and (4.2.2), both (u_k) and $(g(u_k))$ are Cauchy sequences in $L^1(\Omega)$. Thus, there exist $u, v \in L^1(\Omega)$ such that

$$u_k \to u \quad \text{and} \quad g(u_k) \to v \quad \text{in } L^1(\Omega).$$

In particular, $v = g(u)$ a.e. It is then easy to see that

$$-\Delta u + g(u) = \mu \quad \text{in } (C_0^2)^*.$$

Thus μ is a good measure.

We next present a result slightly sharper than Theorem 4.6:

THEOREM 4.6' *Let $\mu \in \mathcal{M}(\Omega)$. The following conditions are equivalent:*

(a) *μ is a good measure;*
(b) *μ^+ is a good measure;*
(c) *μ_c is a good measure;*
(d) *$\mu = f_0 - \Delta v_0$ in $\mathcal{D}'(\Omega)$, for some $f_0 \in L^1$ and some $v_0 \in L^1$ with $g(v_0) \in L^1$.*

Proof. (a) $\Rightarrow$ (b). Since μ and 0 are good measures, it follows from Corollary 4.4 that $\mu^+ = \sup\{\mu, 0\}$ is a good measure.

(b) $\Rightarrow$ (a). Since μ^+ is a good measure and $\mu \leq \mu^+$ in Ω, it follows from Theorem 4.4 that μ is a good measure.

(b) $\Rightarrow$ (c). Note that we always have

$$\mu_c \leq \mu^+. \tag{4.2.3}$$

Indeed, $(\mu^+ - \mu_c)_d = (\mu^+)_d \geq 0$ and $(\mu^+ - \mu_c)_c = \mu_c^+ - \mu_c \geq 0$.
[Here and in the sequel we use the fact that $(\mu^+)_d = (\mu_d)^+$ and $(\mu^+)_c = (\mu_c)^+$ which will be simply denoted μ_d^+ and μ_c^+].

Since μ^+ is a good measure, it follows from (4.2.3) and Theorem 4.4 that μ_c is also a good measure.

(c) $\Rightarrow$ (b). It is easy to see that, for every measure λ,

$$\lambda^+ = \sup\{\lambda_d, \lambda_c\}. \tag{4.2.4}$$

Assume μ_c is a good measure. Since μ_d is diffuse, Corollary 4.2 implies that μ_d is also a good measure. By Corollary 4.4 and (4.2.4), $\mu^+ = \sup\{\mu_d, \mu_c\}$ is a good measure as well.

(a) $\Rightarrow$ (d). Trivial.

(d) $\Rightarrow$ (c). We split the argument into two steps.

Step 1. Proof of (d) $\Rightarrow$ (c) if v_0 has compact support.

Since $\mu = f_0 - \Delta v_0$ in $\mathcal{D}'(\Omega)$ and v_0 has compact support, we have

$$\mu = f_0 - \Delta v_0 \quad \text{in } (C_0^2)^*.$$

Thus, $\mu - f_0 + g(v_0)$ is a good measure. Using the equivalence (a) $\Leftrightarrow$ (c), we conclude that $\mu_c = [\mu - f_0 + g(v_0)]_c$ is a good measure.

Step 2. Proof of (d) $\Rightarrow$ (c) completed.

By assumption,

$$\mu = f_0 - \Delta v_0 \quad \text{in } \mathcal{D}'(\Omega).$$

In particular, we have $\Delta v_0 \in \mathcal{M}(\Omega)$, so that $v_0 \in W_{loc}^{1,p}(\Omega)$, $\forall p < \frac{N}{N-1}$ (see Theorem 4.B.1 below). Let $(\varphi_n) \subset C_c^\infty(\Omega)$ be such that $0 \le \varphi_n \le 1$ in Ω and $\varphi_n(x) = 1$ if $d(x, \partial\Omega) > \frac{1}{n}$. Then

$$\varphi_n \mu = f_n - \Delta(\varphi_n v_0) \quad \text{in } \mathcal{D}'(\Omega),$$

where

$$f_n = \varphi_n f_0 + 2\nabla v_0 \cdot \nabla \varphi_n + v_0 \Delta \varphi_n \in L^1(\Omega).$$

Moreover, since $0 \le g(\varphi_n v_0) \le g(v_0)$ a.e., we have $g(\varphi_n v_0) \in L^1(\Omega)$. Thus, by Step 1,

$$\varphi_n \mu_c = (\varphi_n \mu)_c \in \mathcal{G} \quad \forall n \ge 1.$$

Since $\varphi_n \mu_c \to \mu_c$ strongly in $\mathcal{M}(\Omega)$ and $\mathcal{G}$ is closed with respect to the strong topology in $\mathcal{M}(\Omega)$, we conclude that $\mu \in \mathcal{G}$.

We may now strengthen Corollary 4.7:

Corollary 4.7' *We have*

$$\mathcal{G} + \mathcal{M}_d(\Omega) \subset \mathcal{G},$$

where $\mathcal{M}_d(\Omega)$ denotes the space of diffuse measures.

Proof. Let $\mu \in \mathcal{G}$. By Theorem 4.6', μ_c is a good measure. Thus, for any $\nu \in \mathcal{M}_d$, $(\mu + \nu)_c = \mu_c$ is a good measure. It follows from the equivalence (a) $\Leftrightarrow$ (c) in the theorem above that $\mu + \nu \in \mathcal{G}$.

PROPOSITION 4.3 *Assume*

$$g(2t) \le C(g(t) + 1) \quad \forall t \ge 0. \tag{4.2.5}$$

Then the set of good measures is a convex cone.

Remark 4.7. Assumption (4.2.5) is called in the literature the Δ_2-condition. It holds if $g(t) = t^p$ for $t \ge 0$ (any $p > 1$), but (4.2.5) *fails* for $g(t) = e^t - 1$. In this case, the set of good measures is *not* a cone. As we will see in Section 4.8,

Example 4.5, if $N = 2$, then for any $a \in \Omega$ we have $c\delta_a \in \mathcal{G}$ if $c > 0$ is small, but $c\delta_a \notin \mathcal{G}$ if c is large.

Proof of Proposition 4.3. Assume $\mu \in \mathcal{G}$. Clearly, it suffices to show that $2\mu \in \mathcal{G}$. Let u be the solution of

$$\begin{cases} -\Delta u + g(u) = \mu & \text{in } \Omega, \\ \qquad\qquad\quad u = 0 & \text{on } \partial\Omega. \end{cases}$$

Thus,

$$2\mu = -\Delta(2u) + 2g(u) \quad \text{in } \mathcal{D}'(\Omega).$$

By (4.2.5), $g(2u) \in L^1$. We can now invoke the equivalence (a) $\Leftrightarrow$ (d) in Theorem 4.6′ to conclude that $2\mu \in \mathcal{G}$.

4.3 SOME PROPERTIES OF THE MAPPING $\mu \mapsto \mu^*$

We start with an easy result, which asserts that the mapping $\mu \mapsto \mu^*$ is order preserving:

PROPOSITION 4.4 *Let* $\mu, \nu \in \mathcal{M}(\Omega)$. *If* $\mu \leq \nu$, *then* $\mu^* \leq \nu^*$.

Proof. Since the reduced measure μ^* is a good measure and $\mu^* \leq \mu \leq \nu$, it follows from Theorem 4.1 that $\mu^* \leq \nu^*$.

Next, we have

THEOREM 4.8 *If* $\mu_1, \mu_2 \in \mathcal{M}(\Omega)$ *are mutually singular, then*

$$(\mu_1 + \mu_2)^* = (\mu_1)^* + (\mu_2)^*. \tag{4.3.1}$$

Proof. Since μ_1 and μ_2 are mutually singular, $(\mu_1)^*$ and $(\mu_2)^*$ are also mutually singular (by (4.0.16)). In particular, we have

$$(\mu_1)^* + (\mu_2)^* \leq [(\mu_1)^* + (\mu_2)^*]^+ = \sup\{(\mu_1)^*, (\mu_2)^*\}. \tag{4.3.2}$$

By Corollary 4.4, the right-hand side of (4.3.2) is a good measure. It follows from Theorem 4.4 that $(\mu_1)^* + (\mu_2)^*$ is also a good measure. Since

$$(\mu_1)^* + (\mu_2)^* \leq \mu_1 + \mu_2,$$

we conclude from Theorem 4.1 that

$$(\mu_1)^* + (\mu_2)^* \leq (\mu_1 + \mu_2)^*. \tag{4.3.3}$$

We now establish the reverse inequality. Assume λ is a good measure $\leq (\mu_1 + \mu_2)$. By Radon-Nikodym, we may decompose λ in terms of three measures:

$$\lambda = \lambda_0 + \lambda_1 + \lambda_2,$$

where λ_0 is singular with respect to $|\mu_1| + |\mu_2|$, and, for $i = 1, 2$, λ_i is absolutely continuous with respect to $|\mu_i|$. Since $\lambda_0, \lambda_1, \lambda_2 \leq \lambda^+$, each λ_j, $j = 0, 1, 2$, is a good measure. Moreover, $\lambda \leq \mu_1 + \mu_2$ implies

$$\lambda_0 \leq 0, \quad \lambda_1 \leq \mu_1 \quad \text{and} \quad \lambda_2 \leq \mu_2.$$

Thus, in particular, $\lambda_i \leq (\mu_i)^*$ for $i = 1, 2$. Therefore,

$$\lambda = \lambda_0 + \lambda_1 + \lambda_2 \leq (\mu_1)^* + (\mu_2)^*.$$

Since λ was arbitrary, we have

$$(\mu_1 + \mu_2)^* \leq (\mu_1)^* + (\mu_2)^*. \tag{4.3.4}$$

Combining (4.3.3) and (4.3.4), the result follows.

Here are some consequences of Theorem 4.8:

COROLLARY 4.10 *For every $\mu \in \mathcal{M}(\Omega)$, we have*

$$(\mu^*)_d = (\mu_d)^* = \mu_d \quad \text{and} \quad (\mu^*)_c = (\mu_c)^*. \tag{4.3.5}$$

Also,

$$(\mu^*)^+ = (\mu^+)^* \quad \text{and} \quad (\mu^*)^- = \mu^-. \tag{4.3.6}$$

Proof. Since μ_d is a good measure (see Corollary 4.2), we have $(\mu_d)^* = \mu_d$. By Theorem 4.8,

$$\mu^* = (\mu_d + \mu_c)^* = (\mu_d)^* + (\mu_c)^*.$$

Comparison between the diffuse and concentrated parts gives (4.3.5). Similarly,

$$\mu^* = (\mu^+ - \mu^-)^* = (\mu^+)^* + (-\mu^-)^* = (\mu^+)^* - \mu^-,$$

since every nonpositive measure is good. This identity yields (4.3.6).

More generally, the same argument shows the following:

COROLLARY 4.11 *Let $\mu \in \mathcal{M}(\Omega)$. For every Borel set $E \subset \Omega$, we have*

$$(\mu \lfloor_E)^* = \mu^* \lfloor_E. \tag{4.3.7}$$

Here $\mu \lfloor_E$ denotes the measure defined by $\mu \lfloor_E (A) = \mu(A \cap E)$ for every Borel set $A \subset \Omega$.

For simplicity, from now on we shall write $\mu_d^* = (\mu^*)_d$ and $\mu_c^* = (\mu^*)_c$.

The following result extends Corollary 4.7':

COROLLARY 4.12 *For every $\mu \in \mathcal{M}(\Omega)$ and $\nu \in \mathcal{M}_d(\Omega)$,*

$$(\mu + \nu)^* = \mu^* + \nu.$$

Proof. By Theorem 4.8 and Corollary 4.2, we have

$$(\mu + \nu)^* = \mu_c^* + (\mu_d + \nu)^* = \mu_c^* + \mu_d + \nu = (\mu_c^* + \mu_d^*) + \nu = \mu^* + \nu.$$

Next, we have

THEOREM 4.9 *Given* $\mu, \nu \in \mathcal{M}(\Omega)$, *we have*

$$[\inf \{\mu, \nu\}]^* = \inf \{\mu^*, \nu^*\}, \tag{4.3.8}$$

$$[\sup \{\mu, \nu\}]^* = \sup \{\mu^*, \nu^*\}. \tag{4.3.9}$$

Proof.

Step 1. Proof of (4.3.8).

Clearly,

$$\inf \{\mu^*, \nu^*\} \leq [\inf \{\mu, \nu\}]^*.$$

Assume λ is a good measure $\leq \inf \{\mu, \nu\}$. By Theorem 4.1, $\lambda \leq \mu^*$ and $\lambda \leq \nu^*$. Thus, $\lambda \leq \inf \{\mu^*, \nu^*\}$, whence

$$[\inf \{\mu, \nu\}]^* \leq \inf \{\mu^*, \nu^*\}.$$

Step 2. Proof of (4.3.9).

Applying the Hahn decomposition to $\mu - \nu$, we may write Ω in terms of two disjoint Borel sets $E_1, E_2 \subset \Omega$, $\Omega = E_1 \cup E_2$, so that

$$\mu \geq \nu \text{ in } E_1 \quad \text{and} \quad \nu \geq \mu \text{ in } E_2.$$

By Proposition 4.4 and Corollary 4.11,

$$\mu^* \lfloor_{E_1} = (\mu \lfloor_{E_1})^* \geq (\nu \lfloor_{E_1})^* = \nu^* \lfloor_{E_1}.$$

Thus, $\mu^* \geq \nu^*$ on E_1. Similarly, $\nu^* \geq \mu^*$ on E_2. We then have

$$\sup \{\mu, \nu\} = \mu \lfloor_{E_1} + \nu \lfloor_{E_2} \quad \text{and} \quad \sup \{\mu^*, \nu^*\} = \mu^* \lfloor_{E_1} + \nu^* \lfloor_{E_2}. \tag{4.3.10}$$

On the other hand, by Theorem 4.8 and Corollary 4.11,

$$(\mu \lfloor_{E_1} + \nu \lfloor_{E_2})^* = (\mu \lfloor_{E_1})^* + (\nu \lfloor_{E_2})^* = \mu^* \lfloor_{E_1} + \nu^* \lfloor_{E_2}. \tag{4.3.11}$$

Combining (4.3.10) and (4.3.11), we obtain (4.3.9). $\quad\blacksquare$

We now show that $\mu \mapsto \mu^*$ is nonexpansive:

THEOREM 4.10 *Given* $\mu, \nu \in \mathcal{M}(\Omega)$, *we have*

$$|\mu^* - \nu^*| \leq |\mu - \nu|. \tag{4.3.12}$$

More generally,

$$(\mu^* - \nu^*)^+ \leq (\mu - \nu)^+. \tag{4.3.13}$$

Proof. Clearly, it suffices to show that (4.3.13) holds. We split the proof into two steps.

Step 1. Assume $\nu \leq \mu$. Then we claim that

$$\mu^* - \nu^* \leq \mu - \nu. \tag{4.3.14}$$

Indeed, let ν_n be the solution of (4.0.10) corresponding to the measure ν. Since $\nu \leq \mu$, we have

$$\nu_n \leq u_n \quad \text{a.e.,} \quad \forall n \geq 1.$$

Recall that g_n is nondecreasing; thus,

$$g_n(\nu_n) \leq g_n(u_n) \quad \text{a.e.}$$

Let $n \to \infty$. According to Lemma 4.2, we have

$$g(\nu^*) + (\nu - \nu^*)_c \leq g(u^*) + (\mu - \mu^*)_c.$$

Taking the concentrated part on both sides of this inequality yields

$$(\nu - \nu^*)_c \leq (\mu - \mu^*)_c.$$

Since $\nu_d = \nu_d^*$ and $\mu_d = \mu_d^*$ (by Corollary 4.2), we have

$$\nu - \nu^* \leq \mu - \mu^*,$$

which is (4.3.14).

Step 2. Proof of (4.3.1) completed.

Recall that

$$\sup\{\mu, \nu\} = \nu + (\mu - \nu)^+. \tag{4.3.15}$$

Applying the previous step to the measures ν and $\sup\{\mu, \nu\}$, we have

$$[\sup\{\mu, \nu\}]^* - \nu^* \leq \sup\{\mu, \nu\} - \nu = (\mu - \nu)^+. \tag{4.3.16}$$

By (4.3.9), (4.3.15), and (4.3.16),

$$(\mu - \nu)^+ \geq [\sup\{\mu, \nu\}]^* - \nu^* = \sup\{\mu^*, \nu^*\} - \nu^* = (\mu^* - \nu^*)^+.$$

Therefore, (4.3.13) holds.

4.4 APPROXIMATION OF μ BY $\rho_N * \mu$

Let (ρ_n) be a sequence of mollifiers in $\mathbb{R}^N$ such that supp $\rho_n \subset B_{1/n}$ for every $n \geq 1$. Given $\mu \in \mathcal{M}(\Omega)$, set

$$\mu_n = \rho_n * \mu,$$

that is,

$$\mu_n(x) = \int_\Omega \rho_n(x - y) \, d\mu(y) \quad \forall x \in \mathbb{R}^N. \tag{4.4.1}$$

[The integral in (4.4.1) is well defined in view of Proposition 4.C.1 in Appendix 4C below. Here, we identify μ with $\tilde{\mu} \in [C(\bar{\Omega})]^*$ defined there].
  Let u_n be the solution of

$$\begin{cases} -\Delta u_n + g(u_n) = \mu_n & \text{in } \Omega, \\ \qquad\qquad\quad u_n = 0 & \text{on } \partial\Omega. \end{cases} \tag{4.4.2}$$

THEOREM 4.11 *Assume in addition that g is convex. Then $u_n \to u^*$ in $L^1(\Omega)$, where u^* is given by Proposition 4.1.*

Proof.

Step 1. The conclusion holds if μ is a good measure.

In this case, there exists $u = u^*$ such that

$$\begin{cases} -\Delta u + g(u) = \mu & \text{in } \Omega, \\ \qquad\qquad u = 0 & \text{on } \partial\Omega. \end{cases} \tag{4.4.3}$$

Let $\omega \subset\subset \Omega$. For $n \geq 1$ sufficiently large, we have

$$-\Delta(\rho_n * u) + \rho_n * g(u) = \mu_n \quad \text{in } \omega.$$

Thus, using the convexity of g,

$$\Delta(\rho_n * u - u_n) = \rho_n * g(u) - g(u_n) \geq g(\rho_n * u) - g(u_n) \quad \text{in } \omega.$$

By the standard version of Kato's inequality (see [K]),

$$\Delta(\rho_n * u - u_n)^+ \geq \{g(\rho_n * u) - g(u_n)\}^+ \geq 0 \quad \text{in } \mathcal{D}'(\omega). \tag{4.4.4}$$

Since

$$\int_\Omega |\Delta u_n| \leq 2\|\mu_n\|_{\mathcal{M}} \leq C \quad \forall n \geq 1,$$

we can extract a subsequence (u_{n_k}) such that

$$u_{n_k} \to v \quad \text{in } L^1(\Omega),$$

for some $v \in W_0^{1,1}(\Omega)$. As $n_k \to \infty$ in (4.4.4), we have

$$-\Delta(u - v)^+ \leq 0 \quad \text{in } \mathcal{D}'(\omega).$$

Since $\omega \subset\subset \Omega$ was arbitrary,

$$-\Delta(u - v)^+ \leq 0 \quad \text{in } \mathcal{D}'(\Omega). \tag{4.4.5}$$

On the other hand,

$$(u - v)^+ \in W_0^{1,1}(\Omega). \tag{4.4.6}$$

From (4.4.5), (4.4.6) and the weak form of the maximum principle (see Proposition 4.B.1) we deduce that

$$(u - v)^+ \leq 0 \quad \text{a.e.}$$

Therefore,

$$v \geq u \quad \text{a.e.}$$

By Fatou's lemma, v is a subsolution of (4.0.1); comparison with (4.4.3) yields,

$$v \leq u \quad \text{a.e.}$$

We conclude that

$$v = u \quad \text{a.e.}$$

Since v is independent of the subsequence (u_{n_k}), we must have

$$u_n \to u = u^* \quad \text{in } L^1(\Omega).$$

Step 2.   Proof of Theorem 4.11 completed.

Without loss of generality, we may assume that

$$u_n \to v \quad \text{in } L^1(\Omega).$$

By Fatou, once more, v is a subsolution of (4.0.1). Proposition 4.1 yields

$$v \leq u^* \quad \text{a.e.}$$

Let u_n^* denote the solution of

$$\begin{cases} -\Delta u_n^* + g(u_n^*) = \rho_n * \mu^* & \text{in } \Omega, \\ u_n^* = 0 & \text{on } \partial\Omega. \end{cases}$$

By the previous step,

$$u_n^* \to u^* \quad \text{in } L^1(\Omega).$$

On the other hand, we know from the maximum principle that

$$u_n^* \leq u_n \quad \text{a.e.}$$

Thus, as $n \to \infty$,

$$u^* \leq v \quad \text{a.e.}$$

Since $v \leq u^*$ a.e., the result follows.

Open problem 1.   Does the conclusion of Theorem 4.11 remain valid without the convexity assumption on g?

4.5 FURTHER CONVERGENCE RESULTS

We start with the following

THEOREM 4.12 *Let* $(f_n) \subset L^1(\Omega)$ *and* $f \in L^1(\Omega)$. *Assume*

$$f_n \rightharpoonup f \quad \text{weakly in } L^1. \tag{4.5.1}$$

Let u_n *(resp. u) be the solution of (4.0.1) associated with* f_n *(resp. f). Then* $u_n \to u$ *in* $L^1(\Omega)$.

Proof. By definition,

$$-\Delta u_n + g(u_n) = f_n \quad \text{and} \quad -\Delta u + g(u) = f \quad \text{in } (C_0^2)^*.$$

Using a device introduced by Gallouët-Morel [GM] (see also Proposition 4.B.2 below), we have, for every $M > 0$,

$$\int_{[|u_n|\geq M]} |g(u_n)| \leq \int_{[|u_n|\geq M]} |f_n|.$$

Thus

$$\int_E |g(u_n)| = \int_{\substack{E \\ [|u_n|\geq M]}} + \int_{\substack{E \\ [|u_n|<M]}} \leq \int_{[|u_n|\geq M]} |f_n| + g(M)|E|. \tag{4.5.2}$$

On the other hand, $\|\Delta u_n\|_{L^1} \leq C$ implies $\|u_n\|_{L^1} \leq C$, and thus

$$\text{meas}\,[|u_n| \geq M] \leq \frac{C}{M}.$$

From (4.5.1) and a theorem of Dunford-Pettis (see, e.g., [DS, Corollary IV.8.11]) we infer that (f_n) is equi-integrable. Given $\delta > 0$, fix $M > 0$ such that

$$\int_{[|u_n| \geq M]} |f_n| \leq \delta \quad \forall n \geq 1. \tag{4.5.3}$$

With this fixed M, choose $|E|$ so small that

$$g(M)|E| < \delta. \tag{4.5.4}$$

We deduce from (4.5.2)–(4.5.4) that $g(u_n)$ is equi-integrable.

Passing to a subsequence, we may assume that $u_{n_k} \to v$ in $L^1(\Omega)$ and a.e., for some $v \in L^1(\Omega)$. Then $g(u_{n_k}) \to g(v)$ a.e. By Egorov's lemma, $g(u_{n_k}) \to g(v)$ in $L^1(\Omega)$. It follows that v is a solution of (4.0.1) associated to f. By the uniqueness of the limit, we must have $u_n \to u$ in $L^1(\Omega)$.

Remark 4.8. Theorem 4.12 is no longer true if one replaces the weak convergence $f_n \rightharpoonup f$ in L^1, by the weak* convergence in the sense of measures. Here is an example:

Example 4.1 Assume $N \geq 3$ and let $g(t) = (t^+)^q$ with $q \geq \frac{N}{N-2}$. Let $f \equiv 1$ in Ω. We will construct a sequence (f_k) in $C_c^\infty(\Omega)$ such that

$$f_k \overset{*}{\rightharpoonup} f \quad \text{in } \mathcal{M}(\Omega), \tag{4.5.5}$$

and such that the solutions u_k of (4.0.1) corresponding to f_k converge to 0 in $L^1(\Omega)$. Let (μ_k) be any sequence in $\mathcal{M}(\Omega)$ converging weak* to f, as $k \to \infty$, and such that each measure μ_k is a linear combination of Dirac masses. (For example, each μ_k can be of the form $|\Omega| M^{-1} \sum \delta_{a_i}$, where the M points a_i are uniformly distributed in Ω.) Recall that for $\mu = \delta_a$, the corresponding u^* in Proposition 4.1 is $\equiv 0$ (see [B4] or Theorem 4.B.6 below). Similarly, for each μ_k, the corresponding u^* is $\equiv 0$. Set $h_{n,k} = \rho_n * \mu_k$, with the same notation as in Section 4.4. Let $u_{n,k}$ denote the solution of (4.0.1) relative to $h_{n,k}$. For each *fixed* k we know, by Theorem 4.11, that $u_{n,k} \to 0$ strongly in $L^1(\Omega)$ as $n \to \infty$. For each k, choose $N_k > k$ sufficiently large so that $\|u_{N_k,k}\|_{L^1} < 1/k$. Set $f_k = h_{N_k,k}$, so that $u_k = u_{N_k,k}$ is the corresponding solution of (4.0.1). It is easy to check that, as $k \to \infty$,

$$f_k \overset{*}{\rightharpoonup} f \equiv 1 \quad \text{in } \mathcal{M}(\Omega), \quad \text{but} \quad u_k \to 0 \quad \text{in } L^1(\Omega).$$

Our next result is a refinement of Theorem 4.12 in the spirit of Theorem 4.6. Let $\mu \in \mathcal{M}(\Omega)$ and let (μ_n) be a sequence in $\mathcal{M}(\Omega)$. Assume that

$$\mu = f - \Delta v \quad \text{in } (C_0^2)^*, \tag{4.5.6}$$

$$\mu_n = f_n - \Delta v_n \quad \text{in } (C_0^2)^*, \tag{4.5.7}$$

where $f \in L^1$, $f_n \in L^1$, $v \in L^1$, $v_n \in L^1$, $g(v) \in L^1$, and $g(v_n) \in L^1$.

By Theorem 4.6 we know that there exist u and u_n solutions of

$$-\Delta u + g(u) = \mu \quad \text{in } \Omega, \quad u = 0 \quad \text{on } \partial\Omega, \qquad (4.5.8)$$

$$-\Delta u_n + g(u_n) = \mu_n \quad \text{in } \Omega, \quad u_n = 0 \quad \text{on } \partial\Omega. \qquad (4.5.9)$$

THEOREM 4.13 *Assume (4.5.6)–(4.5.9) and, moreover,*

$$\|\mu_n\|_{\mathcal{M}} \le C, \qquad (4.5.10)$$

$$f_n \rightharpoonup f \quad \text{weakly in } L^1, \qquad (4.5.11)$$

$$v_n \to v \quad \text{in } L^1 \quad \text{and} \quad g(v_n) \to g(v) \quad \text{in } L^1. \qquad (4.5.12)$$

Then $u_n \to u$ in $L^1(\Omega)$.

Proof. We divide the proof into two steps.

Step 1. Fix $0 < \alpha < 1$ and let $u(\alpha)$, $u_n(\alpha)$ be the solutions of

$$-\Delta u(\alpha) + g(u(\alpha)) = \alpha\mu \quad \text{in } \Omega, \quad u(\alpha) = 0 \quad \text{on } \partial\Omega, \qquad (4.5.13)$$

$$-\Delta u_n(\alpha) + g(u_n(\alpha)) = \alpha\mu_n \quad \text{in } \Omega, \quad u_n(\alpha) = 0 \quad \text{on } \partial\Omega. \qquad (4.5.14)$$

Then $u_n(\alpha) \to u(\alpha)$ in $L^1(\Omega)$.

Note that $u(\alpha)$ and $u_n(\alpha)$ exist since $\alpha\mu = \alpha f - \Delta(\alpha v)$ and $g(\alpha v) \le g(v)$, so that $g(\alpha v) \in L^1$, and similarly for $\alpha\mu_n$. We may then apply Theorem 4.6 once more. For simplicity we will omit the dependence in α and we will write $\tilde{u}$, $\tilde{u}_n$ instead of $u(\alpha)$, $u_n(\alpha)$ (recall that in this step α is *fixed*). Since

$$\|\Delta\tilde{u}_n\|_{\mathcal{M}} \le 2\alpha\|\mu_n\|_{\mathcal{M}} \le C,$$

we can extract a subsequence of $(\tilde{u}_n)$ converging strongly in $L^1(\Omega)$ and a.e. Let $w \in W_0^{1,1}(\Omega)$ be such that $\tilde{u}_{n_k} \to w$ in $L^1(\Omega)$ and a.e. We will prove that w satisfies (4.5.13), and therefore, by uniqueness, $w = \tilde{u}$. Since w is independent of the subsequence, we will infer that $(\tilde{u}_n)$ converges to $\tilde{u}$, which is the desired conclusion.

We claim that

$$g(\tilde{u}_n) \quad \text{is equi-integrable.} \qquad (4.5.15)$$

To establish (4.5.15) we argue as in the proof of Theorem 4.12. From (4.5.7) and (4.5.14) we see that

$$-\Delta(\tilde{u}_n - \alpha v_n) + [g(\tilde{u}_n) - g(\alpha v_n)] = h_n \quad \text{in } (C_0^2)^*, \qquad (4.5.16)$$

with

$$h_n = \alpha f_n - g(\alpha v_n). \qquad (4.5.17)$$

Using (4.5.11) and (4.5.12) we see that

$$(h_n) \quad \text{is equi-integrable.} \qquad (4.5.18)$$

From (4.5.16) and Proposition 4.B.2 we obtain (as in the proof of Theorem 4.12) that, for every $M > 0$,

$$\int_{[|\tilde{u}_n - \alpha v_n| \geq M]} |g(\tilde{u}_n) - g(\alpha v_n)| \leq \int_{[|\tilde{u}_n - \alpha v_n| \geq M]} |h_n|. \tag{4.5.19}$$

On the other hand, for any Borel set E of Ω, we have

$$\int_E g(\tilde{u}_n) = \int_{A_n} g(\tilde{u}_n) + \int_{B_n} g(\tilde{u}_n) + \int_{C_n} g(\tilde{u}_n), \tag{4.5.20}$$

where

$$A_n = [\tilde{u}_n \geq v_n] \cap [|\tilde{u}_n - \alpha v_n| \geq M] \cap E,$$
$$B_n = [\tilde{u}_n \geq v_n] \cap [|\tilde{u}_n - \alpha v_n| < M] \cap E,$$
$$C_n = [\tilde{u}_n < v_n] \cap E.$$

To handle the integral on A_n, write

$$\int_{A_n} g(\tilde{u}_n) \leq \int_{[|\tilde{u}_n - \alpha v_n| \geq M]} |g(\tilde{u}_n) - g(\alpha v_n)| + \int_E g(v_n).$$

Thus, by (4.5.19),

$$\int_{A_n} g(\tilde{u}_n) \leq \int_{[|\tilde{u}_n - \alpha v_n| \geq M]} |h_n| + \int_E g(v_n). \tag{4.5.21}$$

Next, on B_n, we have

$$\tilde{u}_n < M + \alpha v_n \leq M + \alpha \tilde{u}_n,$$

and thus

$$\tilde{u}_n < \frac{M}{1 - \alpha}.$$

Therefore,

$$\int_{B_n} g(\tilde{u}_n) \leq g\left(\frac{M}{1 - \alpha}\right)|E|. \tag{4.5.22}$$

Finally, we have

$$\int_{C_n} g(\tilde{u}_n) \leq \int_E g(v_n). \tag{4.5.23}$$

Combining (4.5.20)–(4.5.23) yields

$$\int_E g(\tilde{u}_n) \leq \int_{[|\tilde{u}_n - \alpha v_n| \geq M]} |h_n| + 2\int_E g(v_n) + g\left(\frac{M}{1 - \alpha}\right)|E|. \tag{4.5.24}$$

But $\|\tilde{u}_n - \alpha v_n\|_{L^1} \leq C$ and therefore

$$\text{meas } [|\tilde{u}_n - \alpha v_n| \geq M] \leq \frac{C}{M}. \tag{4.5.25}$$

Given $\delta > 0$, *fix* $M > 0$ sufficiently large such that

$$\int_{[|\tilde{u}_n - \alpha v_n| \geq M]} |h_n| \leq \delta \quad \forall n \geq 1$$

(here we use (4.5.18) and (4.5.25)). With this fixed M, choose $|E|$ so small that

$$2 \int_E g(v_n) + g\left(\frac{M}{1-\alpha}\right) |E| \leq \delta \quad \forall n \geq 1.$$

This finishes the proof of (4.5.15).

Since $g(\tilde{u}_n) \to g(w)$ a.e., we deduce from (4.5.15) and Egorov's lemma that $g(\tilde{u}_n) \to g(w)$ in L^1. We are now able to pass to the limit in (4.5.14) and conclude that w satisfies (4.5.13), which was the goal of Step 1.

Step 2.   Proof of the theorem completed.

Here the dependence on α is important and we return to the notation $u(\alpha)$ and $u_n(\alpha)$. From (4.5.8) and (4.5.13) we deduce that

$$\|\Delta(u(\alpha) - u)\|_{\mathcal{M}} \leq 2(1 - \alpha)\|\mu\|_{\mathcal{M}} \tag{4.5.26}$$

and similarly, from (4.5.9) and (4.5.14), we have

$$\|\Delta(u_n(\alpha) - u_n)\|_{\mathcal{M}} \leq 2(1 - \alpha)\|\mu_n\|_{\mathcal{M}} \leq C(1 - \alpha). \tag{4.5.27}$$

Estimates (4.5.26) and (4.5.27) yield

$$\|u(\alpha) - u\|_{L^1} + \|u_n(\alpha) - u_n\|_{L^1} \leq C(1 - \alpha), \tag{4.5.28}$$

with C independent of n and α. Finally, we write

$$\|u_n - u\|_{L^1} \leq \|u(\alpha) - u\|_{L^1} + \|u_n(\alpha) - u_n\|_{L^1} + \|u_n(\alpha) - u(\alpha)\|_{L^1}. \tag{4.5.29}$$

Given $\varepsilon > 0$, fix $\alpha < 1$ so small that

$$C(1 - \alpha) < \varepsilon, \tag{4.5.30}$$

and then apply Step 1 to assert that

$$\|u_n(\alpha) - u(\alpha)\|_{L^1} < \varepsilon \quad \forall n \geq N, \tag{4.5.31}$$

provided N is sufficiently large. Combining (4.5.28)–(4.5.31) yields

$$\|u_n - u\|_{L^1} \leq 2\varepsilon \quad \forall n \geq N,$$

which is the desired conclusion.

4.6 NONNEGATIVE MEASURES THAT ARE GOOD FOR EVERY g MUST BE DIFFUSE

Let $h : [0, \infty) \to [0, \infty)$ be a continuous nondecreasing function with $h(0) = 0$. Given a compact set $K \subset \Omega$, let

$$\text{cap}_{\Delta,h}(K) = \inf \left\{ \int_\Omega h(|\Delta\varphi|); \varphi \in C_c^\infty(\Omega), 0 \le \varphi \le 1, \text{ and } \varphi = 1 \text{ on } K \right\},$$

where, as usual, $C_c^\infty(\Omega)$ denotes the set of C^∞-functions with compact support in Ω.
  We start with

PROPOSITION 4.5 *Assume*

$$\lim_{t\to\infty} \frac{g(t)}{t} = +\infty \quad \text{and} \quad g^*(s) > 0 \quad \text{for } s > 0. \tag{4.6.1}$$

If μ is a good measure, then $\mu^+(K) = 0$ for every compact set $K \subset \Omega$ such that $\text{cap}_{\Delta,g^*}(K) = 0$.

Here, g^* denotes the convex conjugate of g, which is finite in view of the coercivity of g. Note that if $g'(0) = 0$, then $g^*(s) > 0$ for every $s > 0$.

Proof. Since μ is a good measure, μ^+ is also a good measure. Thus,

$$\mu^+ = -\Delta v + g(v) \quad \text{in } (C_0^2)^*$$

for some $v \in L^1(\Omega)$, $v \ge 0$ a.e., such that $g(v) \in L^1(\Omega)$.
  Let $\varphi_n \in C_c^\infty(\Omega)$ be such that $0 \le \varphi_n \le 1$ in Ω, $\varphi_n = 1$ on K, and

$$\int_\Omega g^*(|\Delta\varphi_n|) \to 0.$$

Passing to a subsequence if necessary, we may assume that

$$g^*(|\Delta\varphi_n|) \to 0 \quad \text{a.e.} \quad \text{and} \quad g^*(|\Delta\varphi_n|) \le G \in L^1(\Omega) \quad \forall n \ge 1.$$

Since $g^*(s) > 0$ if $s > 0$, we also have

$$\varphi_n, |\Delta\varphi_n| \to 0 \quad \text{a.e.}$$

For every $n \ge 1$,

$$\mu^+(K) \le \int_\Omega \varphi_n \, d\mu^+ = \int_\Omega [g(v)\varphi_n - v\Delta\varphi_n]. \tag{4.6.2}$$

Note that

$$|g(v)\varphi_n - v\Delta\varphi_n| \to 0 \quad \text{a.e.}$$

and

$$|g(v)\varphi_n - v\Delta\varphi_n| \le 2g(v) + g^*(|\Delta\varphi_n|) \le 2g(v) + G \in L^1(\Omega).$$

By dominated convergence, the right-hand side of (4.6.2) converges to 0 as $n \to \infty$. We then conclude that $\mu^+(K) = 0$.

As a consequence of Proposition 4.5 we have

THEOREM 4.14 *Given a Borel set* $\Sigma \subset \Omega$ *with zero* H^1-*capacity, there exists g such that*

$$\mu^* = -\mu^- \quad \text{for every measure } \mu \text{ concentrated on } \Sigma.$$

In particular, for every nonnegative $\mu \in \mathcal{M}(\Omega)$ *concentrated on a set of zero* H^1-*capacity, there exists some g such that* $\mu^* = 0$.

Proof. Let $\Sigma \subset \Omega$ be a Borel set of zero H^1-capacity. Let (K_n) be an increasing sequence of compact sets in Σ such that

$$\mu^+\left(\Sigma \setminus \bigcup_n K_n\right) = 0.$$

For each $n \geq 1$, K_n has zero H^1-capacity. By Lemma 4.E.1, one can find $\psi_n \in C_c^\infty(\Omega)$ such that $0 \leq \psi_n \leq 1$ in Ω, $\psi_n = 1$ in some neighborhood of K_n, and

$$\int_\Omega |\Delta \psi_n| \leq \frac{1}{n} \quad \forall n \geq 1.$$

In particular, $\Delta \psi_n \to 0$ in $L^1(\Omega)$. Passing to a subsequence if necessary, we may assume that

$$\Delta \psi_n \to 0 \quad \text{a.e.} \quad \text{and} \quad |\Delta \psi_n| \leq G \in L^1(\Omega) \quad \forall n \geq 1.$$

According to a theorem of De La Vallée-Poussin (see [DVP, Remarque 23] or [DM, Théorème II.22]), there exists a convex function $h : [0, \infty) \to [0, \infty)$ such that $h(0) = 0$, $h(s) > 0$ for $s > 0$,

$$\lim_{t \to \infty} \frac{h(t)}{t} = +\infty, \quad \text{and} \quad h(G) \in L^1(\Omega).$$

By dominated convergence, we then have $h(|\Delta \psi_n|) \to 0$ in $L^1(\Omega)$. Thus,

$$\text{cap}_{\Delta,h}(K_n) = 0 \quad \forall n \geq 1. \tag{4.6.3}$$

Let $g(t) = h^*(t)$ if $t \geq 0$, and $g(t) = 0$ if $t < 0$. By duality, $h = g^*$ on $[0, \infty)$. Let $\mu \in \mathcal{M}(\Omega)$ be any measure concentrated on Σ. By Proposition 4.5, (4.6.3) yields

$$(\mu^*)^+(K_n) = 0 \quad \forall n \geq 1,$$

where the reduced measure μ^* is computed with respect to g. Thus, $(\mu^*)^+(\Sigma) = 0$. Since μ is concentrated on Σ, we have $(\mu^*)^+ = 0$. Applying Corollary 4.10, we then get

$$\mu^* = (\mu^*)^+ - (\mu^*)^- = -\mu^-,$$

which is the desired result.

We may now present the

Proof of Theorem 4.5. Assume $\mu \in \mathcal{M}(\Omega)$ is a good measure for every g. Given a Borel set $\Sigma \subset \Omega$ with zero H^1-capacity, let $\lambda = \mu^+ \lfloor_\Sigma$. In view of Theorem 4.14, there exists $\tilde{g}$ for which $\lambda^* = 0$. On the other hand, by Theorems 4.4 and 4.6', λ is a good measure for $\tilde{g}$. Thus, $\lambda = \lambda^* = 0$. In other words, $\mu^+(\Sigma) = 0$. Since Σ was arbitrary, μ^+ is diffuse. This establishes the theorem.

We conclude this section with the following:

Open problem 2. Let $g : \mathbb{R} \to \mathbb{R}$ be any given continuous, nondecreasing function satisfying (4.0.5). Can one always find some nonnegative $\mu \in \mathcal{M}(\Omega)$ such that μ is good for g, but μ is *not* diffuse?

After this paper was finished, A. C. Ponce [P] gave a positive answer to the above problem.

4.7 SIGNED MEASURES AND GENERAL NONLINEARITIES g

Suppose that $g : \mathbb{R} \to \mathbb{R}$ is a continuous, nondecreasing function, such that $g(0) = 0$. But we will *not* impose in this section that $g(t) = 0$ if $t < 0$. We shall follow the same approximation scheme as in the Introduction. Namely, let (g_n) be a sequence of nondecreasing continuous functions, $g_n : \mathbb{R} \to \mathbb{R}$, $g_n(0) = 0$, satisfying (4.0.8), such that both (g_n^+) and (g_n^-) verify (4.0.7), and

$$g_n^+(t) \uparrow g^+(t), \quad g_n^-(t) \uparrow g^-(t) \quad \forall t \in \mathbb{R} \qquad \text{as } n \uparrow \infty.$$

Let $\mu \in \mathcal{M}(\Omega)$. For each $n \geq 1$, we denote by u_n the unique solution of

$$\begin{cases} -\Delta u_n + g_n(u_n) = \mu & \text{in } \Omega, \\ \quad\quad\quad\quad\quad u_n = 0 & \text{on } \partial\Omega. \end{cases} \tag{4.7.1}$$

First, a simple observation:

LEMMA 4.4 *Assume $\mu \geq 0$ or $\mu \leq 0$. Then there exists $u^* \in L^1(\Omega)$ such that $u_n \to u^*$ in $L^1(\Omega)$. If $\mu \geq 0$, then $u^* \geq 0$ is the largest subsolution of (4.0.1). If $\mu \leq 0$, then $u^* \leq 0$ is the smallest supersolution of (4.0.1). In both cases, we have*

$$\left| \int_\Omega u^* \Delta \zeta \right| \leq 2\|\mu\|_{\mathcal{M}} \|\zeta\|_{L^\infty} \quad \forall \zeta \in C_0^2(\bar{\Omega}) \tag{4.7.2}$$

and

$$\int_\Omega |g(u^*)| \leq \|\mu\|_{\mathcal{M}}. \tag{4.7.3}$$

Proof. If $\mu \geq 0$, then $u_n \geq 0$ a.e. In particular, $g_n(u_n) = g_n^+(u_n)$ for every $n \geq 1$. Since (g_n^+) satisfies the assumptions of Proposition 4.1, we conclude that $u_n \to u^*$ in $L^1(\Omega)$, where $u^* \geq 0$ is the largest subsolution of (4.0.1). If $\mu \leq 0$, then $u_n \leq 0$, so that $w_n = -u_n$ satisfies

$$\begin{cases} -\Delta w_n + \tilde{g}_n(w_n) = -\mu & \text{in } \Omega, \\ \quad\quad\quad\quad\quad w_n = 0 & \text{on } \partial\Omega, \end{cases}$$

where $\tilde{g}_n(t) = g_n^-(-t)$, $\forall t \in \mathbb{R}$. Clearly, the sequence $(\tilde{g}_n)$ satisfies the assumptions of Proposition 4.1. Therefore, $u_n = -w_n \to -w^* = u^*$ in $L^1(\Omega)$. It is easy to see that $u^* \leq 0$ is the smallest supersolution of (4.0.1).

Given $\mu \in \mathcal{M}(\Omega)$ such that $\mu \geq 0$ or $\mu \leq 0$, we *define* $\mu^* \in \mathcal{M}(\Omega)$ by

$$\mu^* = -\Delta u^* + g(u^*) \quad \text{in } (C_0^2)^*. \tag{4.7.4}$$

The reduced measure μ^* is well defined because of (4.7.2) and (4.7.3). It is easy to see that

(a) if $\mu \geq 0$, then $0 \leq \mu^* \leq \mu$;
(b) if $\mu \leq 0$, then $\mu \leq \mu^* \leq 0$.

We now consider the general case of a signed measure $\mu \in \mathcal{M}(\Omega)$. In view of (4.7.4), both measures $(\mu^+)^*$ and $(-\mu^-)^*$ are well defined. Moreover,

$$-\mu^- \leq (-\mu^-)^* \leq 0 \leq (\mu^+)^* \leq \mu^+.$$

The convergence of the approximating sequence (u_n) is governed by the following:

THEOREM 4.15 *Let u_n be given by (4.7.1). Then, $u_n \to u^*$ in $L^1(\Omega)$, where u^* is the unique solution of*

$$\begin{cases} -\Delta u^* + g(u^*) = (\mu^+)^* + (-\mu^-)^* & \text{in } \Omega, \\ u^* = 0 & \text{on } \partial\Omega. \end{cases} \tag{4.7.5}$$

Proof. By standard estimates, $\|\Delta u_n\|_{\mathcal{M}} \leq 2\|\mu\|_{\mathcal{M}}$. Thus, without loss of generality, we may assume that, for a subsequence, still denoted (u_n), $u_n \to u$ in $L^1(\Omega)$ and a.e. We shall show that u satisfies (4.7.5); by uniqueness (see Corollary 4.B.1), this will imply that u is independent of the subsequence.

For each $n \geq 1$, let v_n, $\tilde{v}_n$ be the solutions of

$$\begin{cases} -\Delta v_n + g_n(v_n) = \mu^+ & \text{in } \Omega, \\ v_n = 0 & \text{on } \partial\Omega, \end{cases} \tag{4.7.6}$$

and

$$\begin{cases} -\Delta \tilde{v}_n + g_n^+(\tilde{v}_n) = \mu & \text{in } \Omega, \\ \tilde{v}_n = 0 & \text{on } \partial\Omega, \end{cases} \tag{4.7.7}$$

so that $v_n \geq 0$ a.e., $v_n \downarrow v^*$ and $\tilde{v}_n \downarrow \tilde{v}^*$ in $L^1(\Omega)$. By comparison (see Corollary 4.B.2), we have

$$\tilde{v}_n \leq u_n \leq v_n \quad \text{a.e.}$$

Thus,

$$g_n^+(\tilde{v}_n) \leq g_n^+(u_n) \leq g_n^+(v_n) = g_n(v_n) \quad \text{a.e.} \tag{4.7.8}$$

By Lemma 4.2, we know that

$$g_n^+(\tilde{v}_n) \overset{*}{\rightharpoonup} g^+(\tilde{v}^*) + \mu - \mu^*,$$

$$g_n(v_n) \overset{*}{\rightharpoonup} g(v^*) + \mu^+ - (\mu^+)^*.$$

Here, both reduced measures μ^* and $(\mu^+)^*$ are computed with respect to the non-linearity g^+; in particular (see Corollary 4.10),

$$\mu - \mu^* = \mu^+ - (\mu^+)^*. \qquad (4.7.9)$$

We claim that

$$g_n^+(u_n) \overset{*}{\rightharpoonup} g^+(u) + \mu^+ - (\mu^+)^*. \qquad (4.7.10)$$

This will be a consequence of the following:

LEMMA 4.5 *Let $a_n, b_n, c_n \in L^1(\Omega)$ be such that*

$$a_n \le b_n \le c_n \quad a.e.$$

Assume that $a_n \to a$, $b_n \to b$ and $c_n \to c$ a.e. in Ω for some $a, b, c \in L^1(\Omega)$. If $(c_n - a_n) \overset{*}{\rightharpoonup} (c - a)$ weak* in $\mathcal{M}(\Omega)$, then

$$(c_n - b_n) \overset{*}{\rightharpoonup} (c - b) \quad weak^* \ in \ \mathcal{M}(\Omega). \qquad (4.7.11)$$

Proof. Since

$$0 \le (c_n - b_n) \le (c_n - a_n) \quad a.e., \qquad (4.7.12)$$

the sequence $(c_n - b_n)$ is bounded in $L^1(\Omega)$. Passing to a subsequence if necessary, we may assume that there exists $\lambda \in \mathcal{M}(\Omega)$ such that

$$(c_n - b_n) \overset{*}{\rightharpoonup} \lambda.$$

By (4.7.12), we have $0 \le \lambda \le (c - a)$. Thus, λ is absolutely continuous with respect to the Lebesgue measure. In other words, $\lambda \in L^1(\Omega)$. Given $M > 0$, we denote by S_M the truncation operator $S_M(t) = \min\{M, \max\{t, -M\}\}, \forall t \in \mathbb{R}$. By dominated convergence, we have

$$S_M(a_n) \to S_M(a) \quad \text{strongly in } L^1(\Omega),$$

and similarly for $S_M(b_n)$ and $S_M(c_n)$. Since

$$0 \le [(c_n - S_M(c_n)) - (b_n - S_M(b_n))]$$
$$\le [(c_n - S_M(c_n)) - (a_n - S_M(a_n))] \quad a.e.,$$

as $n \to \infty$ we get

$$0 \le \lambda - (S_M(c) - S_M(b)) \le [(c - S_M(c)) - (a - S_M(a))] \quad a.e.$$

Let $M \to \infty$ in the expression above. We then get $\lambda = (c - b)$. This concludes the proof of the lemma.

We now apply the previous lemma with $a_n = g_n^+(\tilde{v}_n)$, $b_n = g_n^+(u_n)$ and $c_n = g_n(v_n)$. In view of (4.7.8) and (4.7.9), the assumptions of Lemma 4.5 are satisfied. It follows from (4.7.11) that

$$g_n(v_n) - g_n^+(u_n) \overset{*}{\rightharpoonup} g(v^*) - g^+(u). \qquad (4.7.13)$$

Thus,

$$g_n^+(u_n) = g_n(v_n) - [g_n(v_n) - g_n^+(u_n)] \overset{*}{\rightharpoonup} g^+(u) + \mu^+ - (\mu^+)^*,$$

which is precisely (4.7.10). A similar argument shows that

$$g_n^-(u_n) \overset{*}{\rightharpoonup} g^-(u) + \mu^- + (-\mu^-)^*. \qquad (4.7.14)$$

We conclude from (4.7.10) and (4.7.14) that

$$g_n(u_n) \overset{*}{\rightharpoonup} g(u) + \mu - [(\mu^+)^* + (-\mu^-)^*]. \qquad (4.7.15)$$

Therefore, u satisfies (4.7.5), so that (4.7.5) has a solution $u^* = u$. By uniqueness, the whole sequence (u_n) converges to u^* in $L^1(\Omega)$.

Motivated by Theorem 4.15, for any $\mu \in \mathcal{M}(\Omega)$, we *define* the reduced measure μ^* by

$$\mu^* = (\mu^+)^* + (-\mu^-)^*. \qquad (4.7.16)$$

[This definition is coherent if μ is either a positive or a negative measure].

One can derive a number of properties satisfied by μ^*. For instance, the statements of Theorems 4.8–4.10 remain true. Moreover,

THEOREM 4.2′ *There exists a Borel set* $\Sigma \subset \Omega$ *with* cap $(\Sigma) = 0$ *such that*

$$|\mu - \mu^*|(\Omega \backslash \Sigma) = 0.$$

4.8 EXAMPLES

We describe here some simple examples where the measure μ^* can be explicitly identified. Throughout this section, we assume again that $g(t) = 0$ for $t \leq 0$.

Example 4.2 $N = 1$ and g is arbitrary.

This case is very easy since every measure is diffuse (recall that the only set of zero capacity is the empty set). Hence, by Corollary 4.2, every measure is good. Thus, $\mu^* = \mu$ for every μ.

Example 4.3 $N \geq 2$ and $g(t) = t^p$, $t \geq 0$, with $1 < p < \frac{N}{N-2}$.

In this case, we have again $\mu^* = \mu$ since, for *every* measure μ, problem (4.0.1) admits a solution. This result was originally established in 1975 by Ph. Bénilan and H. Brezis (see [BB, Appendix A], [B1], [B2], [B3], [B4] and also Theorem 4.B.5 below). The crucial ingredient is the compactness of the imbedding of the space $\{u \in W_0^{1,1}; \Delta u \in \mathcal{M}\}$, equipped with the norm $\|u\|_{W^{1,1}} + \|\Delta u\|_{\mathcal{M}}$, into L^q for every $q < \frac{N}{N-2}$ (see Theorem 4.B.1 below).

Example 4.4 $N \geq 3$ and $g(t) = t^p$, $t \geq 0$, with $p \geq \frac{N}{N-2}$.

In this case, we have

THEOREM 4.16 *For every measure* μ, *we have*

$$\mu^* = \mu - (\mu_2)^+, \qquad (4.8.1)$$

where $\mu = \mu_1 + \mu_2$ *is the unique decomposition of* μ *(in the sense of Lemma A.1)*
relative to the $W^{2,p'}$-*capacity.*

Proof. By a result of Baras-Pierre [BP] (already mentioned in the Introduction) we
know that a measure $\nu \geq 0$ is a good measure if and only if ν is diffuse with respect
to the $W^{2,p'}$-capacity.
    Set

$$\tilde{\mu} = \mu - (\mu_2)^+ = \mu_1 - (\mu_2)^- \quad \text{and} \quad \tilde{\nu} = (\mu_2)^+. \tag{4.8.2}$$

We claim that

$$(\tilde{\mu})^* = \tilde{\mu} \quad \text{and} \quad (\tilde{\nu})^* = 0. \tag{4.8.3}$$

Clearly, $(\tilde{\mu})^+ = (\mu_1)^+$. From the result of Baras-Pierre [BP], we infer that $(\mu_1)^+$
is a good measure. By Theorem 4.4, $\tilde{\mu}$ is also a good measure. Thus, $(\tilde{\mu})^* = \tilde{\mu}$.
Since $\tilde{\nu}$ is a nonnegative measure concentrated on a set of zero $W^{2,p}$-capacity, it
follows from [BP] that $(\tilde{\nu})^* \leq 0$. Since $(\tilde{\nu})^* \geq 0$, we conclude that (4.8.3) holds.
    Applying Theorem 4.8, we get

$$\mu^* = (\tilde{\mu} + \tilde{\nu})^* = (\tilde{\mu})^* + (\tilde{\nu})^* = \tilde{\mu} = \mu - (\mu_2)^+,$$

which is precisely (4.8.1).

Remark 4.9. In this example we see that the measure $\mu - \mu^*$ is concentrated on a
set Σ whose $W^{2,p'}$-capacity is zero. This is better information than the general fact
that $\mu - \mu^*$ is concentrated on a set Σ whose H^1-capacity is zero.

Example 4.5 $N = 2$ and $g(t) = e^t - 1, t \geq 0$.

    In this case, the identification of μ^* relies heavily on a result of Vázquez [Va].

THEOREM 4.17 *Given any measure* μ, *let*

$$\mu = \mu_1 + \mu_2,$$

where μ_2 *is the purely atomic part of* μ *(this corresponds to the decomposition of*
μ *in the sense of Lemma 4.A.1, where* $\mathcal{Z}$ *consists of countable sets). Write*

$$\mu_2 = \sum_i \alpha_i \delta_{a_i} \tag{4.8.4}$$

with $a_i \in \Omega$ *distinct, and* $\sum |\alpha_i| < \infty$. *Then*

$$\mu^* = \mu - \sum_i (\alpha_i - 4\pi)^+ \delta_{a_i}. \tag{4.8.5}$$

Proof. By a result of Vázquez [Va], we know that a measure ν is a good measure
if and only if $\nu(\{x\}) \leq 4\pi$ for every $x \in \Omega$. (The paper of Vázquez deals with the
equation (4.0.1) in all of $\mathbb{R}^2$, but the conclusion, and the proof, is the same for a
bounded domain.)

Clearly, $\mu_1(\{x\}) = 0$, $\forall x \in \Omega$. From the result of Vázquez [Va] we infer that μ_1 is a good measure. Thus,

$$(\mu_1)^* = \mu_1. \tag{4.8.6}$$

Let $a \in \Omega$ and $\alpha \in \mathbb{R}$. It is easy to see from [Va] that

$$(\alpha \delta_a)^* = \min\{\alpha, 4\pi\}\delta_a. \tag{4.8.7}$$

An induction argument applied to Theorem 4.8 and the continuity of the mapping $\mu \mapsto \mu^*$ show that

$$\mu^* = (\mu_1)^* + (\mu_2)^* = (\mu_1)^* + \sum_i (\alpha_i \delta_{a_i})^*. \tag{4.8.8}$$

By (4.8.6)–(4.8.8), we have

$$\mu^* = \mu_1 + \sum_i \min\{\alpha_i, 4\pi\}\delta_{a_i} = \mu - \sum_i (\alpha_i - 4\pi)^+ \delta_{a_i}. \tag{4.8.9}$$

This establishes (4.8.5).

We conclude this section with two interesting questions:

Open problem 3. Let $N = 2$ and $g(t) = (e^{t^2} - 1)$, $t \ge 0$. Is there an explicit formula for μ^* ?

Open problem 4. Let $N \ge 3$ and $g(t) = (e^t - 1)$, $t \ge 0$. Is there an explicit formula for μ^* ?

A partial answer to Open Problem 4 has been obtained by Bartolucci-Leoni-Orsina-Ponce [BLOP]. More precisely, they have established the following:

THEOREM 4.18 *Any measure μ such that $\mu \le 4\pi \mathcal{H}^{N-2}$ is a good measure.*

Here, $\mathcal{H}^{N-2}$ denotes the $(N-2)$-Hausdorff measure. The converse of Theorem 4.18 is not true. This was suggested by L. Véron in a personal communication; explicit examples are given in [P]. The characterization of good measures is still open; see, however, [MV5].

4.9 FURTHER DIRECTIONS AND OPEN PROBLEMS

4.9.1 Vertical Asymptotes

Let $g : (-\infty, +1) \to \mathbb{R}$ be a continuous, nondecreasing function such that $g(t) = 0$, $\forall t \le 0$, and such that $g(t) \to +\infty$ as $t \to +1$. Let (g_n) be a sequence of functions $g_n : \mathbb{R} \to \mathbb{R}$ which are continuous, nondecreasing, and satisfy the following conditions:

$$0 \le g_1(t) \le g_2(t) \le \cdots \le g(t) \quad \forall t < 1, \tag{4.9.1}$$

$$g_n(t) \to g(t) \quad \forall t < 1 \quad \text{and} \quad g_n(t) \to +\infty \quad \forall t \ge 1. \tag{4.9.2}$$

If $N \geq 2$, then we also assume that

each g_n has subcritical growth, i.e., $g_n(t) \leq C(|t|^p + 1)$ $\forall t \in \mathbb{R}$,   (4.9.3)

for some constant C and some $p < \frac{N}{N-2}$, possibly depending on n.

Given $\mu \in \mathcal{M}(\Omega)$, let u_n be a solution of (4.0.10). Then $u_n \downarrow u^*$ in Ω as $n \uparrow \infty$. Moreover, (4.0.11) and (4.0.12) hold. We may therefore define $\mu^* \in \mathcal{M}(\Omega)$ by (4.0.13).

Open problem 5.   Study the properties of u^* and the reduced measure μ^*.

Clearly, u^* is the largest subsolution. But there are some major differences in this case. When $N \geq 2$, Dupaigne-Ponce-Porretta [DPP] have shown that for any such g one can find a nonnegative measure μ for which the set $\{v \in \mathcal{G}; v \leq \mu\}$ has *no* largest element. In particular, for such measure μ, the reduced measure μ^* cannot be the largest good measure $\leq \mu$. They have also proved that the set of good measures $\mathcal{G}$ is *not* convex for any g. We refer the reader to [DPP] for other results.

Similar questions arise when g is a multivalued graph. For example,

$$g(r) = \begin{cases} 0 & \text{if } r < 1, \\ [0, \infty) & \text{if } r = 1, \\ \emptyset & \text{if } r > 1. \end{cases}$$

This is a simple model of one-sided variational inequality. The objective is to solve in some natural "weak" sense the multivalued equation

$$\begin{cases} -\Delta u + g(u) \ni \mu & \text{in } \Omega, \\ u = 0 & \text{on } \partial\Omega, \end{cases}$$

for any given $\mu \in \mathcal{M}(\Omega)$. This problem has been recently studied by Brezis-Ponce [BP4]. There were some partial results; see, e.g., Baxter [Ba], Dall'Aglio-Dal Maso [DD], Orsina-Prignet [OP], Brezis-Serfaty [BSe], and the references therein.

4.9.2 Nonlinearities Involving ∇u.

Consider the model problem:

$$\begin{cases} -\Delta u + u|\nabla u|^2 = \mu & \text{in } \Omega, \\ u = 0 & \text{on } \partial\Omega, \end{cases} \qquad (4.9.4)$$

where $\mu \in \mathcal{M}(\Omega)$. Problems of this type have been extensively studied, and it is known that they bear some similarities to the problems discussed in this paper. In particular, it has been proved in [BGO2] that (4.9.4) admits a solution if and only if the measure μ is diffuse, i.e., $|\mu|(A) = 0$ for every Borel set $A \subset \Omega$ such that cap $(A) = 0$. Moreover, the solution is unique (see [BM]). When μ is a general measure, not necessarily diffuse, it would be interesting to apply to (4.9.4) the same strategy as in this paper. More precisely, to prove that approximate solutions converge to the solution of (4.9.4), where μ is replaced by its diffuse part μ_d (in the sense of Lemma 4.A.1, relative to the Borel sets whose H^1-capacity are zero):

$$\begin{cases} -\Delta u + u|\nabla u|^2 = \mu_d & \text{in } \Omega, \\ \qquad\qquad\quad u = 0 & \text{on } \partial\Omega, \end{cases} \tag{4.9.5}$$

which possesses a unique solution. There are several "natural" approximations. For example, one may truncate the nonlinearity $g(u, \nabla u) = u|\nabla u|^2$ and replace it by $g_n(u, \nabla u) = \frac{n}{n+|g(u,\nabla u)|} g(u, \nabla u)$. It is easy to see (via a Schauder fixed point argument in $W_0^{1,1}$) that the

$$\begin{cases} -\Delta u_n + g_n(u_n, \nabla u_n) = \mu & \text{in } \Omega, \\ \qquad\qquad\qquad\quad u_n = 0 & \text{on } \partial\Omega, \end{cases} \tag{4.9.6}$$

admits a solution u_n.

Open problem 6. Is it true that (u_n) converges to the solution of (4.9.5)?

Another possible approximation consists of smoothing μ: let u_n be a solution of

$$\begin{cases} -\Delta u_n + u_n|\nabla u_n|^2 = \mu_n & \text{in } \Omega, \\ \qquad\qquad\qquad\quad u_n = 0 & \text{on } \partial\Omega, \end{cases} \tag{4.9.7}$$

where $\mu_n = \rho_n * \mu$, as in Section 4.4. It has been proved by Porretta [Po] that if $\mu \geq 0$, then $u_n \to u$ in $L^1(\Omega)$, where u is the solution of (4.9.5). We have been informed by A. Porretta that the same conclusion holds for any measure μ, by using a substantial modification of the argument in [Po].

4.9.3 Measures as Boundary Data

Consider the problem

$$\begin{cases} -\Delta u + g(u) = 0 & \text{in } \Omega, \\ \qquad\qquad\;\; u = \mu & \text{on } \partial\Omega, \end{cases} \tag{4.9.8}$$

where μ is a measure on $\partial\Omega$ and $g : \mathbb{R} \to \mathbb{R}$ is a continuous, nondecreasing function satisfying (4.0.5). It has been proved by H. Brezis (1972, unpublished) that (4.9.8) admits a unique weak solution when μ is any L^1 function (for a general nonlinearity g). When g is a power, the study of (4.9.8) for measures was initiated by Gmira-Véron [GV], and has vastly expanded in recent years; see the papers of Marcus-Véron [MV1], [MV2], [MV3], [MV4]. Important motivations coming from the theory of probability—and the use of probabilistic methods—have reinvigorated the whole subject; see the pioneering papers of Le Gall [LG1], [LG2], the recent books of Dynkin [D1], [D2], and the numerous references therein. It is known that (4.9.8) has no solution if $g(t) = t^p, t \geq 0$, with $p \geq p_c = \frac{N+1}{N-1}$ and $\mu = \delta_a, a \in \partial\Omega$ (see [GV]). Therefore, it is interesting to develop for (4.9.8) the same program as in this paper. More precisely, let (g_k) be a sequence of functions $g_k : \mathbb{R} \to \mathbb{R}$ which are continuous, nondecreasing, and satisfy (4.0.7) and (4.0.8). Assume, in addition, that each g_k is, e.g., bounded. Then, for every $\mu \in \mathcal{M}(\partial\Omega)$, there exists a unique solution u_k of

$$\begin{cases} -\Delta u_k + g_k(u_k) = 0 & \text{in } \Omega, \\ \qquad\qquad u_k = \mu & \text{on } \partial\Omega, \end{cases} \tag{4.9.9}$$

in the sense that $u_k \in L^1(\Omega)$ and

$$-\int_\Omega u_k \Delta \zeta + \int_\Omega g_k(u_k)\zeta = -\int_{\partial\Omega} \frac{\partial \zeta}{\partial n} d\mu \quad \forall \zeta \in C_0^2(\bar\Omega), \tag{4.9.10}$$

where $\frac{\partial}{\partial n}$ denotes the derivative with respect to the outward normal of $\partial\Omega$. We have the following:

THEOREM 4.19 *As $k \uparrow \infty$, $u_k \downarrow u^*$ in $L^1(\Omega)$, where u^* satisfies*

$$\begin{cases} -\Delta u^* + g(u^*) = 0 & \text{in } \Omega, \\ \qquad\qquad u^* = \mu^* & \text{on } \partial\Omega, \end{cases} \tag{4.9.11}$$

for some $\mu^* \in \mathcal{M}(\partial\Omega)$ such that $\mu^* \leq \mu$. More precisely, $g(u^*)\rho_0 \in L^1(\Omega)$, where $\rho_0(x) = d(x, \partial\Omega)$, and

$$-\int_\Omega u^* \Delta \zeta + \int_\Omega g(u^*)\zeta = -\int_{\partial\Omega} \frac{\partial \zeta}{\partial n} d\mu^* \quad \forall \zeta \in C_0^2(\bar\Omega). \tag{4.9.12}$$

In addition, u^* is the largest subsolution of (4.9.8), i.e., if $v \in L^1(\Omega)$ is any function satisfying $g(v)\rho_0 \in L^1(\Omega)$ and

$$-\int_\Omega v\Delta \zeta + \int_\Omega g(v)\zeta \leq -\int_{\partial\Omega} \frac{\partial \zeta}{\partial n} d\mu \quad \forall \zeta \in C_0^2(\bar\Omega), \zeta \geq 0 \text{ in } \Omega, \tag{4.9.13}$$

then $v \leq u^*$ a.e. in Ω.

Proof. By comparison (see Corollary 4.B.2), we know that (u_k) is nonincreasing. By standard estimates, we have

$$\int_\Omega |u_k| + \int_\Omega g_k(u_k)\rho_0 \leq C\|\mu\|_{\mathcal{M}(\partial\Omega)} \quad \forall k \geq 1.$$

In addition (see [B5, Theorem 3]),

$$\|u_k\|_{C^1(\bar\omega)} \leq C_\omega \quad \forall k \geq 1,$$

for every $\omega \subset\subset \Omega$. Thus, u_k converges in $L^1(\Omega)$ to a limit, say u^*. Moreover,

$$g_k(u_k) \to g(u^*) \quad \text{in } L_{loc}^\infty(\Omega).$$

Let $\zeta_0 \in C_0^2(\bar\Omega)$ be the solution of

$$\begin{cases} -\Delta \zeta_0 = 1 & \text{in } \Omega, \\ \qquad\;\; \zeta_0 = 0 & \text{on } \partial\Omega. \end{cases}$$

Since $(g_k(u_k)\zeta_0)$ is uniformly bounded in $L^1(\Omega)$, then up to a subsequence

$$g_k(u_k)\zeta_0 \overset{*}{\rightharpoonup} g(u^*)\zeta_0 + \lambda \quad \text{in } [C(\bar{\Omega})]^*, \tag{4.9.14}$$

for some $\lambda \in \mathcal{M}(\partial\Omega)$, $\lambda \geq 0$. We claim that

$$\int_{\Omega} g_k(u_k)\zeta \to \int_{\Omega} g(u^*)\zeta + \int_{\partial\Omega} \frac{\partial\zeta}{\partial n}\frac{1}{\frac{\partial\zeta_0}{\partial n}}\,d\lambda \quad \forall \zeta \in C_0^2(\bar{\Omega}). \tag{4.9.15}$$

In fact, given $\zeta \in C_0^2(\bar{\Omega})$, define $\gamma = \zeta/\zeta_0$. It is easy to see that $\gamma \in C(\bar{\Omega})$ and $\gamma = \frac{\partial\zeta}{\partial n}\frac{1}{\frac{\partial\zeta_0}{\partial n}}$ on $\partial\Omega$. Using γ as a test function in (4.9.14), we obtain (4.9.15).

Let $k \to \infty$ in (4.9.10). In view of (4.9.15), we conclude that u^* satisfies (4.9.12), where μ^* is given by

$$\mu^* = \mu + \frac{1}{\frac{\partial\zeta_0}{\partial n}}\lambda \leq \mu. \tag{4.9.16}$$

Finally, it follows from Corollary 4.B.2 that if v is a subsolution of (4.9.8), then $v \leq u_k$ a.e., $\forall k \geq 1$, and thus $v \leq u^*$ a.e.

Some natural questions have been addressed and the following results will be presented in a forthcoming paper (see [BP3]):

(a) the reduced measure μ^* is the largest good measure $\leq \mu$; in other words, if $v \in \mathcal{M}(\partial\Omega)$ is a good measure (i.e., (4.9.8) has a solution with boundary data v) and if $v \leq \mu$, then $v \leq \mu^*$;

(b) $\mu - \mu^*$ is concentrated on a subset of $\partial\Omega$ of zero $\mathcal{H}^{N-1}$-measure (i.e., $(N-1)$-dimensional Lebesgue measure on $\partial\Omega$) and this fact is "optimal", in the sense that any measure $v \geq 0$ which is singular with respect to $\mathcal{H}^{N-1}\lfloor_{\partial\Omega}$ can be written as $v = \mu - \mu^*$ for some $\mu \geq 0$ and some g;

(c) if μ is a measure on $\partial\Omega$ which is good for every g, then $\mu^+ \in L^1(\partial\Omega)$;

(d) given any g, there exists some measure $\mu \geq 0$ on $\partial\Omega$ which is good for g, but $\mu \notin L^1(\partial\Omega)$.

When $g(t) = t^p$, $t \geq 0$, with $p \geq \frac{N+1}{N-1}$, a known result (see, e.g., [MV3]) asserts that $\mu \in \mathcal{M}(\partial\Omega)$ is a good measure if and only if $\mu^+(A) = 0$ for every Borel set $A \subset \partial\Omega$ such that $C_{2/p,p'}(A) = 0$, where $C_{2/p,p'}$ refers to the Bessel capacity on $\partial\Omega$. In this case, we have

(e) the reduced measure μ^* is given by $\mu^* = \mu - (\mu_2)^+$, where $\mu = \mu_1 + \mu_2$ is the decomposition of μ, in the sense of Lemma 4.A.1, relative to $C_{2/p,p'}$.

In contrast with Example 4.5, we do not know what the reduced measure μ^* is when $N = 2$ and $g(t) = e^t - 1$, $t \geq 0$.

Similar issues can be investigated for the parabolic equations

$$\begin{cases} u_t - \Delta u + g(u) = \mu, \\ \quad\quad\quad u(0) = 0, \end{cases} \quad \text{or} \quad \begin{cases} u_t - \Delta u + g(u) = 0, \\ \quad\quad\quad u(0) = \mu. \end{cases}$$

APPENDIX 4A. DECOMPOSITION OF MEASURES
INTO DIFFUSE AND CONCENTRATED PARTS

The following result is taken from [FTS]. We reproduce their proof for the convenience of the reader.

LEMMA 4.A.1 *Let μ be a bounded Borel measure in $\mathbb{R}^N$ and let $\mathcal{Z}$ be a collection of Borel sets such that:*

(a) *$\mathcal{Z}$ is closed with respect to finite or countable unions;*
(b) *$A \in \mathcal{Z}$ and $A' \subset A$ Borel $\Rightarrow A' \in \mathcal{Z}$.*

Then μ can be represented in the form

$$\mu = \mu_1 + \mu_2, \tag{4.A.1}$$

where μ_1 and μ_2 are bounded Borel measures such that

$$\mu_1(A) = 0 \quad \forall A \in \mathcal{Z} \quad \text{and} \quad \mu_2 \text{ vanishes outside a set } A_0 \in \mathcal{Z}.$$

This representation is unique.

Proof. First assume that μ is nonnegative. Denote

$$X_\mu = \sup \{\mu(A); A \in \mathcal{Z}\}.$$

Let $\{A_n\}$ be an increasing sequence of sets in $\mathcal{Z}$ such that

$$\mu(A_n) \to X_\mu.$$

Let $A_0 = \bigcup_n A_n$ and put

$$\mu_1(B) = \mu(B \cap A_0^c), \quad \mu_2(B) = \mu(B \cap A_0),$$

for every Borel set B. Since $A_0 \in \mathcal{Z}$, it remains to verify that μ_1 vanishes on sets of $\mathcal{Z}$. By contradiction, suppose that there exists $E \in \mathcal{Z}$ such that $\mu_1(E) > 0$. Let $E_1 = E \cap A_0^c$. Then $\mu(E_1) > 0$ and $E_1 \in \mathcal{Z}$. It follows that $A_0 \cup E_1 \in \mathcal{Z}$ and $\mu(A_0 \cup E_1) > X_\mu$. Contradiction.

If μ is a signed measure, apply the above to μ^+ and μ^-. The uniqueness is obvious.

APPENDIX 4B. STANDARD EXISTENCE, UNIQUENESS,
AND COMPARISON RESULTS

In this appendix, we collect some well-known results (and a few new ones) that are used throughout this paper. For the convenience of the reader, we shall sketch some of the proofs.

We start with the existence, uniqueness, and regularity of solutions of the linear problem

$$\begin{cases} -\Delta u = \mu & \text{in } \Omega, \\ u = 0 & \text{on } \partial\Omega, \end{cases} \tag{4.B.1}$$

where $\mu \in \mathcal{M}(\Omega)$.

THEOREM 4.B.1 *Given $\mu \in \mathcal{M}(\Omega)$, there exists a unique $u \in L^1(\Omega)$ satisfying*

$$- \int_\Omega u \Delta \zeta = \int_\Omega \zeta \, d\mu \quad \forall \zeta \in C_0^2(\bar{\Omega}). \tag{4.B.2}$$

Moreover, $u \in W_0^{1,q}(\Omega)$ for every $1 \le q < \frac{N}{N-1}$, with the estimates

$$\|u\|_{L^{q^*}} \le C\|\nabla u\|_{L^q} \le C\|\mu\|_{\mathcal{M}}, \tag{4.B.3}$$

where $\frac{1}{q^*} = \frac{1}{q} - \frac{1}{N}$. In particular, $u \in L^P(\Omega)$ for every $1 \le p < \frac{N}{N-2}$, and u satisfies

$$\int_\Omega \nabla u \cdot \nabla \psi = \int_\Omega \psi \, d\mu \quad \forall \psi \in W_0^{1,r}(\Omega), \tag{4.B.4}$$

for any $r > N$.

The proof of Theorem 4.B.1 relies on a standard duality argument and shall be omitted; see [S, Théorème 8.1].

We now establish a weak form of the maximum principle:

PROPOSITION 4.B.1 *Let $v \in W_0^{1,1}(\Omega)$ be such that*

$$- \int_\Omega v \Delta \varphi \le 0 \quad \forall \varphi \in C_c^\infty(\Omega), \varphi \ge 0 \text{ in } \Omega. \tag{4.B.5}$$

Then

$$- \int_\Omega v \Delta \zeta \le 0 \quad \forall \zeta \in C_0^2(\bar{\Omega}), \zeta \ge 0 \text{ in } \Omega, \tag{4.B.6}$$

and, consequently,

$$v \le 0 \quad a.e. \tag{4.B.7}$$

Proof. From (4.B.5) we have

$$\int_\Omega \nabla v \cdot \nabla \varphi \le 0 \quad \forall \varphi \in C_c^\infty(\Omega), \varphi \ge 0 \text{ in } \Omega$$

so that, by density of $C_c^\infty(\Omega)$ in $C_c^2(\Omega)$,

$$\int_\Omega \nabla v \cdot \nabla \varphi \le 0 \quad \forall \varphi \in C_c^2(\Omega), \varphi \ge 0 \text{ in } \Omega.$$

Let (γ_n) be a sequence in $C_c^\infty(\Omega)$ such that $0 \le \gamma_n \le 1$, $\gamma_n(x) = 1$ if $d(x, \partial\Omega) > \frac{1}{n}$, and $|\nabla \zeta_n| \le Cn, \forall n \ge 1$. For any $\zeta \in C_0^2(\bar{\Omega}), \zeta \ge 0$, we have

$$\int_\Omega \nabla v \cdot (\gamma_n \nabla \zeta + \zeta \nabla \gamma_n) = \int_\Omega \nabla v \cdot \nabla(\gamma_n \zeta) \le 0. \tag{4.B.8}$$

Note that

$$\int_\Omega |\nabla v||\nabla \gamma_n|\zeta \leq Cn \int_{d(x,\partial\Omega)\leq\frac{1}{n}} |\nabla v|\zeta \leq C \int_{d(x,\partial\Omega)\leq\frac{1}{n}} |\nabla v| \to 0 \quad \text{as } n \to \infty.$$

Thus, as $n \to \infty$ in (4.B.8), we obtain

$$\int_\Omega \nabla v \cdot \nabla \zeta \leq 0 \quad \forall \zeta \in C_0^2(\bar{\Omega}), \zeta \geq 0 \text{ in } \Omega,$$

which yields (4.B.6) since $v \in W_0^{1,1}(\Omega)$. Inequality (4.B.7) is a trivial consequence of (4.B.6).

LEMMA 4.B.1 *Let $p : \mathbb{R} \to \mathbb{R}$, $p(0) = 0$ be a bounded nondecreasing continuous function. Given $f \in L^1(\Omega)$, let $u \in L^1(\Omega)$ be the unique solution of*

$$-\int_\Omega u\Delta\zeta = \int_\Omega f\zeta \quad \forall \zeta \in C_0^2(\bar{\Omega}). \tag{4.B.9}$$

Then

$$\int_\Omega fp(u) \geq 0. \tag{4.B.10}$$

Proof. Clearly, it suffices to establish the lemma for $p \in C^2(\mathbb{R})$. Assume for the moment $f \in C^\infty(\bar{\Omega})$. In this case, $u \in C_0^2(\bar{\Omega})$. Since $p(0) = 0$, we have $p(u) \in C_0^2(\bar{\Omega})$. Using $p(u)$ as a test function in (4.B.9), we get

$$\int_\Omega fp(u) = \int_\Omega p'(u)|\nabla u|^2 \geq 0.$$

This establishes the lemma for f smooth. The general case when f is just an L^1-function, not necessarily smooth, easily follows by density.

PROPOSITION 4.B.2 *Given $f \in L^1(\Omega)$, let u be the unique solution of (4.B.9). Then, for every $M > 0$, we have*

$$\int_{[u\geq M]} f \geq 0 \quad \text{and} \quad \int_{[u\leq -M]} f \leq 0. \tag{4.B.11}$$

In particular,

$$\int_{[|u|\geq M]} f \operatorname{sgn}(u) \geq 0. \tag{4.B.12}$$

Above, we denote by sgn the function $\operatorname{sgn}(t) = 1$ if $t > 0$, $\operatorname{sgn}(t) = -1$ if $t < 0$, and $\operatorname{sgn}(0) = 0$.

Proof. Clearly, it suffices to establish the first inequality in (4.B.11). Let (p_n) be a sequence of continuous functions in $\mathbb{R}$ such that each p_n is nondecreasing, $p_n(t) = 1$ if $t \geq M$ and $p_n(t) = 0$ if $t \leq M - \frac{1}{n}$. By the previous lemma,

$$\int_\Omega fp_n(u) \geq 0 \quad \forall n \geq 1.$$

As $n \to \infty$, the result follows.

PROPOSITION 4.B.3 *Let* $v \in L^1(\Omega)$, $f \in L^1(\Omega)$ *and* $\nu \in \mathcal{M}(\Omega)$ *satisfy*

$$- \int_\Omega v \Delta \zeta + \int_\Omega f \zeta = \int_\Omega \zeta \, d\nu \quad \forall \zeta \in C_0^2(\bar{\Omega}). \tag{4.B.13}$$

Then

$$\int_{[v>0]} f \le \|\nu^+\|_{\mathcal{M}} \tag{4.B.14}$$

and thus

$$\int_\Omega f \, \mathrm{sgn}\,(v) \le \|\nu\|_{\mathcal{M}}. \tag{4.B.15}$$

Proof. Let $\nu_n = \rho_n * \nu$ (here we use the same notation as in Section 4.4). Let v_n denote the solution of (4.B.13) with ν replaced by ν_n. By Lemma 4.B.1, we have

$$\int_\Omega (v_n - f) \, p(v_n) \ge 0,$$

where p is any function satisfying the assumptions of the lemma. Thus, if $0 \le p(t) \le 1$, $\forall t \in \mathbb{R}$, then we have

$$\int_\Omega f p(v_n) \le \int_\Omega v_n p(v_n) \le \int_\Omega (v_n)^+ \le \|\nu^+\|_{\mathcal{M}}.$$

Let $n \to \infty$ to get

$$\int_\Omega f p(v) \le \|\nu^+\|_{\mathcal{M}}. \tag{4.B.16}$$

Apply (4.B.16) to a sequence of nondecreasing continuous functions (p_n) such that $p_n(t) = 0$ if $t \le 0$ and $p_n(t) = 1$ if $t \ge \frac{1}{n}$. As $n \to \infty$, we obtain (4.B.14). $\blacksquare$

An easy consequence of Proposition 4.B.3 is the following:

COROLLARY 4.B.1 *Let* $g : \mathbb{R} \to \mathbb{R}$ *be a continuous, nondecreasing function such that* $g(0) = 0$. *Given* $\mu \in \mathcal{M}(\Omega)$, *then the equation*

$$\begin{cases} -\Delta u + g(u) = \mu & \text{in } \Omega, \\ u = 0 & \text{on } \partial\Omega, \end{cases} \tag{4.B.17}$$

has at most one solution $u \in L^1(\Omega)$ *with* $g(u) \in L^1(\Omega)$. *Moreover,*

$$\int_\Omega |g(u)| \le \|\mu\|_{\mathcal{M}} \quad \text{and} \quad \int_\Omega |\Delta u| \le 2\|\mu\|_{\mathcal{M}}. \tag{4.B.18}$$

If (B.17) has a solution for $\mu_1, \mu_2 \in \mathcal{M}(\Omega)$, *say* u_1, u_2, *resp., then*

$$\int_\Omega [g(u_1) - g(u_2)]^+ \le \|(\mu_1 - \mu_2)^+\|_{\mathcal{M}}. \tag{4.B.19}$$

In particular,

$$\int_\Omega |g(u_1) - g(u_2)| \le \|\mu_1 - \mu_2\|_{\mathcal{M}}. \tag{4.B.20}$$

We now recall the following unpublished result of H. Brezis from 1972 (see, e.g., [GV]):

PROPOSITION 4.B.4 *Given* $f \in L^1(\Omega; \rho_0 \, dx)$ *and* $h \in L^1(\partial\Omega)$, *there exists a unique* $u \in L^1(\Omega)$ *such that*

$$-\int_\Omega u \Delta \zeta = \int_\Omega f\zeta - \int_{\partial\Omega} h \frac{\partial \zeta}{\partial n} \quad \forall \zeta \in C_0^2(\bar{\Omega}). \tag{4.B.21}$$

In addition, there exists $C > 0$ *such that*

$$\|u\|_{L^1} \le C(\|f\rho_0\|_{L^1(\Omega)} + \|h\|_{L^1(\partial\Omega)}). \tag{4.B.22}$$

We now establish the following:

LEMMA 4.B.2 *Given* $f \in L^1(\Omega; \rho_0 \, dx)$, *let* $u \in L^1(\Omega)$ *be the unique solution of* (4.B.21) *with* $h = 0$. *Then*

$$k \int_{d(x,\partial\Omega) < \frac{1}{k}} |u| \to 0 \quad as \ k \to \infty. \tag{4.B.23}$$

Proof.

Step 1. Proof of the lemma when $f \ge 0$.

Since $f \ge 0$, we have $u \ge 0$. Let $H \in C^2(\mathbb{R})$ be a nondecreasing concave function such that $H(0) = 0$, $H''(t) = -1$ if $t \le 1$ and $H(t) = 1$ if $t \ge 2$. We denote by $\zeta_0 \in C_0^2(\bar{\Omega})$, $\zeta_0 \ge 0$, the solution of

$$\begin{cases} -\Delta \zeta_0 = 1 & \text{in } \Omega, \\ \quad \zeta_0 = 0 & \text{on } \partial\Omega. \end{cases}$$

For any $k \ge 1$, let $w_k = \frac{1}{k} H(k\zeta_0)$. By construction, $w_k \in C_0^2(\bar{\Omega})$ and

$$\Delta w_k = k H''(k\zeta_0)|\nabla \zeta_0|^2 + H'(k\zeta_0)\Delta \zeta_0 \le -k \chi_{[\zeta_0 \le \frac{1}{k}]}|\nabla \zeta_0|^2.$$

Thus,

$$-\int_\Omega u \Delta w_k \ge k \int_{[\zeta_0 \le \frac{1}{k}]} |\nabla \zeta_0|^2 u. \tag{4.B.24}$$

Use w_k as a test function in (4.B.21) (recall that $h = 0$). It follows from (4.B.24) that

$$k \int_{[\zeta_0 \le \frac{1}{k}]} |\nabla \zeta_0|^2 u \le \int_\Omega w_k f. \tag{4.B.25}$$

By Hopf's lemma, we have $|\nabla \zeta_0|^2 \ge \alpha_0 > 0$ in some neighborhood of $\partial\Omega$ in $\bar{\Omega}$. In particular, there exists $c > 0$ such that $c\zeta_0(x) \le d(x, \partial\Omega) \le \frac{1}{c}\zeta_0(x)$ for all $x \in \bar{\Omega}$. Thus, for $k \ge 1$ sufficiently large, we have

$$\alpha_0 k \int_{d(x,\partial\Omega) \le \frac{c}{k}} |\nabla \zeta_0|^2 u \le \int_\Omega w_k f. \tag{4.B.26}$$

Note that the right-hand side of (4.B.26) tends to 0 as $k \to \infty$. In fact, we have $w_k \le C\zeta_0$, $\forall k \ge 1$, and $w_k \le \frac{1}{k}H(k\zeta_0) \to 0$ a.e. Thus, by dominated convergence,

$$\int_\Omega w_k f \to 0 \quad \text{as } k \to \infty. \tag{4.B.27}$$

Combining (4.B.26) and (4.B.27), we obtain (4.B.23).

Step 2. Proof of the lemma completed.

Let $v \in L^1(\Omega)$ denote the unique solution of

$$-\int_\Omega v\Delta\zeta = \int_\Omega |f|\zeta \quad \forall \zeta \in C_0^2(\bar\Omega). \tag{4.B.28}$$

By comparison, we have $|u| \le v$. On the other hand, v satisfies the assumption of Step 1. Thus,

$$k\int_{d(x,\partial\Omega)<\frac1k} |u| \le k\int_{d(x,\partial\Omega)<\frac1k} v \to 0 \quad \text{as } k \to \infty. \tag{4.B.29}$$

This establishes Lemma 4.B.2.

The next result is a new variant of Kato's inequality, where the test function ζ need not have compact support in Ω:

PROPOSITION 4.B.5 *Let $u \in L^1(\Omega)$ and $f \in L^1(\Omega; \rho_0\,dx)$ be such that*

$$-\int_\Omega u\Delta\zeta \le \int_\Omega f\zeta \quad \forall \zeta \in C_0^2(\bar\Omega), \zeta \ge 0 \text{ in } \Omega. \tag{4.B.30}$$

Then

$$-\int_\Omega u^+\Delta\zeta \le \int_{[u\ge0]} f\zeta \quad \forall \zeta \in C_0^2(\bar\Omega), \zeta \ge 0 \text{ in } \Omega. \tag{4.B.31}$$

Proof. We first notice that

$$-\int_\Omega u^+\Delta\varphi \le \int_{[u\ge0]} f\varphi \quad \forall \varphi \in C_c^\infty(\Omega), \varphi \ge 0 \text{ in } \Omega. \tag{4.B.32}$$

In fact, by (4.B.30) we have $-\Delta u \le f$ in $\mathcal{D}'(\Omega)$. Then, Theorem 4.7 yields

$$(-\Delta u^+)_d \le \chi_{[u\ge0]}(-\Delta u)_d \le \chi_{[u\ge0]}f \quad \text{and}$$
$$(-\Delta u^+)_c = (-\Delta u)_c^+ \le (f)_c^+ = 0.$$

Thus,

$$-\Delta u^+ = (-\Delta u^+)_d + (-\Delta u^+)_c \le \chi_{[u\ge0]}f \quad \text{in } \mathcal{D}'(\Omega),$$

which is precisely (4.B.32).

Let $(\gamma_k) \subset C_c^\infty(\Omega)$ be a sequence such that $0 \le \gamma_k \le 1$ in Ω, $\gamma_k(x) = 1$ if $d(x, \partial\Omega) \ge \frac{1}{k}$, $\|\nabla\gamma_k\|_{L^\infty} \le k$, and $\|\Delta\gamma_k\|_{L^\infty} \le Ck^2$. Given $\zeta \in C_0^2(\bar\Omega)$, $\zeta \ge 0$, we apply (4.B.32) with $\varphi = \zeta\gamma_k$ to get

$$-\int_\Omega u^+ \Delta(\zeta\gamma_k) \le \int_{[u \ge 0]} f\zeta\gamma_k. \tag{4.B.33}$$

Consider again the unique solution $v \ge 0$ of (4.B.28). By comparison we have $u \le v$ a.e. and thus $u^+ \le v$ a.e. From Lemma 4.B.2 we see that

$$\int_\Omega u^+ |\nabla\zeta||\nabla\gamma_k| \le Ck \int_{d(x,\partial\Omega)<\frac{1}{k}} u^+ \to 0 \quad \text{as } k \to \infty. \tag{4.B.34}$$

Similarly,

$$\int_\Omega u^+ \zeta |\Delta\gamma_k| \le Ck \int_{d(x,\partial\Omega)<\frac{1}{k}} u^+ \to 0 \quad \text{as } k \to \infty. \tag{4.B.35}$$

Let $k \to \infty$ in (4.B.33). Using (4.B.34) and (4.B.35), we obtain (4.B.31).

Remark 4.B.1. There is an alternative proof of Proposition 4.B.5. First, one shows that (4.B.30) implies that there exist two measures $\mu \le 0, \lambda \le 0$, where $\mu \in \mathcal{M}(\partial\Omega)$ and λ is locally bounded in Ω, with $\int_\Omega \rho_0 \, d|\lambda| < \infty$, satisfying

$$-\int_\Omega u\Delta\zeta = \int_\Omega f\zeta + \int_\Omega \zeta \, d\lambda - \int_{\partial\Omega} \frac{\partial\zeta}{\partial n} d\mu \quad \forall \zeta \in C_0^2(\bar\Omega). \tag{4.B.36}$$

(The existence of λ is fairly straightforward, and the existence of μ is a consequence of Herglotz's theorem concerning positive superharmonic functions.)
Then, inequality (4.B.31) follows from (4.B.36) using the same strategy as in the proof of Lemma 1.5 in [MV2].

As a consequence of Proposition 4.B.5, we have the following:

COROLLARY 4.B.2 *Let $g_1, g_2 : \mathbb{R} \to \mathbb{R}$ be two continuous nondecreasing functions such that $g_1 \le g_2$. Let $u_k \in L^1(\Omega)$, $k = 1, 2$, be such that $g_k(u_k) \in L^1(\Omega; \rho_0 \, dx)$. If*

$$-\int_\Omega (u_2 - u_1)\Delta\zeta + \int_\Omega [g_2(u_2) - g_1(u_1)]\zeta \le 0 \quad \forall \zeta \in C_0^2(\bar\Omega), \zeta \ge 0 \text{ in } \Omega, \tag{4.B.37}$$

then

$$u_2 \le u_1 \quad a.e. \tag{4.B.38}$$

Proof. Applying Proposition 4.B.5 to $u = u_2 - u_1$ and $f = g_1(u_1) - g_2(u_2)$, we have

$$-\int_\Omega (u_2 - u_1)^+ \Delta\zeta \le -\int_\Omega [g_2(u_2) - g_1(u_1)]^+\zeta \le 0$$
$$\forall \zeta \in C_0^2(\bar\Omega), \zeta \ge 0 \text{ in } \Omega.$$

This immediately implies that $u_2 \le u_1$ a.e.

We now present some general existence results for problem (4.B.17). Below, $g : \mathbb{R} \to \mathbb{R}$ denotes a continuous, nondecreasing function, such that $g(0) = 0$.

THEOREM 4.B.2 (Brezis-Strauss [BS]) *For every $f \in L^1(\Omega)$, the equation*

$$\begin{cases} -\Delta u + g(u) = f & \text{in } \Omega, \\ \qquad\qquad u = 0 & \text{on } \partial\Omega \end{cases} \tag{4.B.39}$$

has a unique solution $u \in L^1(\Omega)$ with $g(u) \in L^1(\Omega)$.

Proof. We first observe that if $f \in C^\infty(\bar{\Omega})$, then (4.B.39) always has a solution $u \in C^1(\bar{\Omega})$ (easily obtained via minimization).

For a general $f \in L^1(\Omega)$, let (f_n) be a sequence of smooth functions on $\bar{\Omega}$, converging to f in $L^1(\Omega)$. For each f_n, let u_n denote the corresponding solution of (4.B.39). By (4.B.20), the sequence $(g(u_n))$ is Cauchy in $L^1(\Omega)$. We then conclude from (4.B.3) that (u_n) is also Cauchy in $L^1(\Omega)$, so that

$$u_n \to u \quad \text{and} \quad g(u_n) \to g(u) \quad \text{in } L^1(\Omega).$$

Thus u is a solution of (4.B.39). The uniqueness follows from Corollary 4.B.1.

THEOREM 4.B.3 (Brezis-Browder [BBr]) *For every $T \in H^{-1}(\Omega)$, the equation*

$$\begin{cases} -\Delta u + g(u) = T & \text{in } \Omega, \\ \qquad\qquad u = 0 & \text{on } \partial\Omega, \end{cases} \tag{4.B.40}$$

has a unique solution $u \in H_0^1(\Omega)$ with $g(u) \in L^1(\Omega)$.

Proof. Assume g is uniformly bounded. In this case, the existence of u presents no difficulty, e.g., via a minimization argument in $H_0^1(\Omega)$. In particular, we see that $u \in H_0^1(\Omega)$.

For a general nonlinearity g, let (g_n) be the sequence given by $g_n(t) = g(t)$ if $|t| \le n$, $g_n(t) = g(n)$ if $t > n$, and $g_n(t) = g(-n)$ if $t < -n$. Let $u_n \in H_0^1(\Omega)$ be the solution of (4.B.40) corresponding to g_n. Note that u_n satisfies

$$\int_\Omega \nabla u_n \cdot \nabla v + \int_\Omega g_n(u_n)v = \langle T, v \rangle \quad \forall v \in H_0^1(\Omega).$$

Using $v = u_n$ as a test function, we get

$$\int_\Omega |\nabla u_n|^2 + \int_\Omega g_n(u_n)u_n = \langle T, u_n \rangle \le C \left(\int_\Omega |\nabla u_n|^2 \right)^{1/2}.$$

Thus,

$$\int_\Omega g_n(u_n)u_n \le C \quad \text{and} \quad \int_\Omega |\nabla u_n|^2 \le C, \tag{4.B.41}$$

for some constant $C > 0$ independent of $n \ge 1$. Since (u_n) is uniformly bounded in $H_0^1(\Omega)$, then up to a subsequence we can find $u \in H_0^1(\Omega)$ such that

$$u_n \to u \quad \text{in } L^1 \text{ and a.e.}$$

By (4.B.41), for any $M > 0$, we also have

$$\int_{[|u_n| \geq M]} \int |g_n(u_n)| \leq \frac{1}{M} \int_\Omega g_n(u_n) u_n \leq \frac{C}{M}.$$

We claim that

$$g_n(u_n) \quad \text{is equi-integrable.}$$

In fact, for any Borel set $E \subset \Omega$, we estimate

$$\int_E |g_n(u_n)| = \int_{\substack{E \\ [|u_n| < M]}} |g_n(u_n)| + \int_{\substack{E \\ [|u_n| \geq M]}} |g_n(u_n)| \leq A_M |E| + \frac{C}{M},$$

where $A_M = \max\{g(M), -g(-M)\}$. Given $\varepsilon > 0$, let $M > 0$ be sufficiently large so that $\frac{C}{M} < \varepsilon$. With M fixed, we take $|E|$ small enough so that $A_M |E| < \varepsilon$. We conclude that

$$\int_E |g_n(u_n)| < 2\varepsilon \quad \forall n \geq 1.$$

Thus, $(g_n(u_n))$ is equi-integrable. Since $u_n \to u$ a.e., it follows from Egorov's lemma that $g_n(u_n) \to g(u)$ in $L^1(\Omega)$. Therefore, u satisfies (4.B.40). By Proposition 4.B.3, this solution is unique.

Combining the techniques from both proofs, we have the following:

THEOREM 4.B.4 *For every $f \in L^1(\Omega)$ and every $T \in H^{-1}(\Omega)$, the equation*

$$\begin{cases} -\Delta u + g(u) = f + T & \text{in } \Omega, \\ u = 0 & \text{on } \partial\Omega \end{cases} \tag{4.B.42}$$

has a unique solution $u \in L^1(\Omega)$ with $g(u) \in L^1(\Omega)$.

Proof. Let f_n be a sequence in $C^\infty(\bar{\Omega})$ converging to f in $L^1(\Omega)$. Since $f_n + T \in H^{-1}$, we can apply Theorem 4.B.3 to obtain a solution u_n of (4.B.42) for $f_n + T$. For every $n_1, n_2 \geq 1$, we have

$$-\Delta(u_{n_1} - u_{n_2}) + g(u_{n_1}) - g(u_{n_2}) = f_{n_1} - f_{n_2} \quad \text{in } (C_0^2)^*. \tag{4.B.43}$$

It follows from Proposition 4.B.3 that

$$\int_\Omega |g(u_{n_1}) - g(u_{n_2})| \leq \int_\Omega |f_{n_1} - f_{n_2}|.$$

Thus, $(g(u_n))$ is a Cauchy sequence. Returning to (4.B.43), we conclude from (4.B.3) that (u_n) is Cauchy in $L^1(\Omega)$. Passing to the limit as $n \to \infty$, we find a solution $u \in L^1(\Omega)$ of (4.B.42). By Proposition 4.B.3, the solution is unique. $\quad\blacksquare$

COROLLARY 4.B.3 *Let $\mu \in \mathcal{M}(\Omega)$. If μ is diffuse, then (4.B.17) admits a unique solution $u \in L^1(\Omega)$ with $g(u) \in L^1(\Omega)$.*

Proof. It suffices to observe that, by a result of Boccardo-Gallouët-Orsina [BGO1], every diffuse measure μ belongs to $L^1 + H^{-1}$. $\quad\blacksquare$

Concerning the existence of solutions for *every* measure $\mu \in \mathcal{M}(\Omega)$, we have

THEOREM 4.B.5 (Bénilan-Brezis [BB]) *Assume $N \geq 2$ and*

$$|g(t)| \leq C(|t|^p + 1) \quad \forall t \in \mathbb{R}, \tag{4.B.44}$$

for some $p < \frac{N}{N-2}$. Then, for every $\mu \in \mathcal{M}(\Omega)$, problem (4.B.17) has a unique solution $u \in L^1(\Omega)$.

Assumption (4.B.44) is optimal, in the sense that if $N \geq 3$, $g(t) = |t|^{p-1}t$ and $p \geq \frac{N}{N-2}$, then (4.B.17) has no weak solution for $\mu = \delta_a$, where $a \in \Omega$:

THEOREM 4.B.6 (Bénilan-Brezis [BB]; Brezis-Véron [BV]) *Assume $N \geq 3$. If $p \geq \frac{N}{N-2}$, then, for any $a \in \Omega$, the problem*

$$\begin{cases} -\Delta u + |u|^{p-1}u = \delta_a & in \ \Omega, \\ \qquad\qquad\qquad u = 0 & on \ \partial\Omega \end{cases}$$

has no solution $u \in L^p(\Omega)$.

APPENDIX 4C. CORRESPONDENCE BETWEEN $[C_0(\bar{\Omega})]^*$ AND $[C(\bar{\Omega})]^*$

In this section we establish the following:

PROPOSITION 4.C.1 *Given $\mu \in [C_0(\bar{\Omega})]^*$, there exists a unique $\tilde{\mu} \in [C(\bar{\Omega})]^*$ such that*

$$\tilde{\mu} = \mu \quad on \ C_0(\bar{\Omega}) \quad and \quad |\tilde{\mu}|(\partial\Omega) = 0. \tag{4.C.1}$$

In addition, the map $\mu \mapsto \tilde{\mu}$ is a linear isometry.

In order to prove Proposition 4.C.1, we shall need the following:

LEMMA 4.C.1 *Given $\varepsilon > 0$, there exists $\delta > 0$ such that if $\zeta \in C_0(\bar{\Omega})$, $|\zeta| \leq 1$ in $\bar{\Omega}$, and supp $\zeta \subset \bar{\Omega}\backslash\Omega_\delta$, then*

$$|\langle \mu, \zeta \rangle| \leq \varepsilon.$$

Here, we denote by Ω_δ the set $\{x \in \Omega; d(x, \partial\Omega) > \delta\}$.

Proof. We argue by contradiction. Assume there exist $\varepsilon_0 > 0$ and a sequence $(\zeta_n) \subset C_0(\bar{\Omega})$ such that $|\zeta_n| \leq 1$ in $\bar{\Omega}$, supp $\zeta_n \subset \bar{\Omega}\backslash\Omega_{1/n}$, and

$$\langle \mu, \zeta_n \rangle > \varepsilon_0 \quad \forall n \geq 1.$$

Without loss of generality, we may assume that each ζ_n has compact support in Ω (this is always possible, by density of $C_c^\infty(\Omega)$ in $C_0(\bar{\Omega})$). In particular, we can extract a subsequence (ζ_{n_j}) such that supp ζ_{n_j} are all disjoint. For any $k \geq 1$, let $\tilde{\zeta}_k = \sum_{j=1}^k \zeta_{n_j}$. By construction,

$$\|\tilde{\zeta}_k\|_{L^\infty} \leq 1 \quad and \quad supp \ \tilde{\zeta}_k \subset \Omega.$$

Moreover,

$$k\varepsilon_0 < \langle \mu, \tilde{\zeta}_k \rangle \leq \|\mu\|_{\mathcal{M}}.$$

Since $k \geq 1$ was arbitrary, this gives a contradiction.

Proof of Proposition 4.C.1. Let $\mu \in [C_0(\bar{\Omega})]^*$. Given $\zeta \in C(\bar{\Omega})$, let (ζ_n) be any sequence in $C_0(\bar{\Omega})$ such that

$$\|\zeta_n\|_{L^\infty} \leq C \quad \text{and} \quad \zeta_n \to \zeta \quad \text{in } L_{\text{loc}}^\infty(\Omega).$$

It easily follows from Lemma 4.C.1 that $(\langle \mu, \zeta_n \rangle)$ is Cauchy in $\mathbb{R}$. In particular, the limit $\lim_{n\to\infty} \langle \mu, \zeta_n \rangle$ exists and is independent of the sequence (ζ_n). Set

$$\langle \tilde{\mu}, \zeta \rangle = \lim_{n\to\infty} \langle \mu, \zeta_n \rangle.$$

Clearly, $\tilde{\mu}$ is a continuous linear functional on $C(\bar{\Omega})$ and

$$\langle \tilde{\mu}, \zeta \rangle = \langle \mu, \zeta \rangle \quad \forall \zeta \in C_0(\bar{\Omega}).$$

In addition, Lemma 4.C.1 implies that $|\tilde{\mu}|(\partial\Omega) = 0$; in particular, $\|\tilde{\mu}\|_{C^*} = \|\mu\|_{(C_0)^*}$. The uniqueness of $\tilde{\mu}$ follows immediately from (4.C.1).

APPENDIX 4D. A NEW DECOMPOSITION FOR DIFFUSE MEASURES

The goal of this section is to establish Theorem 4.3. Let G denote the Green function of the Laplacian in Ω. Given $\mu \in \mathcal{M}(\Omega)$, $\mu \geq 0$, set

$$G(\mu)(x) = \int_\Omega G(x, y) \, d\mu(y).$$

Note that $G(\mu)$ is well defined for every $x \in \Omega$, possibly taking values $+\infty$.
    We first present some well-known results in Potential Theory:

LEMMA 4.D.1 *Let* $\mu \in \mathcal{M}(\Omega)$, $\mu \geq 0$ *be such that* $G(\mu) < \infty$ *everywhere in* Ω. *Given* $\varepsilon > 0$, *there exists* $L \subset \Omega$ *compact such that*

$$\mu(\Omega \backslash L) < \varepsilon \quad \text{and} \quad G(\mu \lfloor_L) \in C_0(\bar{\Omega}). \tag{4.D.1}$$

Proof. If μ has compact support in Ω, then Lemma D.1 is precisely Theorem 6.21 in [H]. For an arbitrary $\mu \in \mathcal{M}(\Omega)$, $\mu \geq 0$, such that $G(\mu) < \infty$ in Ω, we proceed as follows. By inner regularity of μ, there exists $K \subset \Omega$ compact such that $\mu(\Omega \backslash K) < \frac{\varepsilon}{2}$. Since $G(\mu \lfloor_K) \leq G(\mu)$, the function $G(\mu \lfloor_K)$ is also finite everywhere in Ω. Then, by Theorem 6.21 in [H], there exists $L \subset \Omega$ compact such that

$$\mu \lfloor_K (\Omega \backslash L) < \frac{\varepsilon}{2} \quad \text{and} \quad G(\mu \lfloor_{K\cap L}) \in C_0(\bar{\Omega}).$$

We conclude that (4.D.1) holds with L replaced by $K \cap L$.

As a consequence of Lemma 4.D.1, we have

PROPOSITION 4.D.1 *Let* $u \in W_0^{1,1}(\Omega)$ *be such that* Δu *is a diffuse measure in* Ω. *Then, there exists a sequence* $(u_n) \subset C_0(\bar{\Omega})$ *such that* $\Delta u_n \in \mathcal{M}(\Omega)$, $\forall n \geq 1$,

$$u = \sum_{n=1}^{\infty} u_n \quad a.e. \text{ in } \Omega \quad and \quad \|\Delta u\|_{\mathcal{M}} = \sum_{n=1}^{\infty} \|\Delta u_n\|_{\mathcal{M}}.$$

Proof. We shall split the proof of Proposition 4.D.1 into three steps.

Step 1. Let $\mu \geq 0$ be a measure such that $G(\mu) < \infty$ everywhere in Ω. Then, there exist disjoint Borel sets $A_n \subset \Omega$ such that

$$\mu\left(\Omega \backslash \bigcup_{n=1}^{\infty} A_n\right) = 0 \quad and \quad G(\mu\lfloor_{A_n}) \in C_0(\bar{\Omega}) \quad \forall n \geq 1. \tag{4.D.2}$$

This result easily follows from Lemma 4.D.1 by an induction argument.

Step 2. Let $\mu \geq 0$ be a diffuse measure in Ω. Then, there exist disjoint Borel sets $A_n \subset \Omega$ such that

$$\mu\left(\Omega \backslash \bigcup_{n=1}^{\infty} A_n\right) = 0 \quad and \quad G(\mu\lfloor_{A_n}) \in C_0(\bar{\Omega}) \quad \forall n \geq 1. \tag{4.D.3}$$

For each $k \geq 1$, let

$$E_k = \{x \in \Omega; G(\mu)(x) \leq k\}.$$

Since $G(\mu)$ is lower semicontinuous (by Fatou), E_k is closed in Ω. Clearly, we have $G(\mu\lfloor_{E_k}) \leq k$ in E_k, and $G(\mu\lfloor_{E_k})$ is harmonic in $\Omega\backslash E_k$. Therefore, by the maximum principle, $G(\mu\lfloor_{E_k}) \leq k$ everywhere in Ω.

Applying the previous step to the measures $\mu\lfloor_{E_k\backslash E_{k-1}}$, one can find disjoint Borel sets $A_n \subset \Omega$ such that

$$\mu\left(F \backslash \bigcup_{n=1}^{\infty} A_n\right) = 0 \quad and \quad G(\mu\lfloor_{A_n}) \in C_0(\bar{\Omega}) \quad \forall n \geq 1,$$

where

$$F = \{x \in \Omega; G(\mu)(x) < \infty\}.$$

Since μ is diffuse and $\Omega\backslash F$ has zero capacity (see, e.g., [H, Theorem 7.33]), we have $\mu(\Omega\backslash F) = 0$. Thus,

$$\mu\left(\Omega \backslash \bigcup_{n=1}^{\infty} A_n\right) = 0, \tag{4.D.4}$$

from which the result follows.

Step 3.   Proof of Proposition 4.D.1 completed.

Set $\mu = -\Delta u$. Applying Step 2 to μ^+, one can find disjoint Borel sets (A_n) such that

$$\mu^+\left(\Omega \backslash \bigcup_{n=1}^{\infty} A_n\right) = 0 \quad \text{and} \quad G(\mu^+ \lfloor A_n) \in C_0(\bar{\Omega}) \quad \forall n \geq 1.$$

Similarly, there exist disjoint Borel sets (B_n) such that

$$\mu^-\left(\Omega \backslash \bigcup_{n=1}^{\infty} B_n\right) = 0 \quad \text{and} \quad G(\mu^- \lfloor B_n) \in C_0(\bar{\Omega}) \quad \forall n \geq 1.$$

Since

$$\mu = \mu^+ - \mu^- = \sum_{n=1}^{\infty} \mu^+ \lfloor A_n - \sum_{n=1}^{\infty} \mu^- \lfloor B_n,$$

we have

$$u = \sum_{n=1}^{\infty} G(\mu^+ \lfloor A_n) - \sum_{n=1}^{\infty} G(\mu^- \lfloor B_n) \quad \text{a.e.}$$

and

$$\|\Delta u\|_{\mathcal{M}} = \sum_{n=1}^{\infty} \|\mu^+ \lfloor A_n\|_{\mathcal{M}} + \sum_{n=1}^{\infty} \|\mu^- \lfloor B_n\|_{\mathcal{M}}.$$

This concludes the proof of the proposition.

We can now present the

Proof of Theorem 4.3.   Let $u \in W_0^{1,1}(\Omega)$ be the unique solution of

$$-\Delta u = \mu \quad \text{in } (C_0^2)^*.$$

Let $(u_n) \subset C_0(\bar{\Omega})$ be the sequence given by Proposition 4.D.1. For $\delta > 0$ fixed, take $w_n \in C_0^2(\bar{\Omega})$ such that

$$\|u_n - w_n\|_{L^\infty} \leq \frac{\delta}{2^n} \quad \text{and} \quad \|\Delta w_n\|_{L^1} \leq \|\Delta u_n\|_{\mathcal{M}}.$$

Let

$$v = \sum_{n=1}^{\infty} (u_n - w_n) \quad \text{and} \quad f = -\sum_{n=1}^{\infty} \Delta w_n.$$

Since

$$\|v\|_{L^\infty} \leq \sum_{n=1}^{\infty} \|u_n - w_n\|_{L^\infty} \leq \delta, \tag{4.D.5}$$

we have $v \in C_0(\bar{\Omega})$ and $\|v\|_{L^\infty} \leq \delta$. Moreover,

$$\|f\|_{L^1} \leq \sum_{n=1}^{\infty} \|\Delta w_n\|_{L^1} \leq \sum_{n=1}^{\infty} \|\Delta u_n\|_{\mathcal{M}} = \|\mu\|_{\mathcal{M}} \tag{4.D.6}$$

implies $f \in L^1(\Omega)$. Finally, by construction, we have

$$\mu = f - \Delta v \quad \text{in } (C_0^2)^*. \tag{4.D.7}$$

In particular, $\Delta v = f - \mu$ is a measure and $\|\Delta v\|_{\mathcal{M}} \leq 2\|\mu\|_{\mathcal{M}}$. Thus,

$$\|\nabla v\|_{L^2}^2 \leq \|v\|_{L^\infty}\|\Delta v\|_{\mathcal{M}} \leq 2\delta\|\mu\|_{\mathcal{M}}. \tag{4.D.8}$$

Since $v \in C_0(\bar{\Omega}) \cap H_0^1$, (4.0.21) immediately follows from (4.D.7). Moreover, replacing δ by $\frac{\delta}{2}\|\mu\|_{\mathcal{M}}$ in (4.D.5) and (4.D.8), we conclude that (4.0.22) holds. The proof of Theorem 4.3 is complete.

Note that our construction of $f \in L^1$ and $v \in L^\infty$ satisfying (4.0.21) is not linear with respect to μ. Here is a natural question:

Open problem 7.   Can one find a bounded linear operator

$$T : \mu \in \mathcal{M}_d(\Omega) \longmapsto (f, v) \in L^1 \times L^\infty$$

such that (4.0.21) and (4.0.22) hold?

After receiving a preprint of our work, A. Ancona [A2] has provided a negative answer to the question above.

APPENDIX 4E. EQUIVALENCE BETWEEN CAP_{H^1} AND $\text{CAP}_{\Delta,1}$

Given a compact set $K \subset \Omega$, let $\text{cap}_{\Delta,1}(K)$ denote the capacity associated to the Laplacian. More precisely,

$$\text{cap}_{\Delta,1}(K) = \inf\left\{\int_\Omega |\Delta\varphi|; \, \varphi \in C_c^\infty(\Omega), \, \varphi \geq 1 \text{ in some neighborhood of } K\right\}.$$

In order to avoid confusion, throughout this section we shall denote by cap_{H^1} the Newtonian capacity with respect to Ω (which we simply denote cap everywhere else in this paper).

The main result in this appendix is the following:

THEOREM 4.E.1 *For every compact set $K \subset \Omega$, we have*

$$\text{cap}_{\Delta,1}(K) = 2\,\text{cap}_{H^1}(K). \tag{4.E.1}$$

Remark 4.E.1.   In an earlier version of this work, we had only established the equivalence between cap_{H^1} and $\text{cap}_{\Delta,1}$. The exact formula (4.E.1) has been suggested to us by A. Ancona.

We first prove the following:

LEMMA 4.E.1 *Let $K \subset \Omega$ be a compact set. Given $\varepsilon > 0$, there exists $\psi \in C_c^\infty(\Omega)$ such that $0 \leq \psi \leq 1$ in Ω, $\psi = 1$ in some neighborhood of K, and*

$$\int_\Omega |\Delta\psi| \leq 2\,\text{cap}_{H^1}(K) + \varepsilon. \tag{4.E.2}$$

Proof. Let $\omega \subset\subset \Omega$ be an open set such that $K \subset \omega$ and

$$\mathrm{cap}_{H^1}(\bar{\omega}) \leq \mathrm{cap}_{H^1}(K) + \frac{\varepsilon}{4}.$$

Let u denote the capacitary potential of $\bar{\omega}$. More precisely, let $u \in H_0^1(\Omega)$ be such that $u = 1$ in $\bar{\omega}$ and

$$\int_\Omega |\nabla u|^2 = \mathrm{cap}_{H^1}(\bar{\omega}).$$

Note that u is superharmonic in Ω and harmonic in $\Omega \setminus \bar{\omega}$. In particular, $0 \leq u \leq 1$. Since supp $\Delta u \subset [u = 1]$, u is continuous (see [H, Theorem 6.20]) and

$$\|\Delta u\|_{\mathcal{M}} = -\int_\Omega \Delta u = -\int_\Omega u \Delta u = \int_\Omega |\nabla u|^2 = \mathrm{cap}_{H^1}(\bar{\omega}).$$

Given $\delta > 0$ small, set

$$v = \frac{(u - \delta)^+}{1 - \delta}.$$

Since v has compact support in Ω, we have

$$\int_\Omega \Delta v = 0. \tag{4.E.3}$$

Moreover, Δv is a diffuse measure (note that $v \in H_0^1(\Omega)$) and

$$\mathrm{supp}\,\Delta v \subset [v = 0] \cup [v = 1]. \tag{4.E.4}$$

Thus, by Corollary 1.3 in [BP2], we have

$$\Delta v \geq 0 \quad \text{in } [v = 0] \quad \text{and} \quad \Delta v \leq 0 \quad \text{in } [v = 1]. \tag{4.E.5}$$

It then follows from (4.E.3)–(4.E.5) that

$$\|\Delta v\|_{\mathcal{M}} = 2 \int_{[v=1]} |\Delta v|.$$

Since $\Delta v = \frac{1}{1-\delta} \Delta u$ in $[v = 1]$, we conclude that

$$\|\Delta v\|_{\mathcal{M}} \leq \frac{2}{1-\delta} \|\Delta u\|_{\mathcal{M}}.$$

Using the same notation as in Section 4.4, we now take $n \geq 1$ sufficiently large so that the function $\psi = \rho_n * v$ has compact support in Ω and $\psi = 1$ in some neighborhood of K. We claim that ψ satisfies all the required properties. In fact, since $0 \leq \psi \leq 1$ in Ω, we only have to show that (4.E.2) holds. Note that

$$\int_\Omega |\Delta \psi| \leq \|\Delta v\|_{\mathcal{M}} \leq \frac{2}{1-\delta} \|\Delta u\|_{\mathcal{M}} = \frac{2}{1-\delta} \mathrm{cap}_{H^1}(\bar{\omega}).$$

Choosing $\delta > 0$ so that

$$\frac{\delta}{1-\delta} \mathrm{cap}_{H^1}(\bar{\omega}) < \frac{\varepsilon}{4},$$

we have

$$\int_\Omega |\Delta\psi| \le 2\left(1 + \frac{\delta}{1-\delta}\right) \operatorname{cap}_{H^1}(\tilde\omega) \le 2\operatorname{cap}_{H^1}(K) + \varepsilon,$$

which is precisely (4.E.2).

We now present the

Proof of Theorem 4.E.1. In view of Lemma 4.E.1, it suffices to show that

$$\operatorname{cap}_{H^1}(K) \le \frac{1}{2}\operatorname{cap}_{\Delta,1}(K). \tag{4.E.6}$$

Let $\varphi \in C_c^\infty(\Omega)$ be such that $\varphi \ge 1$ in some neighborhood of K. Set $\tilde\varphi = \min\{1, \varphi^+\}$. For $n \ge 1$ sufficiently large, the function $\tilde\varphi_n = \rho_n * \tilde\varphi$ belongs to $C_c^\infty(\Omega)$ and $\tilde\varphi_n = 1$ in some neighborhood of K. We then have

$$\operatorname{cap}_{H^1}(K) \le \int_\Omega |\nabla\tilde\varphi_n|^2 \le \int_\Omega |\nabla\tilde\varphi|^2 = \int_\Omega \nabla\tilde\varphi \cdot \nabla\varphi = -\int_\Omega \tilde\varphi \Delta\varphi.$$

Recall that φ has compact support in Ω and $0 \le \tilde\varphi \le 1$. Thus, $\int_\Omega \Delta\varphi = 0$ and we have

$$\operatorname{cap}_{H^1}(K) \le -\int_\Omega \left(\tilde\varphi - \frac{1}{2}\right)\Delta\varphi \le \frac{1}{2}\int_\Omega |\Delta\varphi|.$$

Since φ was arbitrary, we conclude that (4.E.6) holds. This establishes Theorem 4.E.1.

Acknowledgments

We warmly thank A. Ancona for enlightening discussions and suggestions. The first author (H.B.) and the second author (M.M.) are partially sponsored by an E.C. Grant through the RTN Program "Front-Singularities," HPRN-CT-2002-00274. H.B. is also a member of the Institut Universitaire de France. The third author (A.C.P.) is partially supported by CAPES, Brazil, under grant no. BEX1187/99-6.

REFERENCES

[A1]   A. Ancona, *Une propriété d'invariance des ensembles absorbants par perturbation d'un opérateur elliptique*, Comm. Partial Differential Equations 4 (1979), 321–337.

[A2]   ———, *Sur une question de H. Brezis, M. Marcus et A. C. Ponce*, Ann. Inst. H. Poincaré Anal. Non Linéaire 23 (2006), 127–133.

[BP]   P. Baras and M. Pierre, *Singularités éliminables pour des équations semi-linéaires*, Ann. Inst. Fourier (Grenoble) 34 (1984), 185–206.

[BLOP]   D. Bartolucci, F. Leoni, L. Orsina, and A. C. Ponce, *Semilinear equations with exponential nonlinearity and measure data*, Ann. Inst. H. Poincaré Anal. Non Linéaire 22 (2005), 799–815.

[Ba]   J. R. Baxter, *Inequalities for potentials of particle systems*, Illinois J. Math. 24 (1980), 645–652.

[BM]   G. Barles and F. Murat, *Uniqueness and the maximum principle for quasilinear elliptic equations with quadratic growth conditions*, Arch. Rational Mech. Anal. 133 (1995), 77–101.

[BB]   Ph. Bénilan and H. Brezis, *Nonlinear problems related to the Thomas-Fermi equation*, J. Evol. Equ. 3 (2004), 673–770. Dedicated to Ph. Bénilan.

[BGO1] L. Boccardo, T. Gallouët, and L. Orsina, *Existence and uniqueness of entropy solutions for nonlinear elliptic equations with measure data*, Ann. Inst. H. Poincaré Anal. Non Linéaire 13 (1996), 539–551.

[BGO2] ———, *Existence and nonexistence of solutions for some nonlinear elliptic equations*, J. Anal. Math. 73 (1997), 203–223.

[B1]   H. Brezis, *Nonlinear problems related to the Thomas-Fermi equation*. In: Contemporary developments in continuum mechanics and partial differential equations (G. M. de la Penha and L. A. Medeiros, eds.), Proc. Internat. Sympos., Inst. Mat., Univ. Fed. Rio de Janeiro, Rio de Janeiro North Holland Amsterdam, 1978, pp. 74–80.

[B2]   ———, *Some variational problems of the Thomas-Fermi type*. In: Variational inequalities and complementarity problems (R. W. Cottle, F. Giannessi, and J.-L. Lions, eds.), Proc. Internat. School, Erice, 1978, Wiley Chichester, 1980, pp. 53–73.

[B3]   ———, *Problèmes elliptiques et paraboliques non linéaires avec données mesures*. Goulaouic-Meyer-Schwartz Seminar, 1981/1982, École Polytech., Palaiseau, 1982, pp. X.1–X.12.

[B4]   ———, *Nonlinear elliptic equations involving measures*. In: Contributions to nonlinear partial differential equations (C. Bardos, A. Damlamian, J. I. Diaz, and J. Hernandez, eds.), Madrid, 1981, Pitman, Boston, MA, 1983, pp. 82–89.

[B5]   ———, *Semilinear equations in $\mathbb{R}^N$ without condition at infinity*, Appl. Math. Optim. 12 (1984), 271–282.

[BBr]   H. Brezis and F. E. Browder, *Strongly nonlinear elliptic boundary value problems*, Ann. Scuola Norm. Sup. Pisa Cl. Sci. 5 (1978), 587–603.

[BCMR] H. Brezis, T. Cazenave, Y. Martel, and A. Ramiandrisoa, *Blow up for $u_t - \Delta u = g(u)$ revisited*, Adv. Differential Equations 1 (1996), 73–90.

[BMP]   H. Brezis, M. Marcus, and A. C. Ponce, *A new concept of reduced measure for nonlinear elliptic equations*, C. R. Acad. Sci. Paris, Ser. I 339 (2004), 169–174.

[BP1]   H. Brezis and A. C. Ponce, *Remarks on the strong maximum principle*, Differential Integral Equations 16 (2003), 1–12.

[BP2]   ———, *Kato's inequality when Δu is a measure*, C. R. Acad. Sci. Paris, Ser. I 338 (2004), 599–604.

[BP3]   ———, *Reduced measures on the boundary*, J. Funct. Anal. 229 (2005), 95–120.

[BP4]   ———, *Reduced measures for obstacle problems*, Advances in Diff. Eq. 10 (2005), 1201–1234.

[BSe]   H. Brezis and S. Serfaty, *A variational formulation for the two-sided ob-stacle problem with measure data*, Commun. Contemp. Math. 4 (2002), 357–374.

[BS]   H. Brezis and W. A. Strauss, *Semilinear second-order elliptic equations in L^1*, J. Math. Soc. Japan 25 (1973), 565–590.

[BV]   H. Brezis and L. Véron, *Removable singularities for some nonlinear elliptic equations*, Arch. Rational Mech. Anal. 75 (1980/81), 1–6.

[DD]   P. Dall'Aglio and G. Dal Maso, *Some properties of the solutions of obstacle problems with measure data*, Ricerche Mat. 48 (1999), suppl., 99–116. Papers in memory of Ennio De Giorgi.

[DM]   C. Dellacherie and P. -A. Meyer, *Probabilités et potentiel*, Chapitres I à IV, Publications de l'Institut de Mathématique de l'Université de Strasbourg, No. XV, Actualités Scientifiques et Industrielles, No. 1372, Hermann, Paris, 1975.

[DS]   N. Dunford and J. T. Schwartz, *Linear operators. Part I*, Wiley, New York, 1958.

[DP]   L. Dupaigne and A. C. Ponce, *Singularities of positive supersolutions in elliptic PDEs*, Selecta Math. (N.S.) 10 (2004), 341–358.

[DPP]   L. Dupaigne, A. C. Ponce, and A. Porretta, *Elliptic equations with vertical asymptotes in the nonlinear term*, to appear in J. Anal. Math.

[DVP]   C. De La Vallée-Poussin, *Sur l'intégrale de Lebesgue*, Trans. Amer. Math. Soc. 16 (1915), 435–501.

[D1]   E. B. Dynkin, *Diffusions, superdiffusions and partial differential equations*, American Mathematical Society, Providence, RI, 2002.

[D2]   ———, *Superdiffusions and positive solutions of nonlinear partial differential equations*, American Mathematical Society, Providence, RI, 2004.

[FTS]   M. Fukushima, K. Sato, and S. Taniguchi, *On the closable parts of pre-Dirichlet forms and the fine supports of underlying measures*, Osaka J. Math. 28 (1991), 517–535.

[GM]   T. Gallouët and J. -M. Morel, *Resolution of a semilinear equation in L^1*, Proc. Roy. Soc. Edinburgh Sect. A 96 (1984), 275–288; Corrigenda: Proc. Roy. Soc. Edinburgh Sect. A 99 (1985), 399.

[GV]   A. Gmira and L. Véron, *Boundary singularities of solutions of some nonlinear elliptic equations*, Duke Math. J. 64 (1991), 271–324.

[GV]   M. Grillot and L. Véron, *Boundary trace of the solutions of the prescribed Gaussian curvature equation*, Proc. Roy. Soc. Edinburgh Sect. A 130 (2000), 527–560.

[GRe]   M. Grun-Rehomme, *Caractérisation du sous-différentiel d'intégrandes convexes dans les espaces de Sobolev*, J. Math. Pures Appl. 56 (1977), 149–156.

[H]   L. L. Helms, *Introduction to potential theory*, Wiley-Interscience, New York, 1969.

[K]   T. Kato, *Schrödinger operators with singular potentials*, Israel J. Math. 13 (1972), 135–148 (1973).

[LG1]   J. -F. Le Gall, *The Brownian snake and solutions of* $\Delta u = u^2$ *in a domain*, Probab. Theory Related Fields 102 (1995), 393–432.

[LG2]   ———, *A probabilistic Poisson representation for positive solutions of* $\Delta u = u^2$ *in a planar domain*, Comm. Pure Appl. Math. 50 (1997), 69–103.

[MV1]   M. Marcus and L. Véron, *The boundary trace of positive solutions of semilinear elliptic equations: the subcritical case*, Arch. Rational Mech. Anal. 144 (1998), 201–231.

[MV2]   ———, *The boundary trace of positive solutions of semilinear elliptic equations: the supercritical case*, J. Math. Pures Appl. 77 (1998), 481–524.

[MV3]   ———, *Removable singularities and boundary traces*, J. Math. Pures Appl. 80 (2001), 879–900.

[MV4]   ———, *Capacitary estimates of solutions of a class of nonlinear elliptic equations*, C. R. Acad. Sci. Paris, Ser. I 336 (2003), 913–918.

[MV5]   ———, *Nonlinear capacities associated to semilinear elliptic equations*, in preparation.

 [P]   A. C. Ponce, *How to construct good measures*. To appear in Elliptic and parabolic problems (C. Bandle, H. Berestycki, B. Brighi, A. Brillard, M. Chipot, J.-M. Coron, C. Sbordone, I. Shafrir, V. Valente, and G. Vergara-Caffarelli, eds.), Gaeta, 2004, Birkhäuser. A special tribute to the work of Haïm Brezis.

[Po]   A. Porretta, *Absorption effects for some elliptic equations with singularities*, Boll. Unione Mat. Ital. Sez. B Artic. Ric. Mat. (8) 8 (2005), 369–395.

 [S]   G. Stampacchia, *Équations elliptiques du second ordre à coefficients discontinus*, Les Presses de l'Université de Montréal Montréal, 1966.

[Va]   J. L. Vázquez, *On a semilinear equation in* $\mathbb{R}^2$ *involving bounded measures*, Proc. Roy. Soc. Edinburgh Sect. A 95 (1983), 181–202.

Chapter Five

Global Solutions for the Nonlinear Schrödinger Equation
on Three-Dimensional Compact Manifolds

N. Burq, P. Gérard, and N. Tzvetkov

5.1 INTRODUCTION

Let (M, g) be a Riemannian compact manifold of dimension d. In this paper we address global wellposedness of the Cauchy problem for the following nonlinear Schrödinger equation (NLS),

$$i\partial_t u + \Delta u = F(u), \quad u_{|t=0} = u_0. \qquad (5.1.1)$$

In (5.1.1), u is a complex valued function on $\mathbb{R} \times M$ and $u_0 \in H^s(M)$ for s large enough. The nonlinear interaction F is supposed to be of the form

$$F = \frac{\partial V}{\partial \bar{z}}$$

with $V \in C^\infty(\mathbb{C}; \mathbb{R})$ satisfying

$$V(e^{i\theta} z) = V(z), \quad \theta \in \mathbb{R}, z \in \mathbb{C}, \qquad (5.1.2)$$

and, for some $\alpha > 1$,

$$|\partial_z^{k_1} \partial_{\bar{z}}^{k_2} V(z)| \leq C_{k_1, k_2} (1 + |z|)^{1 + \alpha - k_1 - k_2}. \qquad (5.1.3)$$

The number α involved in the second condition on V corresponds to the "degree" of the nonlinearity $F(u)$ in (5.1.1). Moreover, we make the following defocusing assumption:

$$\forall z \in \mathbb{C}, \ V(z) \geq 0. \qquad (5.1.4)$$

The basic question we want to investigate is the global existence of smooth solutions u on $\mathbb{R} \times M$ for smooth data u_0. In view of the well-known properties of the linear Schrödinger flow, it is natural to recast this question in the framework of Sobolev spaces $H^s(M)$. For every $s > d/2$, using the Sobolev embedding $H^s(M) \subset L^\infty(M)$, it is easy to prove that, for every $u_0 \in H^s(M)$, there exists a unique solution $u \in C([-T, T], H^s(M))$ of (5.1.1) for some T depending only on a bound of $\|u_0\|_{H^s}$. Thus our problem reduces to the continuation of u as a global solution of NLS in $C(\mathbb{R}, H^s(M))$. For that purpose, we recall the usual strategy based on the known conservation laws.

It is classical that NLS can be seen as a Hamiltonian equation, associated to the energy functional

$$E(u) = \int_M |\nabla_g u|^2 \, dx + \int_M V(u) \, dx. \tag{5.1.5}$$

It follows from this Hamiltonian structure that smooth solutions of (5.1.1) enjoy the conservation laws

$$\|u(t)\|_{L^2} = \|u_0\|_{L^2}, \quad E(u(t)) = E(u_0). \tag{5.1.6}$$

In view of the defocusing assumption (5.1.4), we infer from the above conservation laws the following a priori estimate

$$\|u(t)\|_{H^1(M)} \leq C(\|u_0\|_{H^1}), \tag{5.1.7}$$

provided that the potential term in the energy is controlled by the H^1 norm, namely,

$$\alpha \leq \frac{d+2}{d-2} \quad \text{if} \quad d \geq 3. \tag{5.1.8}$$

In this situation, global wellposedness for NLS is usually reduced to local wellposedness in $H^1(M)$—namely, that, for every $u_0 \in H^1(M)$, there exists a unique solution u of (5.1.1) in some space $X_T \subset C([-T, T], H^1(M))$ for some T depending only on a bound of $\|u_0\|_{H^1}$, combined with propagation of the regularity. If M is replaced by the Euclidean space $\mathbb{R}^d$, this strategy was achieved successfully for defocusing subcritical nonlinearities ($\alpha < (d+2)/(d-2)$ if $d \geq 3$) and dimension d not too large, thanks to Strichartz estimates, through the contributions of Ginibre-Velo [24] [25] and Kato [28]. The critical case ($d \geq 3$ and $\alpha = (d+2)/(d-2)$) is more involved, since iteration schemes usually give a time T strongly depending on the Cauchy data, except if these data are small (see Cazenave-Weissler [20]). For large data, let us mention that global results for the defocusing critical case $\alpha = 5$ in three-space dimensions are due to Bourgain [5] and Grillakis [27] in the radial case, and more recently to Colliander-Keel-Staffilani-Takaoka-Tao [23] in the general case.

Let us review briefly the state of the art on a compact manifold for dimensions $d = 1, 2, 3$. If $d = 1$, the above strategy applies easily, since $H^1(S^1) \subset L^\infty(S^1)$. In two-space dimensions, the imbedding of H^1 into L^∞ fails; however, the approach of Brezis-Gallouët [8], based on a logarithmic estimate, allows to solve the cubic case $\alpha = 3$. On the torus $\mathbb{T}^2$, Bourgain [1] used new dispersive estimates to prove global existence for every α. The case of an arbitrary compact surface was solved by the authors in [9], appealing to Strichartz estimates with fractional loss of derivatives.

Much less is known in the case $d = 3$. In [1], Bourgain obtained global existence for defocusing subcritical nonlinearities $\alpha < 5$ on the torus $\mathbb{T}^3$. On the other hand, on arbitrary three-manifolds, the use of Strichartz estimates of [9] only yields global existence for cubic defocusing nonlinearities, as we shall recall below. It is therefore natural to look for other examples of three-manifolds for which it is possible to extend global existence to any subquintic defocusing nonlinearity. This is the purpose of this paper.

In Section 5.2 below, we recall the argument of [9] solving the cubic case on an arbitrary three-manifold. In Section 5.3, we observe that this result can be extended to subquintic nonlinearities in the special case of $M = S^3$ by means of bilinear Strichartz estimates. Section 5.4 is devoted to the case of $M = S^2 \times S^1$, where a similar result is obtained through trilinear Strichartz estimates. Finally, in Section 5.5 we describe a recent step to the analysis of critical equation with small energy data on $M = S^3$, namely, that the first iteration of quintic NLS is controlled in the energy space.

Let us mention that Sections 5.3 and 5.4 are essentially taken from [17], sometimes with slight variants, while Section 5.5 is part of a work in progress. Other aspects, related to stability and instability properties of the flow map $u_0 \mapsto u(t)$, are discussed in references [10], [11], and [15], while the more complicated case of boundary problems for NLS is addressed in [12], [13], and [14].

5.2 STRICHARTZ ESTIMATES WITH LOSS AND THE CUBIC CASE

THEOREM 5.2.1 ([9]). *Let M be a three-dimensional compact manifold, and let $s \geq 1$. For every $u_0 \in H^s(M)$, there exists a unique solution $u \in C(\mathbb{R}, H^s(M))$ of (5.1.1) where V satisfies (5.1.2), (5.1.3) with $\alpha = 3$, and (5.1.4).*

Let us recall the main steps of the proof. First, using the WKB approximation and the stationary phase method, one shows the following dispersive inequality for very small times. Given $\chi \in C_0^\infty(\mathbb{R} \setminus \{0\})$, there exists $a > 0$ and $C > 0$ such that, for every dyadic number N, for every time t such that $N|t| \leq a$,

$$\|\chi_N e^{it\Delta}\|_{L^1(M) \to L^\infty(M)} \leq C|t|^{-3/2}, \tag{5.2.1}$$

where χ_N denotes the spectral cutoff $\chi(N^{-2}\Delta)$.

By Keel-Tao's endpoint version of the TT^* lemma [29], we infer a localized Strichartz estimate. Given an interval $I = [t_0, t_1]$ such that $N|I| \lesssim 1$ and a solution v of

$$i\partial_t v + \Delta v = f_1 + f_2, \quad t \in I, \quad u(t_0) = v_0, \tag{5.2.2}$$

we have

$$\|\chi_N v\|_{L^2(I, L^6(M))} \leq$$
$$C \left(\|\chi_N v_0\|_{L^2(M)} + \|\chi_N f_1\|_{L^1(I, L^2(M))} + \|\chi_N f_2\|_{L^2(I, L^{6/5}(M))} \right). \tag{5.2.3}$$

Applying (5.2.3) in the particular case $f_1 = f_2 = 0$, summing on N intervals I, and then summing on all dyadic frequencies N, we obtain, for every solution $v(t) = e^{it\Delta}v_0$ of the homogeneous Schrödinger equation (see also Staffilani-Tataru [34]),

$$\|v\|_{L^2((0,1), L^6(M))} \leq C \|v_0\|_{H^{1/2}(M)}, \tag{5.2.4}$$

and therefore, by the Sobolev inequality,

$$\|v\|_{L^p((0,1), L^\infty(M))} \leq C_s \|v_0\|_{H^s(M)}, \quad s > 1, \quad p = p(s) > 2. \tag{5.2.5}$$

The latter estimate easily implies that NLS with $\alpha = 3$ is locally wellposed in $H^s(M)$ for every $s > 1$. In order to prove that such solutions are global, we derive an a priori

estimate on these solutions by coming back to the localized Strichartz estimate
(5.2.3). Let $T > 0$, $T \lesssim 1$. Given an interval $I = [k/N, (k+3)/N]$ included into
$[0, T]$, we introduce a time cutoff

$$\varphi_N(t) = \varphi(Nt - k),$$

with $\varphi = 1$ on $[1, 2]$, φ compactly supported into $(0, 3)$, and we apply estimate
(5.2.3) to $v = \varphi_N u$ with $v_0 = 0$, $f_1 = i\varphi_N' u$, and $f_2 = \varphi_N F(u)$. We have

$$\|\chi_N f_2\|_{L^2(I, L^{6/5}(M))} \lesssim N^{-3/2} \|\nabla F(u)\|_{L^\infty(I, L^{6/5}(M))} \lesssim N^{-3/2},$$

since $\alpha = 3$ and the H^1 norm of $u(t)$ is $O(1)$ by (5.1.7). On the other hand,

$$\|\chi_N f_1\|_{L^1(I, L^2(M))} \lesssim N^{-1/2} \|\chi_N u\|_{L^2(I, H^1(M))}.$$

Then we square these inequalities and we sum on all such intervals I; we also apply
estimate (5.2.3) on intervals $[0, 1/N]$ and $[T - 1/N, T]$ with nonhomogeneous
Cauchy data. This gives

$$\|\chi_N u\|_{L^2((0,T), L^6(M))} \lesssim N^{-1/2} \|\chi_N u\|_{L^2((0,T), H^1(M))} + N^{-1}.$$

By the Sobolev inequality, this implies

$$\|\chi_N u\|_{L^2((0,T), L^\infty(M))} \lesssim \|\chi_N u\|_{L^2((0,T), H^1(M))} + N^{-1/2}.$$

Finally, we sum on all the dyadic frequencies N, by observing that

$$\sum_{N \leq N_0} \|\chi_N u\|_{L^2((0,T), H^1(M))} \leq \sqrt{\log N_0} \|u\|_{L^2((0,T), H^1(M))}$$

and

$$\sum_{N > N_0} \|\chi_N u\|_{L^2((0,T), H^1(M))} \leq C_s N_0^{-(s-1)} \|u\|_{L^2((0,T), H^s(M))},$$

for every $s > 1$. Choosing

$$N_0 \simeq (2 + \|u\|_{L^\infty((0,T), H^s(M))})^{1/(s-1)}$$

we conclude

$$\|u\|_{L^2((0,T), L^\infty(M))} \lesssim \left(T(1 + \log[2 + \|u\|_{L^\infty((0,T), H^s(M))}])\right)^{1/2}. \qquad (5.2.6)$$

It remains to apply the usual energy estimate in H^s on the interval $[0, T]$, together
with the Gronwall lemma,

$$\|u\|_{L^\infty((0,T), H^s(M))} \leq \|u_0\|_{H^s(M)} e^{C\|u\|^2_{L^2((0,T), L^\infty(M))}}$$

$$\leq C \|u_0\|_{H^s(M)} [2 + \|u\|_{L^\infty((0,T), H^s(M))}]^{DT},$$

where D depends only on a bound of $\|u_0\|_{H^1}$. Therefore, if, for example, $DT = 1/2$,
we get an a priori estimate of $\|u\|_{L^\infty((0,T), H^s(M))}$, which, by a classical iteration,
preserves u from blow-up.

Remark 5.2.1. The logarithmic a priori estimate (5.2.6) can be viewed as a three-
dimensional version of the Brezis-Gallouët inequality [8]. Let us note that it can be
proved avoiding the use of Keel-Tao's endpoint result (see [18]). Moreover, it can
be shown to hold for every subquartic NLS (see [9]). However, we do not know
how to use the information (5.2.6) with a supercubic nonlinearity.

5.3 BILINEAR STRICHARTZ ESTIMATES
AND THE SUBQUINTIC CASE ON S^3

5.3.1 Strichartz Estimates and Sogge's Estimates

Let M be a compact three-manifold. An estimate of the form

$$\|v\|_{L^p((0,1),L^q(M))} \lesssim \|v_0\|_{H^s(M)} \tag{5.3.1}$$

for some p and $q > 2$ and for every solution $v(t) = e^{it\Delta} v_0$ of the linear Schrödinger equation on M implies L^q bounds on eigenfunctions of the Laplace operator on M. Indeed, if

$$\Delta\phi + \lambda\phi = 0, \quad \lambda \geq 1, \tag{5.3.2}$$

the choice $v_0 = \phi$ in (5.3.1) yields

$$\|\phi\|_{L^q(M)} \lesssim \sqrt{\lambda}^s \|\phi\|_{L^2(M)}. \tag{5.3.3}$$

Estimates such as (5.3.3) were obtained by Sogge in [31], [32], [33] for every $q \in [2, \infty]$ and with $s = s(q)$ given by

$$s(q) = \frac{1}{2} - \frac{1}{q} \quad \text{if} \quad 2 \leq q \leq 4, \quad s(q) = 1 - \frac{3}{q} \quad \text{if} \quad 4 \leq q \leq \infty. \tag{5.3.4}$$

Moreover, the value $s = s(q)$ is optimal in the case $M = S^3$. The diagram $1/q \mapsto s(q)$ is represented on fig. 5.1 below.

Note that $s(6) = 1/2$, so that our Strichartz estimate (5.2.4) implies Sogge's estimate for $q = 6$, and so that the loss in (5.2.4) is optimal on S^3.

It is therefore natural to study the optimality of Theorem 5.2.1 on S^3. First, we point an improvement of generalized Strichartz estimates (5.3.1) on S^3 in the specific case $p = q = 4$. Indeed, if one tries to estimate the L^4 norm of the linear solution v on $(0, 1) \times S^3$, the most natural way is to interpolate (5.2.4) with the L^2 estimate provided by the unitarity of the Schrödinger group. This yields

$$\|v\|_{L^4((0,1),L^3(S^3))} \lesssim \|v_0\|_{H^s(S^3)},$$

where the loss s satisfies $s \leq 1/4$ but cannot be smaller than $1/6$ due to the optimality of Sogge's estimates on S^3. Applying the Sobolev inequality, we obtain

$$\|v\|_{L^4((0,1)\times S^3)} \lesssim \|v_0\|_{H^{s+1/4}(S^3)}.$$

However, it turns out that this estimate can be significantly improved by taking into account the properties of the Laplace spectrum on S^3.

PROPOSITION 5.3.1 *If $v(t) = e^{it\Delta} v_0$, then*

$$\forall \varepsilon > 0, \quad \|v\|_{L^4((0,1)\times S^3)} \leq C_\varepsilon \|v_0\|_{H^{1/4+\varepsilon}(S^3)}. \tag{5.3.5}$$

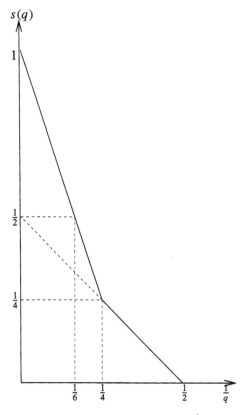

Fig. 5.1. Sogge's diagram on S^3.

Proof. We start with the expansion of $v(t)$ in spherical harmonics,

$$v(t) = \sum_n e^{-it(n^2-1)} H_n,$$

where H_n is a spherical harmonic of degree $n - 1$ on S^3. By the Littlewood-Paley estimate, we may assume that n varies between N and $2N$, where N is a dyadic number. Arguing as in [36], [1], we write

$$\|v\|_{L^4((0,2\pi)\times S^3)}^4 = \|v^2\|_{L^2((0,2\pi)\times S^3)}^2$$

$$= \int_0^{2\pi} \int_{S^3} \left| \sum_{N \le n, \ell \le 2N} e^{-it(n^2+\ell^2)} H_n(x) H_\ell(x) \right|^2 dx\, dt,$$

and we apply the Parseval formula in the t variable. This gives

$$\|v\|_{L^4((0,2\pi)\times S^3)}^4 = 2\pi \sum_{\tau \in \mathbb{Z}} \left\| \sum_{(n,\ell) \in \Lambda_N(\tau)} H_n H_\ell \right\|_{L^2(S^3)}^2,$$

where

$$\Lambda_N(\tau) = \{(n, \ell) : N \le n, \ell \le 2N,\ n^2 + \ell^2 = \tau\}.$$

Combining the classical number-theoretic estimate

$$\forall \varepsilon > 0, \forall \tau, \quad \#\Lambda_N(\tau) \le C_\varepsilon N^\varepsilon$$

with the Schwarz inequality and Sogge's L^4 inequality, we conclude

$$\|v\|_{L^4((0,2\pi)\times S^3)}^4 \le C_\varepsilon N^{1+\varepsilon} \left(\sum_n \|H_n\|_{L^2(S^3)}^2 \right)^2,$$

which is the desired estimate.

Remark 5.3.2. In the case of $\mathbb{R}^3$, the usual Strichartz inequality (see [25]) combined with the Sobolev estimate leads to the scale-invariant inequality

$$\|v\|_{L^4(\mathbb{R}\times\mathbb{R}^3)} \lesssim \|v_0\|_{\dot{H}^{1/4}(\mathbb{R}^3)},$$

while, on the torus $\mathbb{T}^3 = \mathbb{R}^3/\mathbb{Z}^3$, inequality (5.3.5) is due to Bourgain [1]. Moreover, in the case of the sphere, it can be shown that the loss $\varepsilon > 0$ cannot be avoided.

5.3.2 Bilinearization

We now include Proposition 5.3.1 into a bilinear framework. We use the notation

$$P_N = \mathbf{1}_{N \le \sqrt{-\Delta} \le 2N}$$

for the dyadic spectral projector.

DEFINITION 5.3.3 *We shall say that the Schrödinger group satisfies a bilinear Strichartz estimate of loss $\sigma \ge 0$ on M if there exits $C > 0$ such that, for every L^2 functions $f, \tilde{f}$ on M, for every dyadic integers $N, \tilde{N}$, the linear solutions*

$$v_N(t) = e^{it\Delta} P_N f, \quad \tilde{v}_{\tilde{N}}(t) = e^{it\Delta} P_{\tilde{N}} \tilde{f}$$

satisfy

$$\|v_N \tilde{v}_{\tilde{N}}\|_{L^2((0,1)\times M)} \le C \, (\min(N, \tilde{N}))^{2\sigma} \|f\|_{L^2(M)} \|\tilde{f}\|_{L^2(M)}.$$

By choosing $N = \tilde{N}$ and $f(x) = \tilde{f}(x) = \psi(Nx)$ in a coordinate patch, where ψ is smooth and compactly supported, and by describing $v(t)$ for $t \lesssim N^{-2}$, one can prove that no three-dimensional Schrödinger group can satisfy a bilinear Strichartz estimate of loss $\sigma < 1/4$. On the other hand, estimate (5.2.4), combined with the Sobolev inequality for spectrally supported functions, shows that any three-dimensional Schrödinger group satisfies a bilinear Strichartz estimate of loss $\sigma = 1/2$. Hence, the interesting facts concern the range

$$\sigma \in \left[\frac{1}{4}, \frac{1}{2} \right).$$

THEOREM 5.3.1 *For every $\sigma > 1/4$, the Schrödinger group on the sphere S^3 satisfies a bilinear Strichartz estimate of loss σ.*

Proof. Proceeding as in the proof of Proposition 5.3.1, one is reduced to establishing the following bilinear version of Sogge's estimates.

LEMMA 5.3.4 *For any spherical harmonics H_n, $\tilde{H}_\ell$ of degrees n, $\ell \geq 1$ on S^3, for every $\varepsilon > 0$,*

$$\| H_n \, \tilde{H}_\ell \|_{L^2(S^3)} \leq C_\varepsilon (\min(n, \ell))^{\frac{1}{2}+\varepsilon} \| H_n \|_{L^2(S^3)} \| \tilde{H}_\ell \|_{L^2(S^3)}.$$

In fact, Lemma 5.3.4 is valid for eigenfunctions of an elliptic selfadjoint second-order operator on arbitrary compact three-manifolds. A proof of it can be found in [16] or [17], based on dispersive estimates for the oscillatory integrals describing the approximate spectral projectors $\chi(\sqrt{-\Delta} - n)$ for suitable $\chi \in S$. Another proof, which was suggested to us by a discussion with H. Koch and D. Tataru, consists in writing the eigenfunction equation (5.3.2) as a semiclassical evolution equation in two-space dimensions, to which WKB approximation can be applied, leading to dispersive estimates as (5.2.1) (see details in [18]).

5.3.3 The Role of Bilinear Strichartz Estimates

Finally, we recall briefly how to derive global wellposedness for NLS from bilinear Strichartz estimates, following the work of Bourgain [1] on the torus (see also Klainerman-Machedon [30] in the context of hyperbolic equations). For $s \in \mathbb{R}$, $b \in \mathbb{R}$, define

$$X^{s,b}(\mathbb{R} \times M) = \{ v \in S'(\mathbb{R} \times M)$$
$$: (1 + |i\partial_t + \Delta|^2)^{b/2} (1 - \Delta)^{s/2} v \in L^2(\mathbb{R} \times M) \},$$

and introduce the corresponding local spaces

$$X^{s,b}_{\mathrm{loc}}(\mathbb{R} \times M) = \{ v \, : \, \forall \varphi = \varphi(t) \in C_0^\infty(\mathbb{R}), \varphi v \in X^{s,b}(\mathbb{R} \times M) \},$$
$$X^{s,b}((a,b) \times M) = \{ v_{|(a,b) \times M} \, : \, v \in X^{s,b}(\mathbb{R} \times M) \}.$$

It is easy to check that

$$\forall b > \frac{1}{2}, \ X^{s,b}_{\mathrm{loc}}(\mathbb{R} \times M) \subset C(\mathbb{R}, H^s(M)),$$

and that the solutions of the linear Schrödinger equation,

$$v(t) = e^{it\Delta} v_0, \quad v_0 \in H^s,$$

belong to $X^{s,b}_{\mathrm{loc}}(\mathbb{R} \times M)$ for every b. The following result shows how bilinear Strichartz estimates provide an improvement of Theorem 5.2.1.

THEOREM 5.3.2 *Assume the Schrödinger group satisfies a bilinear Strichartz estimate of loss $\sigma \in [\frac{1}{4}, \frac{1}{2})$ on a three-manifold M. Then, for every $s \geq 1$, there exists $b > 1/2$ such that, for every $u_0 \in H^s(M)$, the nonlinear Schrödinger equation (5.1.1) of degree*

$$\alpha < 7 - 8\sigma$$

admits a unique solution u in $X^{s,b}_{\mathrm{loc}}(\mathbb{R} \times M)$. In particular, if $s > 3/2$, the Cauchy problem admits a unique solution $u \in C(\mathbb{R}, H^s(M))$.

Combining Theorem 5.3.1 and Theorem 5.3.2 clearly provides global wellposedness for defocusing subquintic NLS on S^3.

The proof of Theorem 5.3.2 follows the main lines of [1], with necessary adaptations to the case of arbitrary manifolds. Let us briefly review it by focusing on the new case $\alpha > 3$. The main point is to solve the Duhamel equation

$$u(t) = e^{it\Delta}u_0 - i \int_0^t e^{i(t-\tau)\Delta} F(u(\tau))\, d\tau \qquad (5.3.6)$$

by an iteration scheme in $X^{1,b}((0, T) \times M)$ for T small enough depending only on a bound of the H^1 norm of u_0, and to establish $u \in X^{s,b}((0, T) \times M)$ if moreover $u_0 \in H^s$. The globalization is then a consequence of the a priori bound (5.1.7).

In view of the estimate

$$\|w\|_{X^{s,b}((0,T)\times M)} \leq C\, T^{1-b-b'}\|f\|_{X^{s,-b'}((0,T)\times M)}$$

for

$$w(t) = \int_0^t e^{i(t-t')\Delta} f(t')\, dt'$$

and $T \leq 1, 0 \leq b' < 1/2$ and $0 \leq b < 1 - b'$, the key estimates for the convergence of the iteration scheme are, for some power p,

$$\|F(u)\|_{X^{s,-b'}} \leq C_s \left(1 + \|u\|_{X^{1,b}}\right)^p \|u\|_{X^{s,b}},$$
$$\|F(u) - F(v)\|_{X^{s,-b'}} \leq C_s \left(1 + \|u\|_{X^{s,b}} + \|v\|_{X^{s,b}}\right)^p \|u - v\|_{X^{s,b}}, \qquad (5.3.7)$$

for every $s \geq 1$, and for some b, b' such that $0 < b' < \frac{1}{2} < b < 1 - b'$; here the norms are taken in $\mathbb{R} \times M$.

Let us sketch the proof of the first estimate in (5.3.7). By duality, it is equivalent to

$$\left| \int_{\mathbb{R}\times M} F(u)\, w\, dt\, dx \right| \leq C_s \left(1 + \|u\|_{X^{1,b}}\right)^p \|u\|_{X^{s,b}} \|w\|_{X^{-s,b'}}. \qquad (5.3.8)$$

For each pair (N, K) of dyadic integers, we introduce the projector

$$\Delta_{NK} = \mathbf{1}_{N \leq \sqrt{1-\Delta} < 2N}\, \mathbf{1}_{K \leq \langle i\partial_t + \Delta \rangle < 2K},$$

so that every function u on $\mathbb{R} \times M$ can be expanded as

$$u = \sum_{N,K} \Delta_{NK} u$$

with

$$\|u\|_{X^{s,b}}^2 \simeq \sum_{N,K} N^{2s} K^{2b} \|\Delta_{NK} u\|_{L^2}^2.$$

Then the bilinear Strichartz estimate of loss σ can be rephrased as follows. For every $\sigma_1 > \sigma$, there exists $b_1 < 1/2$ such that

$$\|\Delta_{NK} u\, \Delta_{\tilde N \tilde K} \tilde u\|_{L^2} \qquad (5.3.9)$$
$$\leq C \frac{(\min(N, \tilde N))^{2\sigma_1}}{N^s \tilde N^{\tilde s} K^{b-b_1} \tilde K^{\tilde b - b_1}}\, c(N, K)\, \tilde c(\tilde N, \tilde K)\, \|u\|_{X^{s,b}} \|\tilde u\|_{X^{\tilde s, \tilde b}},$$

where the sequences $(c(N, K))$, $(\tilde{c}(\tilde{N}, \tilde{K}))$ belong to the unit ball of ℓ^2. We now expand the integrand in the left-hand side of (5.3.8). Modulo terms of lower order, this integral is a sum of terms such as

$$I_K^N = \int_{\mathbb{R} \times M} \prod_{j=0}^{3} u_{N_j, K_j}^{(j)} H(S_{N_3}(u)) \, dt \, dx,$$

where $N = (N_0, N_1, N_2, N_3)$, $K = (K_0, K_1, K_2, K_3)$, N_j, K_j describe all dyadic integers such that $N_1 \geq N_2 \geq N_3$, $u_{N_0, K_0}^{(0)} = \Delta_{N_0 K_0}(w)$, and, for $j = 1, 2, 3$,

$$u_{N_j, K_j}^{(j)} = \Delta_{N_j K_j}(u) \quad \text{or} \quad \Delta_{N_j K_j}(\overline{u}).$$

As for $S_{N_3}(u)$, it is essentially

$$S_{N_3}(u) = \sum_{N \leq N_3} P_N u,$$

and H is a smooth function built on the third derivatives of F, so that

$$|H(z)| \leq C(1 + |z|)^{\alpha - 3},$$

and

$$\|H(S_{N_3}(u))\|_{L^\infty} \leq C \, N_3^{\frac{\alpha-3}{2}} (1 + \|u\|_{X^{1,b}})^{\alpha - 3}.$$

Using suitable integrations by parts and (5.3.9), it can be shown that, for Λ large enough, there exists $\delta > 0$ and p such that, if $N_0 > \Lambda N_1$,

$$|I_K^N| \lesssim N_0^{-\delta} (K_0 K_1 K_2 K_3)^{-\delta} \|w\|_{X^{-s,b'}} (1 + \|u\|_{X^{1,b}})^p \|u\|_{X^{s,b}},$$

which clearly contributes to the desired right-hand side in (5.3.8).

Let us discuss the main regime $N_0 \leq \Lambda N_1$. Using the Cauchy-Schwarz inequality and (5.3.9), we have

$$|I_K^N| \leq \|u_{N_0, K_0}^{(0)} u_{N_2, K_2}^{(2)}\|_{L^2(\mathbb{R} \times M)} \|u_{N_1, K_1}^{(1)} u_{N_3, K_3}^{(3)}\|_{L^2(\mathbb{R} \times M)} \|H(S_{N_3}(u))\|_{L^\infty(\mathbb{R} \times M)}$$

$$\lesssim \mu_K^N \|w\|_{X^{-s,b'}} \|u\|_{X^{s,b}} \|u\|_{X^{1,b}}^2 (1 + \|u\|_{X^{1,b}})^{\alpha - 3},$$

where

$$\mu_K^N = \left(\frac{N_0}{N_1}\right)^s N_2^{2\sigma_1 - 1} N_3^{2\sigma_1 - 1 + \frac{\alpha-3}{2}} K_0^{b_1 - b'} (K_1 K_2 K_3)^{b_1 - b} \prod_{j=0}^{3} c_j(N_j, K_j),$$

and the sequences $(c_j(N_j, K_j))$ are in the unit ball of ℓ^2. Our assumption on α reads

$$\frac{\alpha - 3}{2} = 2 - 4\sigma - \gamma$$

for some $\gamma > 0$. We then choose $\sigma_1 \in (\sigma, 1/2)$ such that

$$2 - 4\sigma - \gamma \leq 2 - 4\sigma_1,$$

then $b_1 < 1/2$ such that (5.3.9) holds, then $b' \in (b_1, 1/2)$, and finally $b \in (1/2, 1 - b')$. This yields

$$\mu_K^N \leq \left(\frac{N_0}{N_1}\right)^s \left(\frac{N_3}{N_2}\right)^{1-2\sigma_1} K_0^{b_1-b'} (K_1 K_2 K_3)^{b_1-b} \prod_{j=0}^3 c_j(N_j, K_j),$$

leading to the summability of $\sum \mu_K^N$ in view of Schur's lemma.

5.4 TRILINEAR STRICHARTZ ESTIMATES AND THE SUBQUINTIC CASE ON $S^2 \times S^1$

If one tries to apply the previous strategy to $M = S^2 \times S^1$, one observes that the Schrödinger group satisfies a bilinear Strichartz estimate of loss $\sigma = 3/8 + \varepsilon$, which, in view of Theorem 5.3.2, seems to restrict global wellposedness to subquartic NLS. In fact, this "bad" bilinear Strichartz estimate is the trace of a "good" trilinear Strichartz estimate, which will enable us to recover the whole subquintic range.

THEOREM 5.4.1 *For every $\varepsilon > 0$, there exists $C_\varepsilon > 0$ such that, for every $N_1 \geq N_2 \geq N_3$, for every f_1, f_2, f_3 in $L^2(S^2 \times S^1)$, the solutions of the linear Schrödinger equation*

$$v_j(t) = e^{it\Delta} P_{N_j} f_j, \quad j = 1, 2, 3,$$

satisfy

$$\|v_1 v_2 v_3\|_{L^2((0,1) \times S^2 \times S^1)}$$
$$\leq C_\varepsilon N_2^{\frac{3}{4}+\varepsilon} N_3^{\frac{5}{4}} \|f_1\|_{L^2(S^2 \times S^1)} \|f_2\|_{L^2(S^2 \times S^1)} \|f_3\|_{L^2(S^2 \times S^1)}.$$

Remark 5.4.1. For $N_1 = N_2 = N_3$ and $f_1 = f_2 = f_3$, Theorem 5.4.1 provides the following estimate for solutions of the linear Schrödinger equation,

$$\forall \varepsilon > 0, \quad \|v\|_{L^6((0,1) \times S^2 \times S^1)} \leq C_\varepsilon \|v_0\|_{H^{\frac{2}{3}+\varepsilon}(S^2 \times S^1)},$$

while the Strichartz estimates on $\mathbb{R}^3$ combined with the Sobolev inequality give the scale-invariant estimate

$$\|v\|_{L^6(\mathbb{R} \times \mathbb{R}^3)} \leq C \|v_0\|_{\dot{H}^{\frac{2}{3}}(\mathbb{R}^3)}.$$

The proof of Theorem 5.4.1 proceeds as in Theorem 5.3.1, writing

$$v_j(t, x, y) = \sum_{N_j^2 \leq n_j(n_j+1)+p_j^2 \leq 4N_j^2} e^{-it(n_j(n_j+1)+p_j^2)} H_{n_j, p_j}^{(j)}(x) e^{ip_j y},$$

where $H_{n,p}^{(j)}$ denote spherical harmonics of degree n on S^2. The proof is achieved by using the Parseval formula in variables t and y, an elementary number-theoretic counting argument, and the following trilinear version of Sogge's L^6 estimate on S^2.

LEMMA 5.4.2 *If $n_1 \geq n_2 \geq n_3 \geq 1$ are integers and $H_{n_j}^{(j)}$, $j = 1, 2, 3$, is a spherical harmonic of degree n_j on S^2,*

$$\|H_{n_1}^{(1)} H_{n_2}^{(2)} H_{n_3}^{(3)}\|_{L^2(S^2)} \lesssim (n_2 n_3)^{\frac{1}{4}} \|H_{n_1}^{(1)}\|_{L^2(S^2)} \|H_{n_2}^{(2)}\|_{L^2(S^2)} \|H_{n_3}^{(3)}\|_{L^2(S^2)}.$$

As in Subsection 3.3, we prove that three-manifolds that satisfy such trilinear Strichartz estimates enjoy the same global wellposedness properties for subquintic NLS.

THEOREM 5.4.2 *Let M be a compact three-manifold for which there exists $a > 0$ such that, for every $\varepsilon > 0$, solutions of the linear Schrödinger equation*

$$v_j(t) = e^{it\Delta} P_{N_j} f_j, \quad j = 1, 2, 3,$$

satisfy

$$\|v_1 v_2 v_3\|_{L^2((0,1) \times M)} \leq C_\varepsilon \, N_2^{1-a+\varepsilon} \, N_3^{1+a} \, \|f_1\|_{L^2(M)} \|f_2\|_{L^2(M)} \|f_3\|_{L^2(M)}.$$

Then, for every $s \geq 1$, there exists $b > 1/2$ such that, for every $u_0 \in H^s(M)$, the nonlinear Schrödinger equation (5.1.1) of degree $\alpha < 5$ admits a unique solution u in $X_{\mathrm{loc}}^{s,b}(\mathbb{R} \times M)$. In particular, if $s > 3/2$, the Cauchy problem admits a unique solution $u \in C(\mathbb{R}, H^s(M))$.

Let us indicate how Theorem 5.4.2 is obtained by modifying the proof of Theorem 5.3.2 in the previous section. First, in a similar way to (5.3.9), the assumed trilinear Strichartz estimate implies, for every $\varepsilon > 0$, the existence of $\beta = \beta_\varepsilon < 1/2$ such that, for $N_1 \geq N_2 \geq N_3$,

$$\left\| \prod_{j=1}^{3} \Delta_{N_j K_j} u^{(j)} \right\|_{L^2(\mathbb{R} \times M)} \tag{5.4.1}$$

$$\leq C_\varepsilon N_1^{-s_1} N_2^{1-a+\varepsilon-s_2} N_3^{1+a+\varepsilon-s_3} \prod_{j=1}^{3} K_j^{\beta-b_j} c_j(N_j, K_j) \|u^{(j)}\|_{X^{s_j, b_j}},$$

and the sequences $(c_j(N_j, K_j))$ are in the unit ball of ℓ^2.

We now turn to the new ingredients in the proof of estimate (5.3.8), focusing on the hardest case $\alpha > 4$. We perform one more dyadic expansion such that the integral in the left-hand side of (5.3.8) is mainly a sum of terms

$$I_K^N = \int_{\mathbb{R} \times M} \prod_{j=0}^{4} u_{N_j, K_j}^{(j)} G(S_{N_4}(u)) \, dt \, dx,$$

where $N = (N_0, N_1, N_2, N_3, N_4)$, $K = (K_0, K_1, K_2, K_3, K_4)$, N_j, K_j describe all dyadic integers such that $N_1 \geq N_2 \geq N_3 \geq N_4$, $N_0 \leq \Lambda N_1$, $u_{N_0, K_0}^{(0)} = \Delta_{N_0 K_0}(w)$, and, for $j = 1, 2, 3, 4$,

$$u_{N_j, K_j}^{(j)} = \Delta_{N_j K_j}(u) \quad \text{or} \quad \Delta_{N_j K_j}(\overline{u}).$$

G is a smooth function built on the fourth derivatives of F, so that

$$|G(z)| \leq C(1 + |z|)^{\alpha-4},$$

and

$$|G(S_{N_4}(u))| \lesssim \left(1 + \sum_{K_5, N_5: N_5 \leq N_4} |\Delta_{N_5 K_5}(u)| \right)^{\alpha - 4}.$$

We then apply Hölder's inequality to get

$$|I_K^N| \leq [J_K^N]^{1-\gamma} [\tilde{J}_K^N]^\gamma$$

with $\gamma = \alpha - 4 \in (0, 1)$ and

$$J_K^N = \int_{\mathbb{R} \times M} \prod_{j=0}^{4} |u_{N_j, K_j}^{(j)}| \, dt \, dx,$$

$$\tilde{J}_K^N = \int_{\mathbb{R} \times M} \prod_{j=0}^{4} |u_{N_j, K_j}^{(j)}| \, |G|^{1/\gamma} \, dt \, dx \lesssim J_K^N$$

$$+ \sum_{N_5: N_5 \leq N_4} \int_{\mathbb{R} \times M} \prod_{j=0}^{4} |u_{N_j, K_j}^{(j)}| |\Delta_{N_5 K_5}(u)| \, dt \, dx.$$

We estimate J_K^N by the Schwarz inequality and by applying (5.4.1) and its "bad" bilinear trace, which is (5.3.9) with $2\sigma_1 = 1 - a + \varepsilon$. This gives

$$J_K^N \leq \|u_{N_0, K_0}^{(0)} u_{N_2, K_2}^{(2)}\|_{L^2(\mathbb{R} \times M)} \|u_{N_1, K_1}^{(1)} u_{N_3, K_3}^{(3)} u_{N_4, K_4}^{(4)}\|_{L^2(\mathbb{R} \times M)}$$
$$\lesssim \mu_K^N \|w\|_{X^{-s, b'}} \|u\|_{X^{1, b}}^3 \|u\|_{X^{s, b}},$$

where

$$\mu_K^N = \left(\frac{N_0}{N_1} \right)^s N_2^{-a+\varepsilon} N_3^{-a+\varepsilon} N_4^{a+\varepsilon} K_0^{\beta - b'} (K_1 K_2 K_3 K_4)^{\beta - b} \prod_{j=0}^{4} c_j(N_j, K_j),$$

and the sequences $c_j(N_j, K_j)$ belong to the unit ball of ℓ^2.

Then we estimate each term in the sum majorizing $\tilde{J}_K^N$ by Schwarz inequality combined with (5.4.1). Summing on N_5 and K_5, this gives

$$\tilde{J}_K^N \lesssim \mu_K^N N_4^{a+\varepsilon} \|w\|_{X^{-s, b'}} (1 + \|u\|_{X^{1, b}})^4 \|u\|_{X^{s, b}}.$$

We obtained finally

$$|I_K^N| \lesssim \mu_K^N N_4^{\gamma(a+\varepsilon)} \|w\|_{X^{-s, b'}} (1 + \|u\|_{X^{1, b}})^4 \|u\|_{X^{s, b}},$$

with

$$\mu_K^N N_4^{\gamma(a+\varepsilon)}$$
$$= \left(\frac{N_0}{N_1} \right)^s N_2^{-a+\varepsilon} N_3^{-a+\varepsilon} N_4^{(1+\gamma)a+\varepsilon} K_0^{\beta - b'} (K_1 K_2 K_3 K_4)^{\beta - b} \prod_{j=0}^{4} c_j(N_j, K_j).$$

The proof is completed by choosing ε small enough with respect to $(1 - \gamma)a$.

5.5 TOWARD THE CRITICAL CASE

In this last section, we prove some estimates related to the quintic NLS on S^3, which can be seen as a first step to global wellposedness for small energy data. We start with an improvement of the L^6 Strichartz estimate, which, as already observed in Remark 5.4.1, displays the same loss of regularity as on $\mathbb{R}^3$.

PROPOSITION 5.5.1 *If* $v(t) = e^{it\Delta}v_0$, *then*

$$\|v\|_{L^6((0,1)\times S^3)} \leq C \|v_0\|_{H^{2/3}(S^3)}. \tag{5.5.1}$$

Proof. The strategy is similar to Proposition 5.3.1. We write

$$v(t) = \sum_{N\leq n\leq 2N} e^{-it(n^2-1)} H_n,$$

and we compute the L^6 norm in time by using the Fourier series estimate of Proposition 1.10 in Bourgain [7]. For every $x \in S^3$,

$$\int_0^{2\pi} |v(t,x)|^6 \, dt \leq C N \left(\sum_{N\leq n\leq 2N} |H_n(x)|^2 \right)^3.$$

Integrating in x and developing, we obtain

$$\|v\|^6_{L^6((0,2\pi)\times S^3)} \lesssim N \sum_{N\leq n_1,n_2,n_3\leq 2N} \|H_{n_1} H_{n_2} H_{n_3}\|^2_{L^2(S^3)}$$

$$\lesssim N^4 \sum_{N\leq n_1,n_2,n_3\leq 2N} \|H_{n_1}\|^2_{L^2(S^3)} \|H_{n_2}\|^2_{L^2(S^3)} \|H_{n_3}\|^2_{L^2(S^3)},$$

by applying Sogge's estimate (5.3.3), (5.3.4) with $q = 6$. This completes the proof. $\blacksquare$

A trilinearization of Proposition 5.5.1 is in fact possible. This reads

THEOREM 5.5.1 *For every* $\varepsilon > 0$, *there exists* $C_\varepsilon > 0$ *such that, for every* $N_1 \geq N_2 \geq N_3$, *for every* f_1, f_2, f_3 *in* $L^2(S^3)$, *the solutions of the linear Schrödinger equation*

$$v_j(t) = e^{it\Delta} P_{N_j} f_j, \quad j = 1, 2, 3,$$

satisfy

$$\|v_1 v_2 v_3\|_{L^2((0,1)\times S^3)} \leq C_\varepsilon N_2^{\frac{5}{6}+\varepsilon} N_3^{\frac{7}{6}-\varepsilon} \|f_1\|_{L^2(S^3)} \|f_2\|_{L^2(S^3)} \|f_3\|_{L^2(S^3)}.$$

Using the above trilinear theorem, it is easy to derive the following estimate on the first iteration for quintic NLS:

COROLLARY 5.5.2 *For every* $s \geq 1$, *there exists* C_s *such that, for every* $f \in H^s(S^3)$, *the Duhamel term*

$$w(t) = \int_0^t e^{i(t-\tau)\Delta} \left(|e^{i\tau\Delta}f|^4 e^{i\tau\Delta}f \right) d\tau$$

satisfies

$$\sup_{0\leq t\leq 1} \|w(t)\|_{H^s(S^3)} \leq C_s \|f\|^4_{H^1(S^3)} \|f\|_{H^s(S^3)}.$$

Let us first show how Theorem 5.5.1 implies the corollary. As in previous sections, we proceed by duality, and we expand, up to lower-order terms,

$$\int_{S^3} \overline{w}(t, x)\, e^{it\Delta}\, f_0(x)\, dx \simeq \sum_{N_0, N_1 \geq \cdots \geq N_5} I_{N_0, \ldots, N_5}(t),$$

with

$$I_{N_0, \ldots, N_5}(t) = \int_{(0,t) \times S^3} \prod v_j(\tau, x)\, d\tau\, dx,$$

and

$$v_0(\tau) = e^{i\tau\Delta} P_{N_0} f_0, \quad v_j(\tau) = e^{i\tau\Delta} P_{N_j} f, \quad \text{or} \quad \overline{e^{i\tau\Delta} P_{N_j} f}, \ 1 \leq j \leq 5.$$

We fix $\varepsilon \in (0, 1/6)$ and we apply the Schwarz inequality together with Theorem 5.5.1. We obtain

$$|I_{N_0, \ldots, N_5}(t)|$$

$$\lesssim \frac{N_0^s}{N_1^s} \frac{(N_2 N_3)^{\frac{5}{6}+\varepsilon} (N_4 N_5)^{\frac{7}{6}-\varepsilon}}{N_2 N_3 N_4 N_5} \prod_{j=0}^{5} c_j(N_j)\, \|f_0\|_{H^{-s}}\, \|f\|_{H^s}\, \|f\|_{H^1}^4,$$

where the sequences $(c_j(N))$ belong to the unit ball of ℓ^2. This yields

$$|I_{N_0, \ldots, N_5}|$$

$$\lesssim \left(\frac{N_0}{N_1}\right)^s \left(\frac{N_4}{N_2}\right)^{\frac{1}{6}-\varepsilon} \left(\frac{N_5}{N_3}\right)^{\frac{1}{6}-\varepsilon} \prod_{j=0}^{5} c_j(N_j)\, \|f_0\|_{H^{-s}}\, \|f\|_{H^s}\, \|f\|_{H^1}^4,$$

which completes the proof by the usual Schur lemma.

We close this section by describing the main steps of the proof of Theorem 5.5.1. A first lemma is the following slight extension of Bourgain's result in [7], which we already used in the proof of Proposition 5.5.1.

LEMMA 5.5.3 *Let $\psi \in C_0^\infty(\mathbb{R})$ be supported in an interval I of length $\lesssim N$, with*

$$\sup_{\tau \in \mathbb{R}} |\psi''(\tau)| \lesssim N^{-2}.$$

Then, for every sequence (c_n),

$$\left\| \sum_{n \in \mathbb{Z}} \psi(n)\, c_n\, e^{itn^2} \right\|_{L^6(\mathbb{R}/2\pi\mathbb{Z})} \lesssim N^{\frac{1}{6}} \left(\sum_{n \in I} |c_n|^2 \right)^{\frac{1}{2}}.$$

This lemma enables us to trilinearize the L^6 estimate of [7], in the spirit of [1].

LEMMA 5.5.4 *Let $\varphi \in C_0^\infty(\mathbb{R})$. Then, for $N_1 \geq N_2 \geq N_3$, the Fourier series*

$$F_j(t) = \sum_{n_j} \varphi\left(\frac{n_j}{N_j}\right) c_j(n_j)\, e^{itn_j^2}, \quad j = 1, 2, 3,$$

satisfy

$$\|F_1 F_2 F_3\|_{L^2(\mathbb{R}/2\pi\mathbb{Z})} \lesssim N_2^{\frac{1}{3}} N_3^{\frac{1}{6}} \prod_{j=1}^{3} \|F_j\|_{L^2(\mathbb{R}/2\pi\mathbb{Z})}.$$

Proof. Introduce $\chi \in C_0^\infty(\mathbb{R})$ supported into $(-1, 1)$, such that

$$\sum_{\alpha \in \mathbb{Z}} \chi(\tau - \alpha) = 1,$$

and write

$$F_1 = \sum_{\alpha \in \mathbb{Z}} F_{1,\alpha}$$

with

$$F_{1,\alpha}(t) = \sum_{n_1} \varphi\left(\frac{n_1}{N_1}\right) \chi\left(\frac{n_1}{N_2} - \alpha\right) c_1(n_1) e^{itn_1^2}.$$

It is clear that the functions $F_{1,\alpha} F_2 F_3$ are almost orthogonal as α varies. Hence,

$$\|F_1 F_2 F_3\|_{L^2}^2 \simeq \sum_\alpha \|F_{1,\alpha} F_2 F_3\|_{L^2}^2,$$

and each term in the right-hand side can be estimated by means of the Hölder inequality and of Lemma 5.5.3, observing that

$$\psi_\alpha(\tau) = \chi\left(\frac{\tau}{N_2} - \alpha\right) \varphi\left(\frac{\tau}{N_1}\right)$$

satisfies the assumption of Lemma 5.5.3 with $N = N_2$. This yields

$$\|F_{1,\alpha} F_2 F_3\|_{L^2}^2 \lesssim N_3^{\frac{1}{3}} N_2^{\frac{2}{3}} \|F_2\|_{L^2}^2 \|F_3\|_{L^2}^2 \left(\sum_{(\alpha-1)N_2 \leq n_1 \leq (\alpha+1)N_2} |c_1(n_1)|^2 \right)$$

and the summation on α completes the proof.

The second ingredient is once again a trilinear Sogge estimate (see [17]).

LEMMA 5.5.5 *If $n_1 \geq n_2 \geq n_3 \geq 1$ are integers and $H_{n_j}^{(j)}$, $j = 1, 2, 3$, is a spherical harmonic of degree $n_j - 1$ on S^3, then, for every $\varepsilon > 0$,*

$$\|H_{n_1}^{(1)} H_{n_2}^{(2)} H_{n_3}^{(3)}\|_{L^2(S^3)}$$

$$\leq C_\varepsilon \, n_2^{\frac{1}{2}+\varepsilon} \, n_3^{1-\varepsilon} \|H_{n_1}^{(1)}\|_{L^2(S^3)} \|H_{n_2}^{(2)}\|_{L^2(S^3)} \|H_{n_3}^{(3)}\|_{L^2(S^3)}.$$

At this stage it is easy to conclude the proof of Theorem 5.5.1, similarly to Proposition 5.5.1. We write, for a suitable $\varphi \in C_0^\infty(\mathbb{R})$,

$$v_j(t) = \sum_{n_j} e^{-it(n_j^2-1)} \varphi(n_j/N_j) H_{n_j}^{(j)},$$

and we apply Lemma 5.5.4 to $F_j(t) = v_j(t, x)$ for fixed x. It remains to integrate in x and to apply Lemma 5.5.5.

5.6 CONCLUSION

Sections 3 and 4 provided two different proofs of the global wellposedness of defocusing subquintic NLS on S^3 and $S^2 \times S^1$, which add to the proofs for $\mathbb{R}^3$ and for $\mathbb{T}^3$. Moreover, J. Bourgain [4] recently gave another proof of this result for irrational tori. It is of course tempting to conjecture that this fact is true for every three-manifold, though a general argument seems out of reach for the moment.

As for critical NLS, we still have to introduce a convenient space of functions where an iterative scheme can converge with small energy data.

Note that arguments of Section 5 cannot be extended to cubic NLS on S^4. Indeed, in this case, it is known [9] that the Strichartz inequality

$$\|v\|_{L^4((0,1)\times S^4)} \lesssim \|v_0\|_{H^{\frac{1}{2}}(S^4)}$$

fails for the solutions v of the linear Schrödinger equation. Therefore, the question of wellposedness of this four-dimensional critical NLS seems even more intricate.

REFERENCES

[1] J. Bourgain. Fourier transform restriction phenomena for certain lattice subsets and application to nonlinear evolution equations I. Schrödinger equations. *Geom. and Funct. Anal.*, 3: 107–156, 1993.

[2] J. Bourgain. Exponential sums and nonlinear Schrödinger equations. *Geom. and Funct. Anal.*, 3: 157–178, 1993.

[3] J. Bourgain. Eigenfunction bounds for the Laplacian on the n-torus. *Internat. Math. Res. Notices*, 3: 61–66, 1993.

[4] J. Bourgain. On Strichartz's inequalities and the Nonlinear Schrödinger Equation on irrational tori (chapter 1 of this book).

[5] J. Bourgain. Global wellposedness of defocusing 3D critical NLS in the radial case. *Jounal of the A.M.S.*, 12: 145–171, 1999.

[6] J. Bourgain. Global solutions of nonlinear Schrödinger equations. *Colloq. Publications, Amer. Math. Soc.*, 1999.

[7] J. Bourgain. On $\Lambda(p)$-subsets of squares. *Israël Journal of Mathematics*, 67: 291–311, 1989.

[8] H. Brezis and T. Gallouët. Nonlinear Schrödinger evolution equations. *Nonlinear Analysis, Theory, Methods and Applications*, 4: 677–681, 1980.

[9] N. Burq, P. Gérard, and N. Tzvetkov. Strichartz inequalities and the nonlinear Schrödinger equation on compact manifolds. *Amer. J. Math.*, 126, no. 3: 569–605, 2004.

[10] N. Burq, P. Gérard, and N. Tzvetkov. An instability property of the nonlinear Schrödinger equation on S^d. *Math. Res. Lett.*, 9(2–3): 323–335, 2002.

[11] N. Burq, P. Gérard, and N. Tzvetkov. The Cauchy problem for the nonlinear Schrödinger equation on compact manifolds. *J. Nonlinear Math. Physics*, 10: 12–27, 2003.

[12] N. Burq, P. Gérard, and N. Tzvetkov. Two singular dynamics of the nonlinear Schrödinger equation on a plane domain. *Geom. Funct. Anal.*, 13: 1–19, 2003.

[13] N. Burq, P. Gérard, and N. Tzvetkov. An example of singular dynamics for the nonlinear Schrödinger equation on bounded domains. *Hyperbolic Problems and Related Topics*, F. Colombini and T. Nishitani, editors, Graduate Series in Analysis, International Press, 2003.

[14] N. Burq, P. Gérard, and N. Tzvetkov. On nonlinear Schrödinger equations in exterior domains. *Ann. I. H. Poincaré-AN*, 21: 295–318, 2004.

[15] N. Burq, P. Gérard, and N. Tzvetkov. Bilinear eigenfunction estimates and the nonlinear Schrödinger equation on surfaces. *Invent. math.* 159: 187–223, 2005.

[16] N. Burq, P. Gérard, and N. Tzvetkov. Multilinear estimates for Laplace spectral projectors on compact manifolds. *C. R. Acad. Sci. Paris*, Ser. I 338: 359–364, 2004.

[17] N. Burq, P. Gérard, and N. Tzvetkov. Multilinear eigenfunction estimates and global existence for the three dimensional nonlinear Schrödinger equations. *Ann. Scient. Éc. Norm. Sup.* 38: 255–301, 2005.

[18] N. Burq, P. Gérard, and N. Tzvetkov. The Cauchy Problem for the nonlinear Schrödinger equation on compact manifolds. In *Phase Space Analysis of Partial Differential Equations* (ed. by F. Colombini and L. Pernazza), vol.I. Centro di Ricerca Matematica Ennio de Giorgi, Scuola Normale Superiore, Pisa, 2004, 21–52.

[19] T. Cazenave. Semilinear Schrödinger equations. Courant Lecture Notes in Mathematics, 10. New York University. American Mathematical Society, Providence, RI, 2003.

[20] T. Cazenave and F. Weissler. The Cauchy problem for the critical nonlinear Schrödinger equation in H^s. *Nonlinear Analysis, Theory, Methods and Applications*, 14, no. 10: 807–836, 1990.

[21] M. Christ, J. Colliander, and T. Tao. Asymptotics, modulation and low regularity ill-posedness for canonical defocusing equations. *Amer. J. Math.*, 125: 1225–1293, 2003.

[22] M. Christ, J. Colliander, and T. Tao. Ill-posedness for nonlinear Schrödinger and wave equations, math.AP/0311048. To appear in *Ann. I. H. Poincaré-Analyse non linéaire*.

[23] J. Colliander, M. Keel, G. Staffilani, H. Takaoka, and T. Tao. Global wellposedness and scattering in the energy space for the critical nonlinear Schrödinger equation in $\mathbb{R}^3$, math.AP/0402129. To appear in *Annals of Math*.

[24] J. Ginibre and G. Velo. On a class of nonlinear Schrödinger equations. *J. Funct. Anal.*, 32: 1–71, 1979.

[25] J. Ginibre and G. Velo. The global Cauchy problem for the nonlinear Schrödinger equation. *Ann. Inst. H. Poincaré-AN*, 2: 309–327, 1985.

[26] J. Ginibre. Le problème de Cauchy pour des EDP semi-linéaires périodiques en variables d'espace (d'après Bourgain). Séminaire Bourbaki, Exp. 796, *Astérisque*, 237: 163–187, 1996.

[27] M. Grillakis. On nonlinear Schrödinger equations. *Comm. Partial Differential Equations*, 25: 1827–1844, 2000.

[28] T. Kato. On nonlinear Schrödinger equations. *Ann. Inst. H. Poincaré, Physique théorique*, 46: 113–129, 1987.

[29] M. Keel and T. Tao. Endpoint Strichartz estimates. *Amer. J. Math.*, 120: 955–980, 1998.

[30] S. Klainerman and M. Machedon. Finite energy solutions of the Yang-Mills equations in $\mathbb{R}^{3+1}$ *Ann. of Math. (2)*, 142 (1): 39–119, 1995.

[31] C. Sogge. Oscillatory integrals and spherical harmonics. *Duke Math. Jour.*, 53: 43–65, 1986.

[32] C. Sogge. Concerning the L^p norm of spectral clusters for second order elliptic operators on compact manifolds. *J. Funct. Anal.*, 77: 123–138, 1988.

[33] C. Sogge. Fourier integrals in classical analysis. *Cambridge Tracts in Mathematics*, 1993.

[34] G. Staffilani and D. Tataru. Strichartz estimates for a Schrödinger operator with nonsmooth coefficients. *Comm. Partial Differential Equations*, 27(7–8): 1337–1372, 2002.

[35] V. Yudovich. Non-stationary flows of an ideal incompressible fluid. *Journal Vytchisl. Mat. i Mat. Fis.*, 3: 1032–1066, 1963.

[36] A. Zygmund. On Fourier coefficients and transforms of functions of two variables. *Studia Math.*, 50:189–201, 1974.

Chapter Six

Power series solution of a nonlinear
Schrödinger equation

M. Christ

6.1 INTRODUCTION

6.1.1 The NLS Cauchy Problem

The Cauchy problem for the one-dimensional periodic cubic nonlinear Schrödinger equation is

$$\begin{cases} iu_t + u_{xx} + \omega|u|^2 u = 0 \\ u(0, x) = u_0(x), \end{cases} \tag{NLS}$$

where $x \in \mathbb{T} = \mathbb{R}/2\pi\mathbb{Z}$, $t \in \mathbb{R}$, and the parameter ω equals ± 1. Bourgain [2] has shown this problem to be wellposed in the Sobolev space H^s for all $s \geq 0$, in the sense of uniformly continuous dependence on the initial datum. In H^0 it is wellposed globally in time, and as is typical in this subject, the uniqueness aspect of wellposedness is formulated in a certain auxiliary space more restricted than $C^0([0, T], H^s(\mathbb{T}))$, in which existence is also established. For $s < 0$ it is illposed in the sense of uniformly continuous dependence [3], and is illposed in stronger senses [5] as well. The objectives of this chapter are twofold: to establish the existence of solutions for wider classes of initial data than H^0, and to develop an alternative method of solution.

The spaces of initial data considered here are the spaces $\mathcal{F}L^{s,p}$ for $s \geq 0$ and $p \in [1, \infty]$, defined as follows:

DEFINITION 6.1.1 $\mathcal{F}L^{s,p}(\mathbb{T}) = \{f \in \mathcal{D}'(\mathbb{T}) : \langle \cdot \rangle^s \widehat{f}(\cdot) \in \ell^p\}$.

Here $\mathcal{D}'(\mathbb{T})$ is the usual space of distributions, and $\mathcal{F}L^{s,p}$ is equipped with the norm $\|f\|_{\mathcal{F}L^{s,p}} = \|\widehat{f}\|_{\ell^{s,p}(\mathbb{Z})} = \left(\sum_{n\in\mathbb{Z}}\langle n\rangle^{ps}|\widehat{f}(n)|^p\right)^{1/p}$. We write $\mathcal{F}L^p = \mathcal{F}L^{0,p}$, and are mainly interested in these spaces since, for $p > 2$, they are larger function spaces than the borderline Sobolev space H^0 in which (NLS) is already known to be wellposed.

The author was supported by NSF grant DMS-040126.

6.1.2 Motivations

At least four considerations motivate analysis of the Cauchy problem in these particular function spaces. The first is the desire for existence theorems for initial data in function spaces that scale like the Sobolev spaces H^s, for negative s. $\mathcal{F}L^p$ scales like $H^{s(p)}$ where $s(p) = -\frac{1}{2} + \frac{1}{p} \downarrow -\frac{1}{2}$ as $p \uparrow \infty$, thus spanning the gap between the optimal exponent $s = 0$ for Sobolev space wellposedness, and the scaling exponent $-\frac{1}{2}$. Moreover, $\mathcal{F}L^p$ is invariant under the Galilean symmetries of the equation.

Some existence results are already known in spaces scaling like H^s for certain negative exponents, for the nonperiodic one-dimensional setting. Vargas and Vega [12] proved existence of solutions for arbitrary initial data in certain such spaces for a certain range of strictly negative exponents. In particular, for the local in time existence theory, their spaces contain $\mathcal{F}L^p$ for all $p < 3$, and scale like $\mathcal{F}L^p$ for a still larger though bounded range of p. Grünrock [7] has proved wellposedness for the cubic nonlinear Schrödinger equation in the real line analogues of $\mathcal{F}L^{s,p}$, and for other PDE in these function spaces, as well.

A second motivation is the work of Kappeler and Topalov [9], [10], who showed via an inverse scattering analysis that the periodic KdV and mKdV equations are wellposed for wider ranges of Sobolev spaces H^s than had previously been known. It is reasonable to seek a corresponding improvement for (NLS). We obtain here such an improvement, but with $\mathcal{F}L^p$ with $p > 2$ substituted for H^s with $s < 0$.

Third, Christ and Erdoğan, in unpublished work, have investigated the conserved quantities in the inverse scattering theory relevant to (NLS), and have found that for any distribution in $\mathcal{F}L^p(\mathbb{T})$ with small norm, the sequence of gap lengths for the associated Dirac operator belongs to ℓ^p and has comparable norm.[1] Thus $\mathcal{F}L^p$ for $2 < p < \infty$ may be a natural setting for the Dirac operator inverse scattering theory relevant to the periodic cubic nonlinear Schrödinger equation.

For $p = 2$, the existing proof [2] of wellposedness via a contraction mapping argument implies that the mapping from initial datum to solution has a convergent power series expansion; that is, certain multilinear operators are well defined and satisfy appropriate inequalities. Our fourth motivation is the hope of understanding more about the structure of these operators.

6.1.3 Modified Equation

In order for the Cauchy problem to make any sense in $\mathcal{F}L^p$ for $p > 2$, it seems to be essential to modify the differential equation. We consider

$$\begin{cases} iu_t + u_{xx} + \omega\big(|u|^2 - 2\mu(|u|^2)\big)u = 0 \\ u(0, x) = u_0(x), \end{cases} \tag{NLS*}$$

---

[1] Having slightly better than bounded Fourier coefficients seems to be a minimal condition for the applicability of this machinery, since the eigenvalues for the free periodic Dirac system are equally spaced, and gap lengths for perturbations are to leading order proportional to absolute values of Fourier coefficients of the perturbing potential.

where

$$\mu(|f|^2) = (2\pi)^{-1} \int_{\mathbb{T}} |f(x)|^2 \, dx \qquad (6.1.1)$$

equals the mean value of the absolute value squared of f. In (NLS*), $\mu(|u|^2)$ is short-hand for $\mu(|u(t, \cdot)|^2) = \|u(t, \cdot)\|_{L^2}^2$, which is independent of t for all sufficiently smooth solutions; modifying the equation in this way merely introduces a unimodular scalar factor $e^{2i\mu t}$, where $\mu = \mu(|u_0|^2)$. For parameters p, s such that $\mathcal{F}L^{s,p}$ is not embedded in H^0, $\mu(|u_0|^2)$ is not defined for typical $u_0 \in \mathcal{F}L^{s,p}$, but of course the same goes for the function $|u_0(x)|^2$, and we will nonetheless prove that the equation makes reasonable sense for such initial data.

The coefficient 2 in front of $\mu(|u|^2)$ is the unique one for which solutions depend continuously on initial data in $\mathcal{F}L^p$ for $p > 2$.

6.1.4 Conclusions

Our main result is as follows. Recall that there exists a unique mapping $u_0 \mapsto Su_0(t, x)$, defined for $u_0 \in C^\infty$, which for all sufficiently large s extends to a uniformly continuous mapping from $H^s(\mathbb{T})$ to $C^0([0, \infty), H^s(\mathbb{T})) \cap C^1([0, \infty), H^{s-2}(\mathbb{T}))$, such that Su_0 is a solution of the modified Cauchy problem (NLS*). $C^\infty(\mathbb{T})$ is of course a dense subset of $\mathcal{F}L^{s,p}$ for any $p \in [1, \infty)$.

THEOREM 6.1.1 *For any $p \in [1, \infty)$, any $s \geq 0$, and any $R < \infty$, there exists $\tau > 0$ for which the solution mapping S extends by continuity to a uniformly continuous mapping from the ball centered at 0 of radius R in $\mathcal{F}L^{s,p}(\mathbb{T})$ to $C^0([0, \tau], \mathcal{F}L^{s,p}(\mathbb{T}))$.*

For the unmodified equation this has the following obvious consequence. Denote by $H_c^0 = H_c^0(\mathbb{T})$ the set of all $f \in H^0$ such that $\|f\|_{L^2} = c$. Denote by $S'u_0$ the usual solution [2] of the unmodified Cauchy problem (NLS) with initial datum u_0, for $u_0 \in H^0$.

COROLLARY 6.1.2 *Let $p \in [1, \infty)$ and $s \geq 0$. For any $R < \infty$ there exists $\tau > 0$ such that for any finite constant $c > 0$, the mapping $H_c^0 \ni u_0 \mapsto S'u_0$ is uniformly continuous as a mapping from H_c^0 intersected with the ball centered at 0 of radius R in $\mathcal{F}L^{s,p}$, equipped with the $\mathcal{F}L^{s,p}$ norm, to $C^0([0, \tau], \mathcal{F}L^{s,p}(\mathbb{T}))$.*

The unpublished result of the author and Erdoğan says that for smooth initial data, if $\|u_0\|_{\mathcal{F}L^p}$ is sufficiently small then $\|u(t)\|_{\mathcal{F}L^p}$ remains bounded uniformly for all $t \in [0, \infty)$. This result in combination with Theorem 6.1.1 would yield global wellposedness for sufficiently small initial data.

The following result concerns the discrepancy between the nonlinear evolution (NLS*) and the corresponding linear Cauchy problem

$$\begin{cases} iv_t + v_{xx} = 0 \\ v(0, x) = u_0(x). \end{cases} \qquad (6.1.2)$$

PROPOSITION 6.1.3 *Let $R < \infty$ and $p \in [1, \infty)$. Let $q > p/3$ also satisfy $q \geq 1$. Then there exist $\tau, \varepsilon > 0$ and $C < \infty$ such that for any initial datum u_0 satisfying $\|u_0\|_{\mathcal{F}L^p} \leq R$, the solutions $u = Su_0$ of (NLS*) and v of (6.1.2) satisfy*

$$\|u(t, \cdot) - v(t, \cdot)\|_{\mathcal{F}L^q} \leq Ct^\varepsilon \text{ for all } t \in [0, \tau]. \tag{6.1.3}$$

Here u is the solution defined by approximating u_0 by elements of C^∞ and passing to the limit. Thus for $p > 1$ the linear evolution approximates the nonlinear evolution, modulo correction terms that are smoother in the $\mathcal{F}L^q$ scale.

Our next result indicates that the function $u(t, x)$ defined by the limiting procedure of Theorem 6.1.1 is a solution of the differential equation in a more natural sense than merely being a limit of smooth solutions. Define Fourier truncation operators T_N, acting on $\mathcal{F}L^{s,p}(\mathbb{T})$, by $\widehat{T_N f}(n) = 0$ for all $|n| > N$, and $= \widehat{f}(n)$ whenever $|n| \leq N$. T_N acts also on functions $v(t, x)$ by acting on $v(t, \cdot)$ for each time t separately. We denote by $S(u_0)$ the limiting function whose existence, for nonsmooth u_0, is established by Theorem 6.1.1.

PROPOSITION 6.1.4 *Let $p \in [1, \infty)$, $s \geq 0$, and $u_0 \in \mathcal{F}L^{s,p}$. Write $u = S(u_0)$. Then for any $R < \infty$ there exists $\tau > 0$ such that whenever $\|u_0\|_{\mathcal{F}L^{s,p}} \leq R$, $\mathcal{N}u(t, x) = (|u|^2 - 2\mu(|u|^2))u$ exists in the sense that*

$$\lim_{N \to \infty} \mathcal{N}(T_N u)(t, x) \text{ exists in the sense of distributions in } C^0([0, \tau], \mathcal{D}'(\mathbb{T})).$$
$$\tag{6.1.4}$$

Moreover if $\mathcal{N}(u)$ is interpreted as this limit, then $u = S(u_0)$ satisfies (NLS*) in the sense of distributions in $(0, \tau) \times \mathbb{T}$.

More generally, the same holds for any sequence of Fourier multipliers of the form $\widehat{T_\nu f}(n) = m_\nu(n)\widehat{f}(n)$, where each sequence m_ν is finitely supported, $\sup_\nu \|m_\nu\|_{\ell^\infty} < \infty$, and $m_\nu(n) \to 1$ as $\nu \to \infty$ for each $n \in \mathbb{Z}$; the limit is of course independent of the sequence (m_ν). Making sense of the nonlinearity via this limiting procedure is connected with general theories of multiplication of distributions [1], [6], but the existence here of the limit over all sequences (m_ν) gives u stronger claim to the title of solution than in the general theory.

Unlike the fixed-point method, our proof yields no uniqueness statement corresponding to these existence results. For any $p > 2$, solutions of the Cauchy problem in the class $C^0([0, \tau], \mathcal{F}L^p)$, in the sense of Proposition 6.1.4, are in fact not unique [4].

6.1.5 Method

Define the partial Fourier transform

$$\widehat{u}(t, n) = (2\pi)^{-1} \int_{\mathbb{T}} e^{-inx} u(t, x) \, dx. \tag{6.1.5}$$

Our approach is to regard the partial differential equation as an infinite coupled nonlinear system of ordinary differential equations for these Fourier coefficients, to express the solution as a power series in the initial datum

$$\widehat{u}(t, n) = \sum_{k=0}^{\infty} \hat{A}_k(t)(\widehat{u_0}, \ldots, \widehat{u_0}) \tag{6.1.6}$$

where each $\hat{A}_k(t)$ is a bounded multilinear operator[2] from a product of k copies of $\mathcal{F}L^{s,p}$ to $\mathcal{F}L^{s,p}$, to show that the individual terms $\hat{A}_k(t)(\widehat{u_0}, \ldots, \widehat{u_0})$ are well defined, and to show that the formal series converges absolutely in $C^0(\mathbb{R}, \mathcal{F}L^{s,p})$ to a solution in the sense of (6.1.4). The case $s \geq 0$ follows from a very small modification of the analysis for $s = 0$, so we discuss primarily $s = 0$, indicating the necessary modifications for $s > 0$ at the end of the paper.

The analysis is rather elementary, much of the chapter being devoted to setting up the definitions and notation required to describe the operators $\hat{A}_k(t)$. A single number theoretic fact enters the discussion: the number of factorizations of an integer n as a product of two integer factors is $O(n^\delta)$, for all $\delta > 0$; this same fact was used by Bourgain [2].

The author is grateful to J. Bourgain, C. Kenig, H. Koch, and D. Tataru for invitations to conferences that stimulated this work, and to Betsy Stovall for proofreading a draft of the manuscript.

6.2 A SYSTEM OF COUPLED ORDINARY DIFFERENTIAL EQUATIONS

6.2.1 General Discussion

Define

$$\sigma(j, k, l, n) = n^2 - j^2 + k^2 - l^2. \tag{6.2.1}$$

It factors as

$$\sigma(j, k, l, n) = 2(n - j)(n - l) = 2(k - l)(k - j) \tag{6.2.2}$$

provided that $j - k + l = n$.

Written in terms of Fourier coefficients $\widehat{u}_n(t) = \widehat{u}(t, n)$, the equation $iu_t + u_{xx} + \omega(|u|^2 - 2\mu(|u|^2))u = 0$ becomes

$$i\frac{d\widehat{u}_n}{dt} - n^2\widehat{u}_n + \omega \sum_{j-k+l=n} \widehat{u}_j\overline{\widehat{u}_k}\widehat{u}_l - 2\omega \sum_m |\widehat{u}_m|^2\widehat{u}_n = 0. \tag{6.2.3}$$

Here the first summation is taken over all $(j, k, l) \in \mathbb{Z}^3$, satisfying the indicated identity, and the second over all $m \in \mathbb{Z}$. Substituting

$$a_n(t) = e^{in^2t}\widehat{u}(t, n), \tag{6.2.4}$$

---

[2] Throughout the discussion we allow multilinear operators to be either conjugate linear or linear in each of their arguments, independently.

(6.2.3) becomes

$$\frac{da_n}{dt} = i\omega \sum_{j-k+l=n}^{*} a_j\bar{a}_k a_l e^{i\sigma(j,k,l,n)t} - i\omega|a_n|^2 a_n, \tag{6.2.5}$$

where the notation $\sum_{j-k+l=n}^{*}$ means that the sum is taken over all $(j, k, l) \in \mathbb{Z}^3$ for which neither $j = n$ nor $l = n$. This notational convention will be used throughout the discussion. The effect of the term $-2\omega\mu(|u|^2)u$ in the modified differential equation (NLS*) is to cancel out a term $2i\omega(\sum_m |a_m|^2)a_n$, which would otherwise appear on the right-hand side of (6.2.5).

Reformulated as an integral equation, (6.2.5) becomes

$$a_n(t) = a_n(0) + i\omega \sum_{j-k+l=n}^{*} \int_0^t a_j(s)\bar{a}_k(s)a_l(s)e^{i\sigma(j,k,l,n)s}\,ds$$

$$- i\omega \int_0^t |a_n(s)|^2 a_n(s)\,ds. \tag{6.2.6}$$

However, in deriving (6.2.6) from (6.2.5), we have interchanged the integral over $[0, t]$ with the summation over j, k, l without any justification.

In terms of Fourier coefficients, (6.2.6) is restated as

$$\hat{u}(t, n) = \hat{u}_0(n) - in^2 \int_0^t \hat{u}(s, n)\,ds$$

$$+ i\omega \sum_{j-k+l=n}^{*} \int_0^t \hat{u}(s, j)\overline{\hat{u}(s, k)}\hat{u}(s, l)\,ds - i\omega \int_0^t |\hat{u}(s, n)|^2\hat{u}(s, n)\,ds. \tag{6.2.7}$$

Substituting for $a_j(s)$, $a_k(s)$, $a_l(s)$ in the right-hand side of (6.2.6) by means of (6.2.6) itself yields

$$a_n(t) = a_n(0) + i\omega \sum_{j-k+l=n}^{*} a_j(0)\bar{a}_k(0)\bar{a}_l(0) \int_0^t e^{i\sigma(j,k,l,n)s}\,ds$$

$$- i\omega|a_n(0)|^2 a_n(0) \int_0^t 1\,ds$$

$$+ \text{additional terms}$$

$$= a_n(0)\left(1 - i\omega t|a_n(0)|^2\right) + \tfrac{1}{2}\omega \sum_{j-k+l=n}^{*} \frac{a_j(0)\bar{a}_k(0)a_l(0)}{(n-j)(n-l)}\left(e^{i(n^2-j^2+k^2-l^2)t} - 1\right)$$

$$+ \text{additional terms}. \tag{6.2.8}$$

These additional terms involve the functions a_m, not only the initial data $a_m(0)$. The right-hand side of the integral equation (6.2.6) can then be substituted for each function a_n, replacing it by $a_n(0)$ but producing still more complex additional terms. Repeating this process indefinitely produces an infinite series, whose convergence certainly requires justification. Each substitution by means of (6.2.6) results in multilinear expressions of increased complexity in terms of functions $a_n(t)$ and initial data $a_n(0)$.

We recognize $1 - i\omega t |a_n(0)|^2$ as a Taylor polynomial for $\exp(-i|a_n(0)|^2 t)$, but for our purposes it will not be necessary to exploit this by recombining terms. In particular, we will not exploit the coefficient i which makes this exponential unimodular.

6.2.2 A Sample Term

One of the very simplest additional terms arises when (6.2.6) is substituted into itself twice:

$$(i\omega)^4 \sum_{j_1-j_2+j_3=n}^{*} \sum_{m_1^1-m_2^1+m_3^1=j_1}^{*} \sum_{m_1^2-m_2^2+m_3^2=j_2}^{*} \sum_{m_1^3-m_2^3+m_3^3=j_3}^{*}$$

$$\int_{0\leq r_1,r_2,r_3\leq s\leq t} a_{m_1^1}(r_1)\bar{a}_{m_2^1}(r_1)a_{m_3^1}(r_1)\bar{a}_{m_1^2}(r_2)a_{m_2^2}(r_2)\bar{a}_{m_3^2}(r_2)a_{m_1^3}(r_3)\bar{a}_{m_2^3}(r_3)a_{m_3^3}(r_3)$$

$$e^{i\sigma(j_1,j_2,j_3,n)s}e^{i\sigma(m_1^1,m_2^1,m_3^1,j_1)r_1}e^{-i\sigma(m_1^2,m_2^2,m_3^2,j_2)r_2}e^{i\sigma(m_1^3,m_2^3,m_3^3,j_3)r_3} \, dr_1 \, dr_2 \, dr_2 \, ds. \tag{6.2.9}$$

Substituting once more via (6.2.6) for each function $a_n(r_j)$ in (6.2.9) yields a main term

$$(i\omega)^4 \sum_{(m_k^i)_{1\leq i,k\leq 3}}^{*} \mathcal{I}(t, (m_k^i)_{1\leq i,k\leq 3}) \prod_{i,j=1}^{3} a^*_{m_j^i}(0), \tag{6.2.10}$$

which arises when $a_n(r_j)$ is replaced by $a_n(0)$, plus higher-degree terms. Here the superscript $*$ indicates that the sum is taken over only certain $(m_k^i)_{1\leq i,k\leq 3} \in \mathbb{Z}^9$, where $a^*_{m_j^i}(0) = a_{m_j^i}(0)$ if $i + j$ is even and $= \overline{a_{m_j^i}(0)}$ if $i + j$ is odd, and where

$$\mathcal{I}(t, (m_k^i)_{1\leq i,k\leq 3}) = \int_{0\leq r_1,r_2,r_3\leq s\leq t} e^{i\phi(s,r_1,r_2,r_3,\{m_j^i:1\leq i,j\leq 3\})} \, dr_1 \, dr_2 \, dr_2 \, ds, \tag{6.2.11}$$

with

$$\phi(s, r_1, r_2, r_3, (m_j^i)_{1\leq i,j\leq 3}) \tag{6.2.12}$$

$$= \sigma(j_1, j_2, j_3, n)s + \sum_{i=1}^{3}(-1)^{i+1}\sigma(m_1^i, m_2^i, m_3^i, j_i)r_i;$$

and j_1, j_2, j_3, n are defined as functions of (m_j^i) by the equations governing the sums in (6.2.9). Continuing in this way yields formally an infinite expansion for the sequence $(a_n(t))_{n\in\mathbb{Z}}$ in terms of multilinear expressions in the initial datum $(a_n(0))$. This expansion is doubly infinite; the single (and relatively simple) term (6.2.10) is for instance an infinite sum over most elements of an eight-dimensional free $\mathbb{Z}$-module for each n.

The discussion up to this point has been purely formal, with no justification of convergence. In the next section we will describe the terms in this expansion systematically. The main work will be to show that each multilinear operator is well defined on ℓ^p initial data, and then that the resulting fully nonlinear infinite series is convergent.

6.3 TREES AND OPERATORS INDEXED BY TREES

6.3.1 Trees

On a formal level $a(t) = (a_n(t))_{n \in \mathbb{Z}}$ equals an infinite sum

$$\sum_{k=1}^{\infty} A_k(t)(a(0), a(0), a(0), \ldots),\tag{6.3.1}$$

where each $A_k(t)$ is a sum of finitely many multilinear operators, each of degree k. Throughout the chapter, by a multilinear operator we mean one which is either linear or conjugate linear with respect to each argument; for instance, $(f, g) \mapsto f\bar{g}$ is considered to be multilinear. We now describe a class of trees that will be used both to name, and to analyze, these multilinear operators.

In a partially ordered set with partial order $\leq$, w is said to be a child of v if $w \leq v$, $w \neq v$, and if $w \leq u \leq v$ implies that either $u = w$, or $u = v$.

The word "tree" in this chapter will always refer to a special subclass of what are usually called trees, equipped with additional structure.

DEFINITION 6.3.1 *A tree T is a finite partially ordered set with the following properties:*

(1) *Whenever $v_1, v_2, v_3, v_4 \in T$ and $v_4 \leq v_2 \leq v_1$ and $v_4 \leq v_3 \leq v_1$, then either $v_2 \leq v_3$ or $v_3 \leq v_2$.*
(2) *There exists a unique element $\mathbf{r} \in T$ satisfying $v \leq \mathbf{r}$ for all $v \in T$.*
(3) *T equals the disjoint union of two subsets T^0, T^∞, where each element of T^∞ has zero children, and each element of T^0 has three children.*
(4) *For each $v \in T$ there is given a number in $\{\pm 1\}$, denoted $\pm_v$.*
(5) *There is given a partition of the set of all nonterminal nodes of T into two disjoint classes, called simple nodes and ordinary nodes.*

Terminal nodes are neither simple nor ordinary. The distinction between ordinary and simple nodes will encode the distinction between the two types of nonlinear terms on the right-hand side of (6.2.6).

DEFINITION 6.3.2 *Elements of T are called nodes. A terminal node is one with zero children. The maximal element of T is called its root node and will usually be denoted by $\mathbf{r}$. T^∞ denotes the set of all terminal nodes of T, while $T^0 = T \setminus T^\infty$ denotes the set of all nonterminal nodes. The three children of any $v \in T^0$ are denoted by $(v, 1), (v, 2), (v, 3)$.*

For any $u \in T$, $T_u = \{v \in T : v \leq u\}$ is a tree, with root node u. The number $|T|$ of nodes of a tree is of the form $1 + 3k$ for some nonnegative integer k.

$$|T^\infty| = 1 + 2k \text{ and } |T^0| = k \tag{6.3.2}$$

so that T, T^∞, T^0 have uniformly comparable cardinalities, except in the trivial case $k = 0$ where $T = \{\mathbf{r}\}$.

Given a tree T, we will work with the auxiliary space $\mathbb{Z}^T$; the latter symbol T denotes the set of all nodes of the tree with the same name. Elements of $\mathbb{Z}^T$ will be denoted by $\mathbf{j} = (j_v)_{v \in T} \in \mathbb{Z}^T$ with each coordinate $j_v \in \mathbb{Z}$.

DEFINITION 6.3.3 *Let T be any tree. A function $\sigma_w : \mathbb{Z}^T \to \mathbb{Z}$ is defined by*

$$\sigma_w(\mathbf{j}) = \begin{cases} 0 & \text{if } w \text{ is terminal,} \\ j_w^2 - j_{(w,1)}^2 + j_{(w,2)}^2 - j_{(w,3)}^2 & \text{if } w \text{ is nonterminal.} \end{cases} \tag{6.3.3}$$

$\sigma_v(\mathbf{j})$ depends only on the four coordinates $j_v, j_{(v,1)}, j_{(v,2)}, j_{(v,3)}$ of $\mathbf{j}$.

DEFINITION 6.3.4 *An ornamented tree is a tree T, together with a coefficient $\varepsilon_{v,i} \in \{-1, 0, 1\}$ for each nonterminal node $v \in T^0$, and for each $i \in \{1, 2, 3\}$.*

DEFINITION 6.3.5 *Let T be an ornamented tree. The function $\rho : \mathbb{Z}^T \to \mathbb{Z}$ is defined recursively by*

$$\rho_v(\mathbf{j}) = 0 \text{ if } v \in T^\infty \tag{6.3.4}$$

and

$$\rho_v(\mathbf{j}) = \sigma(j_{(v,1)}, j_{(v,2)}, j_{(v,3)}, j_v) + \sum_{i=1}^{3} \varepsilon_{v,i} \rho_{(v,i)}(\mathbf{j}) \text{ if } v \in T^0. \tag{6.3.5}$$

Whenever all children of v are terminal, $\rho_v(\mathbf{j}) = \sigma_v(\mathbf{j})$. But if T has many elements, then for typical $v \in T^0$, ρ_v will be a quadratic polynomial in many variables, which will admit no factorization like that enjoyed by σ_v. $\rho_v(\mathbf{j})$ depends only on $\{j_u, \varepsilon_{u,i} : u \leq v\}$. To simplify notation and language, we will use the symbol T to denote the ornamented tree, the underlying tree, and the underlying set.

DEFINITION 6.3.6 *Let T be a tree. $\mathcal{J}(T) \subset \mathbb{Z}^T$ denotes the set of all $\mathbf{j} = (j_v)_{v \in T}$ satisfying the restrictions*

$$j_v = j_{(v,1)} - j_{(v,2)} + j_{(v,3)} \text{ for every } v \in T^0 \tag{6.3.6}$$

$$\{j_v, j_{(v,2)}\} \cap \{j_{(v,1)}, j_{(v,3)}\} = \emptyset \text{ for every ordinary node } v \in T^0 \tag{6.3.7}$$

$$j_v = j_{(v,i)} \text{ for all } i \in \{1, 2, 3\} \text{ for every simple node } v \in T^0. \tag{6.3.8}$$

(6.3.6) implies that for any $v \in T^0$, j_v can be expressed as a linear combination, with coefficients in $\{\pm 1\}$, of $\{j_w : w \in T^\infty\}$.

Let $\delta, c_0 > 0$ be sufficiently small positive numbers, to be specified later. The following key definition involves these quantities.

DEFINITION 6.3.7 *Let T be an ornamented tree. If $\mathbf{j} \in \mathcal{J}(T)$ and $v \in T$, we say that the ordered pair $(v, \mathbf{j})$ is nearly resonant if v is nonterminal and*

$$|\rho_v(\mathbf{j})| \leq c_0 |\sigma_v(\mathbf{j})|^{1-\delta}. \tag{6.3.9}$$

$(v, \mathbf{j})$ *is said to be exceptional if $v \in T^0$ and $\rho_v(\mathbf{j}) = 0$.*

Whether $(v, \mathbf{j})$ is nearly resonant depends on the values of j_u for all $u \le v$.

Exceptional pairs $(v, \mathbf{j})$ are of course nearly resonant. If $v \in T^0$ is an ordinary node all three of whose children are terminal, then $(v, \mathbf{j})$ cannot be exceptional, for $\rho_v(\mathbf{j}) = \sigma(j_{(v,1)}, j_{(v,2)}, j_{(v,3)}, j_v) = 2(j_v - j_{(v,1)})(j_v - j_{(v,3)})$ cannot vanish, by (6.3.7). But if v has at least one nonterminal child, then nothing prevents $\rho_v(\mathbf{j})$ from vanishing, and if v is a simple node all of whose children are terminal, then any pair $(v, \mathbf{j})$ is certainly exceptional.

6.3.2 Multilinear Operators Associated to Trees

DEFINITION 6.3.8 *Let T be any tree, and let t be any real number. If T is not the trivial tree $\{\mathbf{r}\}$ with only element, then the associated interaction amplitudes are*

$$\mathcal{I}_T(t, \mathbf{j}) = \int_{\mathcal{R}(T,t)} \prod_{u \in T^0} e^{\pm u i \omega \sigma_u(\mathbf{j}) t_u} \, dt_u \qquad (6.3.10)$$

where $\mathcal{R}(T, t) \subset [0, t]^{T^0}$ is defined to be

$$\mathcal{R}(T, t) = \{(t_u)_{u \in T^0} : 0 \le t_u \le t_{u'} \le t \text{ whenever } u, u' \in T^0 \text{ satisfy } u \le u'\}. \qquad (6.3.11)$$

When $T = \{\mathbf{r}\}$ has a single element, $\mathcal{J}(T) = \mathbb{Z}$, and $\mathcal{I}_T(t, \mathbf{j})$ is defined to be 1 for all $t, \mathbf{j}$.

The following upper bounds for the interaction amplitudes $\mathcal{I}_T(t, \mathbf{j})$ are the only information concerning them that will be used in the analysis.

LEMMA 6.3.1 *Let T be any tree, and let $\mathbf{j} \in \mathcal{J}(T)$. Then for all $t \in [0, 1]$,*

$$|\mathcal{I}_T(t, \mathbf{j})| \le t^{|T^0|} \qquad (6.3.12)$$

and

$$|\mathcal{I}_T(t, \mathbf{j})| \le 2^{|T|} \sum_{(\varepsilon_{u,i})} \prod_{w \in T^0} \langle \rho_w(\mathbf{j}) \rangle^{-1}. \qquad (6.3.13)$$

The notation $\langle x \rangle$ means $(1 + |x|^2)^{1/2}$. The sum in (6.3.13) is taken over all of the $3^{|T^0|}$ possible choices of $\varepsilon_{u,i} \in \{0, 1, -1\}$; these choices in turn determine the functions ρ_w. Lemma 6.3.1 will be proved in Section 6.5.

DEFINITION 6.3.9 *Let T be any tree, and let $t \in \mathbb{R}$. The tree operator $\mathfrak{S}_T(t)$ associated to T, t is the multilinear operator that maps the $|T^\infty|$ sequences $(x_v)_{v \in T^\infty}$ of complex numbers to the sequence of complex numbers*

$$\mathfrak{S}_T(t)\big((x_v)_{v \in T^\infty}\big)(n) = \sum_{\mathbf{j} \in \mathcal{J}(T): j_{\mathbf{r}} = n} \mathcal{I}_T(t, \mathbf{j}) \prod_{w \in T^\infty} x_w(j_w) \qquad (6.3.14)$$

indexed by $n \in \mathbb{Z}$.

$\mathfrak{S}_T(t)$ takes as input $|T^\infty|$ complex sequences, each belonging to a Banach space $\ell^p(\mathbb{Z})$, and outputs a single complex sequence, which will be shown to belong to some $\ell^q(\mathbb{Z})$.

When T is the trivial tree $\{\mathbf{r}\}$ having only one element, $\mathfrak{S}_T(t)$ is the identity operator for every time t, mapping any sequence $(x_n(0))_{n\in\mathbb{Z}}$ to itself. This corresponds to the linear Schrödinger evolution; it is independent of t because we are dealing with twisted Fourier coefficients (6.2.4).

6.4 FORMALITIES

With all these definitions and notations in place, we can finally formulate the conclusion of the discussion in Section 6.2.

PROPOSITION 6.4.1 *The recursive procedure indicated in Section 6.2 yields a formal expansion*

$$a(t) = \sum_{k=1}^{\infty} A_k(t)(a_{T,1}^*(0), a_{T,2}^*(0), \ldots), \tag{6.4.1}$$

where each $A_k(t)$ is a multilinear operator of the form

$$A_k(t) = \sum_{|T|=3k+1} c_T \mathfrak{S}_T(t), \tag{6.4.2}$$

each sequence $a_{T,n}^*(0)$ equals either $a(0)$ or $\bar{a}(0)$, the scalars $c_T \in \mathbb{C}$ satisfy $|c_T| \leq C^{1+|T|}$, and for each index k, the sum in (6.4.2) is taken over a finite collection of $O(C^k)$ ornamented trees T of the indicated cardinalities.

This asserts that the outcome of the repeated substitution of (6.2.6) into itself, as described in Section 6.2, is accurately encoded in the definitions in Section 6.3. This proposition and the following result will be proved later in the chapter.

PROPOSITION 6.4.2 *There exists a finite positive constant c_0 such that whenever $a(0) \in \ell^1$, the multiply infinite series $\sum_k A_k(t)(a^*(0), \ldots)$ converges absolutely to a function in $C^0([0, \tau], \ell^1)$ provided that $\tau\|a(0)\|_{\ell^1} \leq c_0$.*

Conversely, if $u \in C^0([0, \tau], \ell^1)$, then for such τ, the sequence $a_n(t) = e^{in^2 t}\widehat{u}(t, n)$ equals the sum of this series, for $t \in [0, \tau]$.

By the first statement we mean that $\sum_{\mathbf{j}\in\mathcal{J}(T)} |\mathcal{I}_T(t, \mathbf{j})| \prod_{w\in T^\infty} |a(0)(j_w)|$ converges absolutely for each ornamented tree T, and that if its sum is denoted by $\mathfrak{S}_T^*(a(0), a(0), \ldots)(t)$, then the resulting series $\sum_{k=1}^{\infty} \sum_{|T|=3k+1} c_T \mathfrak{S}_T^*(a(0), a(0), \ldots)(t)$ likewise converges.

The operators $\mathfrak{S}_T$ and coefficients c_T were defined so that the following holds automatically.

LEMMA 6.4.3 *There exists $c > 0$ with the following property. Let $\widehat{u}_0$ be any numerical sequence and define $a(0)(n) = \widehat{u}_0(n)$. Suppose that the infinite series defining $\mathfrak{S}_T^*(a^\star(0), a^\star(0), \dots)(t)$ converges absolutely and uniformly for all $t \in [0, \tau]$ and that its sum is $O(c^{|T|})$, uniformly for every ornamented tree T. Define $a(t)$ to be the sequence $\sum_{k=1}^\infty A_k(t)(a^\star(0), a^\star(0), \dots)$. Then a satisfies the integral equation (6.2.6) for $t \in [0, \tau]$. Moreover, the function $u(t, x)$ defined by $\widehat{u}(t, n) = e^{-in^2 t} a(t, n)$ is a solution of the modified Cauchy problem (NLS*) in the corresponding sense (6.2.7).*

The main estimate in our analysis is as follows.

PROPOSITION 6.4.4 *Let $p \in (1, \infty)$. Then for any exponent $q > \frac{p}{|T^\infty|}$ satisfying also $q \geq 1$, there exist $\varepsilon > 0$ and $C < \infty$ such that for all trees T and all sequences $x_v \in \ell^1$,*

$$\|\mathfrak{S}_T(t)\big((x_v)_{v \in T^\infty}\big)\|_{\ell^q} \leq (Ct^\varepsilon)^{|T^\infty|} \prod_{v \in T^\infty} \|x_v\|_{\ell^p}. \tag{6.4.3}$$

Proposition 6.4.4 and Lemma 6.4.3 will be proved in subsequent sections. Together, they give:

COROLLARY 6.4.5 *Let $p \in [1, \infty)$. For any $R < \infty$ there exists $\tau > 0$ such that the solution mapping $u_0 \mapsto u(t, \cdot)$ for the modified Cauchy problem (NLS*), initially defined for all sufficiently smooth u_0, extends by uniform continuity to a real analytic mapping from $\{u_0 \in \mathcal{F}L^p : \|u_0\|_{\mathcal{F}L^p} \leq R\}$ to $C^0([0, \tau], \mathcal{F}L^p(\mathbb{T}))$.*

We emphasize that analytic dependence on t is not asserted; solutions are Hölder continuous with respect to time.

6.5 BOUND FOR THE INTERACTION AMPLITUDES $\mathcal{I}_T(T, \mathbf{j})$

Proof of Lemma 6.3.1. Let $\mathbf{j} \in \mathbb{Z}^T$ be given; it will remain constant throughout the proof. The first bound of the lemma holds simply because $|\mathcal{I}_T(t, \mathbf{j})| \leq |\mathcal{R}(T, t)|$. The proof of the second bound (6.3.13) proceeds recursively in steps. In each step we integrate with respect to t_v for certain nodes v in the integral defining $\mathcal{R}(T, t)$, holding certain other coordinates t_w fixed. Once integration has been performed with respect to some coordinate, that coordinate is of course removed from later steps.

In Step 1, we hold t_v fixed whenever at least one child of v is not terminal. We also fix t_v for every simple node v having only terminal children. The former coordinates t_v, and underlying nodes v, are said to be temporarily fixed; the latter coordinates and nodes are said to be permanently fixed. We integrate with respect to all nonfixed coordinates t_w.

When $|T| = 1$ there is nothing to prove. Otherwise there must always exist at least one node, all of whose children are terminal. If there exists such a node which is also ordinary, then at least one coordinate t_v is not fixed. The subset, or slice, of $\mathcal{R}(T, t)$ defined by setting each of the fixed coordinates equal to some constant is either empty, or takes the product form $\times_{u \text{ not fixed}}[0, t_{u^*}]$, where u^* denotes the parent of u. The integrand is likewise a product, of simple exponentials. Integrating over this slice with respect to all of the nonfixed coordinates thus yields

$$\prod_w e^{\pm w i \sigma_w t_w} \prod_u \int_0^{t_{u^*}} e^{\pm u i \sigma_u t_u} \, dt_u,$$

where the first product is taken over all fixed $w \in T^0$, and the second over all remaining nonfixed $u \in T^0$.

None of the quantities σ_u can vanish in Step 1, since an ordinary node having only terminal children can never be exceptional, by (6.3.7). Therefore, the preceding expression equals

$$\prod_w e^{\pm w i \sigma_w t_w} \prod_u (\pm_u i \sigma_u)^{-1} \left(e^{\pm u i \sigma_u t_{u^*}} - 1 \right).$$

This may be expanded as a sum of 2^N terms, where N is the number of nonfixed nodes in T^0. Each of these terms has the form

$$\pm \prod_w e^{\pm w i \sigma_w t_w} \prod_u (i \sigma_u)^{-1} e^{\varepsilon_u i \sigma_u t_{u^*}} \qquad (6.5.1)$$

for some numbers $\varepsilon_u \in \{0, 1, -1\}$.

The other possibility in Step 1 is that $|T| > 1$, but every nonterminal node that has only terminal children is simple. In that case all coordinates t_v are fixed at Step 1, no integration is performed, and we move on to Step 2.

Any node v that is permanently fixed at any step of the construction remains fixed through all subsequent steps; we never integrate with respect to t_v. On the other hand, once we've integrated with respect to some t_w, then the node w is also removed from further consideration.

We now carry out Step 2. The set T_1 of all nodes temporarily fixed during Step 1 is itself a tree. There is an associated subset $\mathcal{R}_{T_1}$ of $\{(t_w : w \in T_1)\}$, defined by the inequalities $0 \le t_w \le t_{w'} \le t$ whenever $w \le w'$, and also by $t_u \le t_w$ if $u \le w$ and u was permanently fixed in Step 1. To each node $w \in T_1$ is associated a modified phase $\sigma_w^{(2)}$, defined to be $\sigma_w + \sum_i \varepsilon_{(w,i)} \sigma_{(w,i)}$, where the sum is taken over all $i \in \{1, 2, 3\}$ such that we integrated with respect to $t_{(w,i)}$ in the first step. Thus, the product of exponentials in (6.5.1) can be rewritten as

$$\prod_w e^{\pm w i \sigma_w t_w} \prod_u e^{\varepsilon_u i \sigma_u t_{u^*}} = \prod_{v \in T_1} e^{\pm v i \sigma_v^{(2)} t_v}, \qquad (6.5.2)$$

which takes the same general form as the original integrand.

A node w is permanently fixed at Step 2 if it was permanently fixed at Step 1, or if w is terminal in T_1 and satisfies $\sigma_w^{(2)} = 0$. A node $w \in T_1$ is temporarily fixed at Step 2 if w is not terminal in T_1. We now integrate $\prod_{w \in T_1} e^{\pm i \sigma_w^{(2)}(t_w)}$ with respect to t_u for all $u \in T_1$ that are neither temporarily nor permanently fixed at Step 2. As in Step 1, this integral has a product structure $\times_u [t_{u,*}, t_{u^*}]$, where the product is taken over all nodes u not fixed at this step, u^* is the parent of u, and the lower limit $t_{u,*}$ is either zero, or equals t_w for some child w of u that has been permanently fixed. Now 2^{N_2} terms are obtained after integration, where N_2 is the number of variables with respect to which we integrate.

In Step 3 we consider the tree T_2 consisting of all $w \in T_1$ that were temporarily fixed in Step 2. Associated to T_2 is a set $\mathcal{R}_{T_2}$, and associated to each node $v \in T_2$ is a modified phase $\sigma_w^{(3)} = \sigma_w^{(2)} + \sum_i \varepsilon_{(w,i)} \sigma_{(w,i)}^{(2)}$, the sum being taken over all $i \in \{1, 2, 3\}$ such that (w, i) was not fixed in Step 2. A node $v \in T_2$ is then permanently fixed if it is terminal in T_2 and $\sigma_v^{(3)} = 0$. $v \in T_2$ is temporarily fixed if it is not terminal in T_2. We then integrate with respect to t_v for all $v \in T_2$ that are neither temporarily nor permanently fixed.

This procedure terminates after finitely many steps, when for each node $v \in T^0$, either v has become permanently fixed, or we have integrated with respect to t_v. This yields a sum of at most $2^{|T^0|}$ terms. Each term arises from some particular choice of the parameters $\varepsilon_{u,i}$, and is expressed as an integral with respect to t_v for all nodes $v \in T^0$ that were permanently fixed at some step; the vector (t_v) indexed by all such v varies over a subset of $[0, t]^M$ where M is the number of such v. At step n, each integration with respect to some t_u yields a factor of $(\sigma_u^{(n)})^{-1}$, multiplied by some unimodular factor; $\sigma_u^{(n)}$ is nonzero, since u would otherwise have been permanently fixed.

Thus, for each term we obtain an upper bound of $\prod_u |\rho_u(\mathbf{j})|^{-1}$, where the product is taken over all nonexceptional nodes u; this bound must still be integrated with respect to all t_w where w ranges over all the exceptional nodes. Each such coordinate t_w is restricted to $[0, t]$. Thus we obtain a total bound

$$|\mathcal{I}(t, \mathbf{j})| \leq \sum_{(\varepsilon_{u,i})} t^M \prod_{w \in T^0}^* |\rho_w(\mathbf{j})|^{-1}, \tag{6.5.3}$$

where for each $(\varepsilon_{u,i})$, $M = M((\varepsilon_{u,i}))$ is the number of exceptional nodes encountered in this procedure, that is, the number of permanently fixed nodes, and where for each $(\varepsilon_{u,i})$, $\prod_{w \in T^0}^*$ denotes the product over all nodes $w \in T^0$ that are nonexceptional with respect to the parameters $(\varepsilon_{u,i})$ and $\mathbf{j}$. Since $t \in [0, 1]$, the stated result follows. $\qquad \square$

6.6 A SIMPLE ℓ^1 BOUND

This section is devoted to a preliminary bound for simplified multilinear operators. For any tree T and any sequences $y_v \in \ell^1$, define

$$\tilde{S}_T\big((y_v)_{v \in T^\infty}\big)(n) = \sum_{\mathbf{j}: j_\mathbf{r} = n}^{**} \prod_{u \in T^\infty} y_u(j_u). \tag{6.6.1}$$

The notation $\sum_{\mathbf{j}: j_\mathbf{r} = n}^{**}$ indicates that the sum is taken over all indices $\mathbf{j} \in \mathbb{Z}^T$ satisfying (6.3.6) as well as $j_\mathbf{r} = n$; the restrictions (6.3.7) and (6.3.8) are not imposed here. $\tilde{S}_T$

has the same general structure as $\mathfrak{S}_T$, except that the important interaction ampli-tudes $\mathcal{I}_T(t, \mathbf{j})$ have been omitted.

LEMMA 6.6.1 *For any tree T and any sequences $\{(y_v) : v \in T^\infty\}$*

$$\|\tilde{S}_T((y_v)_{v \in T^\infty})\|_{\ell^1} \leq \prod_{w \in T^\infty} \|y_w\|_{\ell^1}, \tag{6.6.2}$$

with equality when all $y_v(j_v)$ are nonnegative.

Proof. There exists a nonnegative integer k for which $|T| = 3k+1$, $|T^\infty| = 2k+1$, and $|T^0| = k$. Consider the set $B \subset T$ whose elements are the root node $\mathbf{r}$ together with all (v, i) such that $v \in T^0$ and $i \in \{1, 3\}$. Thus $|B| = 1 + 2k = |T^\infty|$. Define

$$k_{v,i} = j_v - j_{(v,i)} \text{ for } v \in T^0 \text{ and } i \in \{1, 3\}. \tag{6.6.3}$$

Consider the $\mathbb{Z}$-linear mapping L from $\mathbb{Z}^{T^\infty}$ to $\mathbb{Z}^B$ defined so that $L(\mathbf{j})$ has coordi-nates $j_\mathbf{r}$ and all $k_{v,i}$. The definition of $k_{v,i}$ makes sense for $i = 2$, but that quantity is redundant; $k_{v,1} - k_{v,2} + k_{v,3} \equiv 0$.

j_v and $j_{(v,i)}$ are well-defined linear functionals of $\mathbf{j} \in \mathbb{Z}^{T^\infty}$, because given the quantities j_w for all $w \in T^\infty$, j_v can be recovered for all other $v \in T$ via the relations (6.3.6), by ascending induction on v. We claim that L is invertible. Indeed, from the quantities $j_\mathbf{r}$ and all $j_v - j_{(v,i)}$ with $v \in T^0$ and $i \in \{1, 3\}$, j_u can be recovered for all $u \in T$ by descending induction on u, using again (6.3.6) at each stage. For instance, at the initial step, $j_{(\mathbf{r},i)} = j_\mathbf{r} + k_{\mathbf{r},i}$ for $i = 1, 3$, and then $j_{(\mathbf{r},2)}$ can be recovered via (6.3.6). Thus L is injective, hence invertible.

By descending induction on nodes it follows in the same way from (6.3.6) that $\mathbf{j} = (j_w)_{w \in T^\infty}$ satisfies a certain linear relation of the form

$$j_\mathbf{r} = \sum_{w \in T^\infty} \pm_w j_w, \tag{6.6.4}$$

where each coefficient $\pm_w$ equals ± 1. By the conclusion of the preceding paragraph, $(j_w)_{w \in T^\infty}$ is subject to no other relation; the sum defining $\tilde{S}_T((y_w)_{w \in T^\infty})(j_\mathbf{r})$ is taken over all $\mathbf{j}$ satisfying this relation. Therefore, $\sum_{j_\mathbf{r}} \tilde{S}_T(j_\mathbf{r})$ equals the summation over all $w \in T^\infty$ and all $j_w \in \mathbb{Z}$, without restriction, of $\prod_{w \in T^\infty} y_w(j_w)$. The lemma follows. $\square$

COROLLARY 6.6.2 *For any tree, the sum defining $\mathfrak{S}_T((y_v)_{v \in T^\infty})(n)$ converges ab-solutely for all $n \in \mathbb{Z}$ whenever all $y_v \in \ell^1$, and the resulting sequence satisfies*

$$\|\mathfrak{S}_T((y_v)_{v \in T^\infty})\|_{\ell^1} \leq \prod_{v \in T^\infty} \|y_v\|_{\ell^1}. \tag{6.6.5}$$

Proof. This is a direct consequence of the preceding lemma together with the simple bound $|\mathcal{I}_T(t, \mathbf{j})| \leq t^{|T^0|}$ of Lemma 6.3.1. $\square$

Estimates in ℓ^p for $p > 1$ are less simple; there is no bound for $\tilde{S}_T$ in terms of the quantities $\|y_w\|_{\ell^p}$ for $p > 1$. The additional factors $\langle \rho_u \rangle^{-1}$ in the second interaction amplitude bound (6.3.13), reflecting the dispersive character of the partial differential equation, are essential for estimates in terms of weaker ℓ^p norms.

Proof of Propositions 6.4.1 and 6.4.2. The first conclusion of Proposition 6.4.2 follows directly from the preceding corollary. To establish Proposition 6.4.1, let $y = y_n(t) = y(t, n) \in C^0([0, \tau], \ell^1)$ be any sequence-valued solution of the integral equation

$$y(t, n) = y(0, n) - i\omega \int_0^t |y(s, n)|^2 y(s, n) \, ds$$

$$+ i\omega \sum_{j-k+l=n}^* \int_0^t y(s, j)\bar{y}(s, k)y(s, l)e^{i\sigma(j,k,l,n)s} \, ds. \qquad (6.6.6)$$

Consider any tree T, and let each node $v \in T^\infty$ be designated as either finished or unfinished. Consider the associated function

$$\int_{\mathcal{R}(T,t)} \sum_{j \in \mathcal{J}(T)} \prod_{v \in T^0} e^{\pm v i \sigma_v t_v} \prod_{u \in T^\infty} y_u(t_u, j_u) \, dt_u \qquad (6.6.7)$$

for $0 \le t \le \tau$, with $t_r \equiv t$, where for each $u \in T^\infty$, $y_u(t, \cdot)$ is identically equal to one of $y(t, \cdot)$, $\bar{y}(t, \cdot)$ of u is unfinished, and to one of $y(0, \cdot)$, $\bar{y}(0, \cdot)$ if u is finished. The simplest such expression, associated to the tree $T = \{r\}$ having only one element, is any constant sequence $y_r(0, j_r)$.

For each unfinished node u, substitute the right-hand side of (6.6.6) or its complex conjugate, as appropriate, for $y_u(t_u, j_u)$ in (6.6.7). The $C^0(\ell^1)$ hypothesis guarantees that an absolutely convergent integral and sum are produced. Thus, we may interchange the outer integral with the sums. What results is a finite linear combination of expressions of the same character as (6.6.7), associated to trees $T^\sharp$. At most $3^{|T^\infty|}$ such expressions are obtained, and each is multiplied by a unimodular numerical coefficient.

Each nonterminal node of T is a nonterminal node of $T^\sharp$, and each finished node of T^∞ remains a terminal node of $T^\sharp$. When the first term on the right in (6.6.6) is substituted for $y_u(t_u, j_u)$, then the unfinished node u becomes a finished terminal node. When the second or third terms on the right are substituted, new unfinished terminal nodes are added to create $T^\sharp$, in which u is a nonterminal simple node or an ordinary node, respectively. Each child of u in $T^\sharp$ is a terminal node of $T^\sharp$, and is (consequently) unfinished.

When $T = \{r\}$, we have simply $y(t)$. Repeatedly substituting as above produces an infinite sum of expressions as described in Proposition 6.4.1. Thus the proof of that result is complete.

To prove that any solution y in $C^0([0, \tau], \ell^1)$ must agree with the sum of our power series for sufficiently small τ, regard y as being the function associated as above to $T = \{r\}$ and apply the substitution procedure a large finite number of times, N. If M is given and N is chosen sufficiently large in terms of N, then what results is an expression for y as a sum of some terms of the power series, including all terms associated to trees of orders $\le M$, together with certain error terms. There are at most C^N error terms, and each is $O(\tau^{cN})$ in $C^0(\ell^1)$ norm, where the constants depend on the $C^0(\ell^1)$ norm of y. Therefore, these expressions converge, as $N \to \infty$, to the sum of the power series in $C^0([0, \tau], \ell^1)$ norm provided that τ is sufficiently small relative to the $C^0(\ell^1)$ norm of y. $\qquad\qquad\qquad\qquad\qquad\qquad\qquad\qquad$ □

6.7 TREE SUM MAJORANTS

In this section we introduce majorizing operators that are the essence of the problem, and decompose them into suboperators.

6.7.1 Majorant Operators Associated to Ornamented Trees

DEFINITION 6.7.1 *Let T be an ornamented tree. The tree sum majorant associated to T is the multilinear operator*

$$S_T\big((y_w)_{w \in T^\infty}\big)(n) = \sum_{\mathbf{j} \in \mathcal{J}(T): j_r = n} \prod_{u \in T^0} \langle \rho_u(\mathbf{j}) \rangle^{-1} \prod_{w \in T^\infty} y_w(j_w). \qquad (6.7.1)$$

S_T is initially defined when all $y_w \in \ell^1$, in order to ensure absolute convergence of the sum.

LEMMA 6.7.1 *Let $p \in [1, \infty)$ and suppose that $q > |T^\infty|^{-1} p$ and $q \geq 1$. Then there exists $C < \infty$ such that for all ornamented trees,*

$$\big\| S_T\big((x_v)_{v \in T^\infty}\big) \big\|_{\ell^q} \leq C^{|T|} \prod_{v \in T^\infty} \|x_v\|_{\ell^p} \qquad (6.7.2)$$

for all sequences $x_v \in \ell^1$.

Assuming this for the present, we show how it implies Proposition 6.4.4.

Proof of Proposition 6.4.4. Let T be any tree. We already have

$$\big\| \mathfrak{S}_T(t)\big((x_v)_{v \in T^\infty}\big) \big\|_{\ell^1} \leq t^{|T^0|} \prod_{v \in T^\infty} \|x_v\|_{\ell^1} \qquad (6.7.3)$$

for all sequences $x_v \in \ell^1$ by Lemma 6.6.1 together with the first bound for the interaction amplitudes $\mathcal{I}_T(t, \mathbf{j})$ provided by Lemma 6.3.1.

On the other hand, to T are associated at most $3^{|T|}$ ornamented trees $\tilde{T}$, defined by specifying coefficients $\varepsilon_{v,i}$. According to the second conclusion (6.3.13) of Lemma 6.3.1, $\|\mathfrak{S}_T(t)((x_v)_{v \in T^\infty})\|_{\ell^q}$ is majorized by $C^{|T|}$ times the sum over these $\tilde{T}$ of $\|S_{\tilde{T}}((x_v)_{v \in T^\infty})\|_{\ell^q}$. This bound holds uniformly in t, provided that t is restricted to a bounded interval. Thus (6.7.2) implies that

$$\big\| \mathfrak{S}_T(t)\big((x_v)_{v \in T^\infty}\big) \big\|_{\ell^q} \leq C^{|T|} \prod_{v \in T^\infty} \|x_v\|_{\ell^p} \qquad (6.7.4)$$

under the indicated assumptions on p, q. Interpolating this with the bound for $p = q = 1$ yields

$$\left\| \mathfrak{S}_T(t)\big((x_v)_{v \in T^\infty}\big) \right\|_{\ell^q} \leq (Ct^\varepsilon)^{|T|} \prod_{v \in T^\infty} \|x_v\|_{\ell^p} \tag{6.7.5}$$

for some $\varepsilon > 0$. $\qquad\qquad\qquad\qquad\qquad\qquad\qquad\qquad\qquad\qquad\qquad\qquad$ $\square$

6.7.2 Marked Ornamented Trees and Associated Operators

The analysis of S_T will rely on several further decompositions.

DEFINITION 6.7.2 *A marked ornamented tree* (T, T') *is an ornamented tree* T *together with a subset* $T' \subset T^0$, *the set of marked nodes, and the collection*

$$\mathcal{J}(T, T') = \{ \mathbf{j} \in \mathcal{J}(T) : \{ v \in T : (v, \mathbf{j}) \text{ is nearly resonant} \} = T' \}. \tag{6.7.6}$$

DEFINITION 6.7.3 *Let* (T, T') *be a marked ornamented tree. The associated tree sum majorant is the multilinear operator*

$$S_{(T,T')}\big((y_w)_{w \in T^\infty}\big)(n) = \sum_{\mathbf{j} \in \mathcal{J}(T,T'):\, j_r = n} \prod_{u \in T^0} \langle \rho_u(\mathbf{j}) \rangle^{-1} \prod_{w \in T^\infty} y_w(j_w). \tag{6.7.7}$$

Now for any ornamented tree T,

$$S_T = \sum_{T' \subset T^0} S_{(T,T')}, \tag{6.7.8}$$

the sum being taken over all subsets $T' \subset T^0$. The total number of such subsets is $2^{|T^0|} \leq 2^{|T|} \leq 2^{3|T^\infty|/2} = C^{|T^\infty|}$. Therefore, in order to establish the bound stated in Lemma 6.7.1 for the operator S_T associated to an ornamented tree T, it suffices to prove that same bound for $S_{(T,T')}$, for all subsets $T' \subset T^0$.

6.7.3 A Further Decomposition

Let (T, T') be any marked ornamented tree, which will remain fixed for the remainder of the analysis. To avoid having to write absolute value signs, we assume that y_v are all sequences of nonnegative real numbers.

We seek an upper bound for the associated tree sum operator $S_{(T,T')}$. The factors $\langle \rho_v \rangle^{-1}$ in the definition of $S_{(T,T')}$ are favorable when $|\rho_v|$ is large; nearly resonant pairs are those for which $|\rho_v(\mathbf{j})|$ is relatively small, and hence these require special attention.

Denote by $\Gamma = (\gamma_u)_{u \in T'}$ any element of $\mathbb{Z}^{T'}$. Let

$$\mathcal{J}(T, T', \Gamma) = \{ \mathbf{j} \in \mathcal{J}(T, T') : \rho_u(\mathbf{j}) = \gamma_u \text{ for all } u \in T' \}. \tag{6.7.9}$$

T' is the set of all nearly resonant nodes, so by its definition we have

$$|\gamma_u| = |\rho_u(\mathbf{j})| \leq c_0 |\sigma_u(\mathbf{j})|^{1-\delta} \ \forall u \in T'. \tag{6.7.10}$$

This leads to a further decomposition and majorization

$$
S_{(T,T')}\big((y_v)_{v \in T^\infty}\big)(n) = \sum_{\Gamma \in \mathbb{Z}^{T'}} \sum_{\mathbf{j} \in \mathcal{J}(T,T',\Gamma):\, j_r = n} \prod_{u \in T^0} \langle \rho_u(\mathbf{j}) \rangle^{-1} \prod_{w \in T^\infty} y_w(j_w)
$$

$$
\leq C^{|T|} \sum_{\mathbf{N}} \prod_{v \in T'} 2^{-N_v} \sum_{\mathbf{M}} \prod_{u \in T^0 \setminus T'} 2^{-(1-\delta)M_u} \sum_{\Gamma} \sum_{\mathbf{j} \in \mathcal{J}(T,T',\Gamma):\, j_r = n} \prod_{w \in T^\infty} y_w(j_w),
$$

$$(6.7.11)$$

where $\mathbf{N} = (N_v)_{v \in T'}$ and $\mathbf{M} = (M_u)_{u \in T^0 \setminus T'}$. The notation in the last line means that the first two sums are taken over all nonnegative integers N_v, M_u as v ranges over T' and u over $T^0 \setminus T'$; the third sum is taken over all $\Gamma = (\gamma_u)_{u \in T'}$ such that

$$
\langle \gamma_v \rangle \in [2^{N_v}, 2^{1+N_v}) \quad \text{for all } v \in T'; \tag{6.7.12}
$$

and the sum with respect to $\mathbf{j}$ is taken over all $\mathbf{j} \in \mathcal{J}(T, T', \Gamma)$, satisfying $j_r = n$ together with the additional restrictions

$$
|\sigma_u(j_{(u,1)}, j_{(u,2)}, j_{(u,3)}, j_u)| \sim 2^{M_u} \quad \text{for all } u \in T^0 \setminus T' \tag{6.7.13}
$$

$$
\rho_v(\mathbf{j}) = \gamma_v \quad \text{for all } v \in T'. \tag{6.7.14}
$$

Thus there is an upper bound $2^{N_v} \leq C c_0 |\sigma_v(\mathbf{j})|^{1-\delta}$ for all $v \in T'$.

6.7.4 Rarity of Near Resonances

Let δ_1 be a small constant, to be chosen later. Recall that for any positive integer n, there are at most $C_{\delta_1} n^{\delta_1}$ pairs (n', n'') of integers for which n can be factored as $n = n'n''$. This fact was exploited by Bourgain [2] in his proof of H^0 wellposedness.

The key to the control of near resonances is a strong limitation on the number of $\mathbf{j}$ satisfying (6.7.14), for any fixed Γ. Given $v \in T'$ any parameter γ_v, and any $\mathbf{j} \in \mathcal{J}(T, T', \Gamma)$, the equation (6.7.14) can be written as

$$
\sigma_v(\mathbf{j}) = \gamma_v - \sum_{i=1}^{3} \varepsilon_{v,i} \rho_{(v,i)}(\mathbf{j}),
$$

and $\rho_{(v,i)}(\mathbf{j})$ depends only on $\{j_w - j_{(w,i)} : w < v, i \in \{1, 2, 3\}\}$. Since the quantity σ_v on the left-hand side of this rewritten equation can be factored as $2(j_v - j_{(v,1)})(j_v - j_{(v,3)})$, we conclude that for any $\{j_w - j_{(w,l)} : w < v, l \in \{1, 2, 3\}\}$ and any γ_v there are at most $C_{\delta_1} |\gamma_v - \sum_{i=1}^{3} \varepsilon_{v,i} \rho_{(v,i)}(\mathbf{j})|^{\delta_1}$ ordered pairs $\big(j_v - j_{(v,1)}, j_v - j_{(v,3)}\big)$ satisfying (6.7.14).

For any nearly resonant node $v \in T'$, $|\gamma_v|$ is small relative to $\sum_{i=1}^{3} |\rho_{(v,i)}(\mathbf{j})|^{1-\delta}$, provided that the constant c_0 is chosen to be sufficiently small in the definition of a nearly resonant node. Therefore, we can choose for each $\mathbf{N}$, $\mathbf{M}$ a family $\mathcal{F} = \mathcal{F}_{\mathbf{N},\mathbf{M}}$ of vector-valued functions $F = (f_{v,i} : v \in T', i \in \{1, 3\})$ such that for any Γ satisfying (6.7.12) and any $\mathbf{j} \in \mathcal{J}(T, T', \Gamma)$, there exists $F \in \mathcal{F}_{\mathbf{N},\mathbf{M}}$ such that for each $v \in T'$ and each $i \in \{1, 3\}$,

$$
k_{v,i} = f_{v,i}(\gamma_v, (k_{w,i} : w < v)), \tag{6.7.15}
$$

where $k_{u,i} = j_u - j_{(u,i)}$.

The number of such functions is strongly restricted:

$$|\mathcal{F}_{N,M}| \leq C_{\delta_1}^{|T'|} \prod_{v \in T'} 2^{\delta_1 \max_i K_{(v,i)}} \qquad (6.7.16)$$

where $K_u = N_u$ for $u \in T'$ and $K_u = M_u$ for $u \in T^0 \setminus T'$, and the maximum is taken over $i \in \{1, 3\}$. Powers of $2^{\delta_1 N_{(v,i)}}$ are undesirable; we will show in Lemma 6.8.2 below that the product on the right-hand side of (6.7.16) satisfies a better bound in which N does not appear.

6.7.5 A Final Decomposition

For M, N as above, we set $|M| = \sum_{u \in T^0 \setminus T'} M_u$ and $|N| = \sum_{v \in T'} N_v$.

DEFINITION 6.7.4 *To any M, N, Γ and any function $F \in \mathcal{F}_{N,M}$ is associated the multilinear operator*

$$S_{T,T',N,M,\Gamma,F}\big((y_w)_{w \in T^\infty}\big)(n) = \sum_{\substack{\mathbf{j} \in \mathcal{J}(T,T',\Gamma):j_r=n}} \prod_{w \in T^\infty} y_w(j_w), \qquad (6.7.17)$$

where the sum in (6.7.17) is taken over all $\mathbf{j} \in \mathcal{J}(T, T', \Gamma)$ satisfying $j_r = n$, (6.7.13), (6.7.14), and the additional restriction (6.7.15).

The multilinear operators $S_{T,T',N,M,\Gamma,F}$ are our basic building blocks. We have shown so far that for all nonnegative sequences y_w and all $n \in \mathbb{Z}$,

$$\big|S_{(T,T')}\big((y_w)_{w \in T^\infty}\big)(n)\big|$$
$$\leq C^{|T|} \sum_{N,M} 2^{-|N|} 2^{-(1-\delta)|M|} \sum_{\Gamma} \sum_{F \in \mathcal{F}_{N,M}} \big|S_{T,T',N,M,\Gamma,F}\big((y_w)_{w \in T^\infty}\big)(n)\big|, \quad (6.7.18)$$

where the second summation in (6.7.17) is taken over all $\Gamma = (\gamma_u)_{u \in T'}$ satisfying both (6.7.12) and (6.7.10). The factor of $2^{-(1-\delta)|M|}$ arises because for each $u \in T^0 \setminus T'$, we have by virtue of Lemma 6.3.1 a factor of $\langle \rho_u(\mathbf{j}) \rangle^{-1}$, and this factor is $\leq C 2^{-(1-\delta)M_u}$ because u is not nearly resonant.

6.8 BOUNDS FOR THE MOST BASIC MULTILINEAR OPERATORS

LEMMA 6.8.1 *Let $p \in [1, \infty)$ and $\delta_1 > 0$. Then for every exponent $q \geq \max(1, p/|T^\infty|)$, there exists $C < \infty$ such that for every T, T', N, M, Γ, F and for every sequence y_v,*

$$\big\|S_{T,T',N,M,\Gamma,F}\big((y_v)_{v \in T^\infty}\big)\big\|_{\ell^q} \leq C^{|T|} 2^{(1+\delta_1)|M|} \prod_{v \in T^\infty} \|y_v\|_{\ell^p}. \qquad (6.8.1)$$

This involves no positive power of $2^{|N|}$, and thus improves on (6.7.16).

Proof. As was shown in the proof of Lemma 6.6.1, each quantity j_v in the summation defining $S_{T,T',N,M,\Gamma,F}\big((y_w)_{w \in T^\infty}\big)(j_r)$ can be expressed as a function, depending on

Γ and on F, of j_r together with all $k_{w,i} = j_w - j_{(w,i)}$, where w varies over the set T^0 and i varies over $\{1, 3\}$. The equation (6.7.15) can then be used by descending induction on T to eliminate $k_{w,i}$ for all $w \in T'$ so long as F, Γ are given. More precisely, j_v equals $j_r + g_v$, where g_v is some function of all $k_{w,i}$ with $w \in T^0 \setminus T'$ and $i \in \{1, 3\}$.

$\prod_{v \in T^\infty} y_v(j_v)$ can thus be rewritten as $\prod_{v \in T^\infty} y_v(j_r + g_v)$. If every $k_{w,i}$ is held fixed, then as a function of j_r, this product belongs to ℓ^q for $q = p/|T^\infty|$ with bound $\prod_{v \in T^\infty} \|y_v\|_{\ell^p}$, by Hölder's inequality.

The total number of terms in the sum defining $S_{T,T',\mathbf{N},\mathbf{M},\Gamma,F}$ is the total possible number of vectors $(k_{w,i})$ where w ranges over $T^0 \setminus T'$ and i over $\{1, 3\}$. The number of such pairs for a given w is $\leq C_{\delta_1} 2^{(1+\delta_1)M_w}$, since $|2k_{w,1}k_{w,3}| = |\sigma_w(\mathbf{j})| \leq 2^{M_w+1}$. Thus in all there are at most $C_{\delta_1}^{|T|} 2^{(1+\delta_1)|\mathbf{M}|}$ terms. Minkowski's inequality thus gives the stated bound. $\square$

A difficulty now appears. For each $v \in T'$ we have a compensating factor of $\langle \rho_v(\mathbf{j}) \rangle^{-1} = \langle \gamma_v(\mathbf{j}) \rangle^{-1} \sim 2^{-N_v}$, but no upper bound whatsoever is available for the ratio of $\max_i |\rho_{(v,i)}(\mathbf{j})|^{\delta_1}$ to $\langle \gamma_v(\mathbf{j}) \rangle$. Thus for any particular $v \in T'$, the factor lost through the nonuniqueness of F need not be counterbalanced by the favorable factor ρ_v^{-1}. Nonetheless, the product of all these favorable factors does compensate for the product of all those factors lost, as will now be shown.

LEMMA 6.8.2 *For any $\varepsilon > 0$ there exists $C_\varepsilon < \infty$ such that uniformly for all $T, T', \mathbf{N}, \mathbf{M}$,*

$$|\mathcal{F}_{\mathbf{N},\mathbf{M}}| \leq C_\varepsilon^{|T|} 2^{\varepsilon |\mathbf{M}|}. \tag{6.8.2}$$

Proof. Let $\mathbf{j} \in \mathcal{J}(T, T', \Gamma)$ satisfy $\rho_v(\mathbf{j}) = \gamma_v$ for all $v \in T'$ but be otherwise arbitrary. Throughout this argument, $\mathbf{j}$ will remain fixed, and ρ_v will be written as shorthand for $\rho_v(\mathbf{j})$.

If the constant c_0 in the definition (6.3.9) of a nearly resonant node is chosen to be sufficiently small, then any nearly resonant node u has a child (u, i) such that $|\rho_u| \leq \frac{1}{2}|\rho_{(u,i)}|^{1-\delta}$. Consider any chain $v = u_h \geq u_{h-1} \geq \cdots \geq u_1$ of nodes such that u_{k+1} is the parent of u_k for each $1 \leq k < h$ (u_k is called the $(k-1)$-th generation ancestor of u_1), u_k is nearly resonant for all $k > 1$, u_1 is either not nearly resonant or is terminal, and $|\rho_{u_k}| \leq \frac{1}{2}|\rho_{u_{k-1}}|^{1-\delta}$. Then

$$|\rho_{u_k}| \leq |\rho_{u_1}|^{(1-\delta)^{k-1}}; \tag{6.8.3}$$

hence

$$2^{K_{u_k}} = 2^{N_{u_k}} \leq C 2^{(1-\delta)^{k-1} M_{u_1}}. \tag{6.8.4}$$

If u_1 is terminal, then $\rho_{u_1} = 0$ by definition, whence the inequality $|\rho_{u_k}| \leq \rho_{u_1}|^{(1-\delta)^{k-1}}$ forces $\rho_{u_k} = 0$ for all u_k, as well. This means that $2^{\max_i K_{(u_k,i)}} \sim 1$. In particular, this holds for $u_k = v$, so the factor $2^{\max_i K_{(v,i)}}$ will be harmless in our estimates. We say that a node v is negligible if there exists such a chain, with $v = u_h$ for some $h \geq 1$.

Recall that $|\mathcal{F}_{\mathbf{N},\mathbf{M}}| \leq C_{\delta_1}^{|T|} \prod_{v \in T'} 2^{\max_i K_{(v,i)}\delta_1}$. For each nonnegligible nearly resonant node v, choose one such chain with $u_h = v$, thus uniquely specifying h and u_1 as functions of v; we then write $(u_1, h) = A(v)$. Any node has at most one h-th generation ancestor; therefore, given both u_1 and h, there can be at most one v such that $(u_1, h) = A(v)$. Consequently,

$$\prod_{\substack{v \in T' \text{ nonnegligible}}} 2^{\max_i K_{(v,i)}\delta_1} \leq \prod_{w \in T^0 \setminus T'} \prod_{h=1}^{\infty} 2^{(1-\delta)^{h-1}\delta_1 M_w}$$

$$= \prod_{w \in T^0 \setminus T'} 2^{M_w \delta_1/\delta}, \tag{6.8.5}$$

since each factor $2^{\max_i K_{(v,i)}\delta_1}$ in the first product is majorized by $2^{(1-\delta)^{h-1}\delta_1 M_w}$ in the second product, where $(w, h) = A(v)$. Forming the product with respect to h for each fixed v yields the desired inequality, since the series $\sum_{h=0}^{\infty}(1 - \delta)^{h-1}\delta_1$ is convergent. The exponent $1 - \delta < 1$ in the definition (6.3.9) of a nearly resonant node was introduced solely for this purpose. If negligible nodes are also allowed in the product on the left-hand side of (6.8.5), then they contribute a factor bounded by $C^{|T|}$, so the conclusion remains valid for the full product.

The desired bound now follows by choosing δ_1 so that $\delta_1/\delta = \varepsilon$. □

Conclusion of proof of Lemma 6.7.1. As already noted, it suffices to establish (6.7.2) with S_T replaced by $S_{(T,T')}$. Combining the preceding two lemmas gives

$$\sum_{F \in \mathcal{F}_{\mathbf{N},\mathbf{M}}} \left\| S_{T,T',\mathbf{N},\mathbf{M},\Gamma,F}\left((y_v)_{v \in T^\infty}\right) \right\|_{\ell^q} \leq C_\varepsilon^{|T|} 2^{(1+\varepsilon)|\mathbf{M}|} \prod_{v \in T^\infty} \|y_v\|_{\ell^p} \tag{6.8.6}$$

for arbitrarily small $\varepsilon > 0$, provided $q \geq \max(1, \frac{p}{|T^\infty|})$. Since $|\Gamma| \leq C^{|T|}2^{|\mathbf{N}|}$,

$$\sum_{\Gamma} \sum_{F \in \mathcal{F}_{\mathbf{N},\mathbf{M}}} \left\| S_{T,T',\mathbf{N},\mathbf{M},\Gamma,F}\left((y_v)_{v \in T^\infty}\right) \right\|_{\ell^q} \leq C_\varepsilon^{|T|}2^{|\mathbf{N}|}2^{(1+\varepsilon)|\mathbf{M}|} \prod_{v \in T^\infty} \|y_v\|_{\ell^p}. \tag{6.8.7}$$

On the other hand, Lemma 6.6.1 gives

$$\sum_{\Gamma} \sum_{F \in \mathcal{F}_{\mathbf{N},\mathbf{M}}} \left\| S_{T,T',\mathbf{N},\mathbf{M},\Gamma,F}\left((y_v)_{v \in T^\infty}\right) \right\|_{\ell^1} \leq C^{|T|} \prod_{v \in T^\infty} \|y_v\|_{\ell^1}. \tag{6.8.8}$$

Thus, if $q > \frac{p}{|T^\infty|}$ and $q \geq 1$, we may interpolate to find that there exists $\eta > 0$ depending on $q - \frac{p}{|T^\infty|}$ but not on δ such that

$$\sum_{\Gamma} \sum_{F \in \mathcal{F}_{\mathbf{N},\mathbf{M}}} \left\| S_{T,T',\mathbf{N},\mathbf{M},\Gamma,F}\left((y_v)_{v \in T^\infty}\right) \right\|_{\ell^q}$$

$$\leq C_\eta^{|T|}2^{(1-\eta)|\mathbf{N}|+(1-\eta)|\mathbf{M}|} \prod_{v \in T^\infty} \|y_v\|_{\ell^p}. \tag{6.8.9}$$

Taking into account the factors $2^{-|\mathbf{N}|}2^{-(1-\delta)|\mathbf{M}|}$ in (6.7.18), and summing over $\mathbf{N}, \mathbf{M}$ as well as over all subsets $T' \subset T^0$, completes the proof of Lemma 6.7.1. □

6.9 LOOSE ENDS

We may reinterpret the sum of our power series (6.4.1),(6.4.2) as a function via the relation $\widehat{u}(t, n) = e^{in^2 t} a_n(t)$ with $a(0)$ defined by $\widehat{u_0}(n) = a_n(0)$, and will do so consistently without further comment, abusing notation mildly by writing $u(t, x) = S(t)u_0(x)$.

LEMMA 6.9.1 *Let $p \in [1, \infty)$. For any $R > 0$ there exists $\tau > 0$ such that for any $u_0 \in \mathcal{F}L^p$ with norm $\leq R$, the element $u(t, x) \in C^0([0, \tau], \mathcal{F}L^p)$ defined by (6.4.1),(6.4.2) is a limit, in $C^0([0, \tau], \mathcal{F}L^p)$ norm, of smooth solutions of (NLS*).*

Proof. All of our estimates apply also in the spaces $\mathcal{F}L^{s,p}$ defined by the condition that $(\langle n \rangle^s \widehat{f}(n))_{n \in \mathbb{Z}} \in \ell^p$, provided that $1 \leq p < \infty$ and $s > 0$. This follows from the proof given for $s = 0$ above, for the effect of working in $\mathcal{F}L^{s,p}$ is to introduce a factor of $\prod_{v \in T^0} \frac{\langle j_v \rangle^s}{\prod_{i=1}^3 \langle j_{(v,i)} \rangle^s}$ in the definition of the tree operator. The relation $j_v = j_{(v,1)} - j_{(v,2)} + j_{(v,3)}$ ensures that $\max_i |j_{(v,i)}| \geq \frac{1}{3}|j_v|$, whence $\frac{\langle j_v \rangle^s}{\prod_{i=1}^3 \langle j_{(v,i)} \rangle^s} \lesssim 1$, so the estimates for $s = 0$ apply directly to all $s > 0$.

More generally, if $\mathcal{F}L^{s,p}$ is equipped with the norm

$$\|f\|_{\mathcal{F}L^{s,p}_\varepsilon} = \|(1 + |\varepsilon \cdot|^{2s})^{1/2} \widehat{f}(\cdot)\|_{\ell^p},$$

then all estimates hold uniformly in $\varepsilon \in [0, 1]$ and $s \geq 0$. This follows from the same reasoning.

Fix a sufficiently large positive exponent s. Given any initial datum u_0 satisfying $\|u_0\|_{\mathcal{F}L^p} \leq R$ with the additional property that $\widehat{u_0}(n) = 0$ whenever $|n|$ exceeds some large quantity N, we may choose $\varepsilon > 0$ so that $\|u_0\|_{\mathcal{F}L^{s,p}_\varepsilon} \leq 2R$. This ε depends on N, but not on R. Thus the infinite series converges absolutely and uniformly in $C^0([0, \tau], H^{s - \frac{1}{2} + \frac{1}{p}})$ if $p \geq 2$ and in $C^0([0, \tau], H^s)$ if $p \leq 2$, where τ depends only on R, not on s. By Lemma 6.4.3, the series sums to a solution of (NLS*) in the sense of (6.2.7); but since the sum is very smooth as a function of x (that is, its Fourier coefficients decay rapidly), this implies that it is a solution in the classical sense. Given an arbitrary u_0 satisfying $\|u_0\|_{\mathcal{F}L^p} \leq R$, we can thus approximate it by such special initial data to conclude that $S(t)u_0$ is indeed a limit, in $C^0([0, \tau], \mathcal{F}L^p)$, of smooth solutions. □

Proof of Proposition 6.1.4. Let $u_0 \in \mathcal{F}L^p$ be given, let $u(t, x) = S(t)(u_0) \in C^0([0, \tau], \mathcal{F}L^p)$. We aim to prove that the nonlinear expression $|u|^2 u$ has an intrinsic meaning as the limit as $N \to \infty$ of $|T_N u|^2 T_N u$ in the sense of distributions in $(0, \tau) \times \mathbb{T}$. Forming $T_N S(t)(u_0)$ is of course not the same thing as forming $S(t)(T_N u_0)$.

Define $a_n(t) = e^{in^2 t} \widehat{u}(t, n)$. Denote also by T_N the operator that maps a sequence-valued function $(b_n(t))$ to $(T_N b_n(t))$, where $T_N b_n = b_n$ if $|n| \leq N$, and $= 0$ otherwise. It suffices to prove that

$$\int_0^t \sum_{j-k+l=n}^* T_N a_j(s) \overline{T_N a_k(s)} T_N a_l(s) e^{i\sigma(j,k,l,n)s} \, ds \tag{6.9.1}$$

$$- \int_0^t |T_N a_n(s)|^2 T_N a_n(s) \, ds$$

converges in ℓ^p norm as $N \to \infty$, uniformly for all $t \in [0, \tau]$, to

$$\sum_{j-k+l=n}^{*} \int_0^t a_j(s)\overline{a_k(s)}a_l(s)e^{i\sigma(j,k,l,n)s}\,ds - \int_0^t |a_n(s)|^2 a_n(s)\,ds.$$

Convergence in the distribution sense follows easily from this by expressing any sufficiently smooth function of the time t as a superposition of characteristic functions
of intervals $[0, t]$.

Now, in the term $\int_0^t \sum_{j-k+l=n}^{*} T_N a_j(s)\overline{T_N a_k(s)}T_N a_l(s)e^{i\sigma(j,k,l,n)s}\,ds$, the integral
may be interchanged with the sum since the truncation operators restrict the sum-
mation to finitely many terms. Expanding a_j, a_k, a_l out as infinite series of tree
operators applied to $a(0)$, we obtain finally an infinite series of the general form
$\sum_{k=1}^{\infty} B_k(t)(a(0), \ldots, a(0))$, where $B_k(t)$ is a finite linear combination of $O(C^k)$
tree sum operators, with coefficients $O(C^k)$, applied to $a(0)$ just as before, with the
sole change that the extra restriction $|j_{(\tau,i)}| \le N$ for $i \in \{1, 2, 3\}$ is placed on $\mathbf{j}$ in
the summation defining $\mathfrak{S}_T$ for each tree T.

Since we have shown that all bounds hold for the sums of the absolute values of
the terms in the tree sum, it follows immediately that this trilinear term converges as
$N \to \infty$. Convergence for the other nonlinear term is, of course, trivial. Likewise,
it is trivial that $(T_N u)_t \to u_t$ and $(T_N u)_{xx} \to u_{xx}$, by linearity. $\square$

This reasoning shows that the limit of each term equals the sum of a convergent
power series, taking values in $C^0([0, \tau], \mathcal{F}L^p)$, in u_0.

Given $R > 0$, there exists $\tau > 0$ for which we have shown that for any
$a(0) \in \ell^p$ satisfying $\|a(0)\|_{\mathcal{F}L^p} \le R$, our power series expansion defines $a(t) \in$
$C^0([0, \tau], \ell^p)$, as an ℓ^p-valued analytic function of $a(0)$. Moreover, for any $t \in$
$[0, \tau]$, both cubic terms in the integral equation (6.2.6) are well defined as limits
obtained by replacing $a(s)$ by $T_N a(s)$, evaluating the resulting cubic expressions,
and passing to the limit $N \to \infty$.

LEMMA 6.9.2 *Whenever $\|a(0)\|_{\ell^p} \le R$, the function $a(t) \in C^0([0, \tau], \ell^p)$ defined
as the sum of the power series expansion (6.4.1) satisfies the integral equation (6.2.7)
when the nonlinear terms in (6.2.6) are defined by the limiting procedure described
in the preceding paragraph.*

Proof. This follows by combining Lemma 6.4.3 with the result just proved. $\square$

Proof of Proposition 6.1.3. Let $u_0 \in \mathcal{F}L^p$. If $u = Su_0$, and if v is the solution
of the Cauchy problem (NLS*) for the modified linear Schrödinger equation with
initial datum u_0, then $u_0 - v$ is expressed as $\sum_{k=1}^{\infty} B_k(t)(u_0, \ldots, u_0)$, where the n-th
Fourier coefficient of the function $B_k(t)(u_0, \ldots)(t)$ equals $e^{-in^2 t}A_k(t)(a^\star(0), \ldots)$
with $a_n(0) = \widehat{u_0}(n)$. According to Proposition 6.4.4,

$$\|A_k(t)(a^\star(0), \ldots)\|_{\ell^q} = O(t^{k\varepsilon}\|a(0)\|_{\ell^p}^k)$$

whenever $q > \frac{p}{3}$ and $q \ge 1$. Summation with respect to k yields the conclusion. $\square$

REFERENCES

[1] H. A. Biagioni, *A nonlinear theory of generalized functions*, second edition. Lecture Notes in Mathematics, 1421. Springer-Verlag, Berlin, 1990.

[2] J. Bourgain, *Fourier transform restriction phenomena for certain lattice subsets and applications to nonlinear evolution equations. I. Schrödinger equations*, Geom. Funct. Anal. 3 (1993), no. 2, 107–156.

[3] N. Burq, P. Gérard, and N. Tzvetkov, *An instability property of the nonlinear Schrödinger equation on S^d*, Math. Res. Lett. 9 (2002), nos. 2-3, 323–335.

[4] M. Christ, *Nonuniqueness of weak solutions of the nonlinear Schrödinger equation*, preprint February 2005, math.AP/0503366.

[5] M. Christ, J. Colliander, and T. Tao, *Instability of the periodic nonlinear Schrödinger equation*, preprint, math.AP/0311227.

[6] J.-F. Colombeau, *Multiplication of distributions. A tool in mathematics, numerical engineering and theoretical physics*, Lecture Notes in Mathematics, 1532. Springer-Verlag, Berlin, 1992.

[7] A. Grünrock, *Bi- and trilinear Schrödinger estimates in one space dimension with applications to cubic NLS and DNLS*, Int. Math. Res. Not. 2005, no. 41, 2525–2558.

[8] ———, *An improved local wellposedness result for the modified KdV equation*, Int. Math. Res. Not. 2004, no. 61, 3287–3308.

[9] T. Kappeler and P. Topalov, *Global well-posedness of KdV in $H^{-1}(T, R)$*, Preprint Series, Institute of Mathematics, University of Zurich.

[10] ———, *Global well-posedness of mKDV in $L^2(T, R)$*, Comm. Partial Differential Equations 30 (2005), nos. 1–3, 435–449.

[11] Y. Tsutsumi, *L^2-solutions for nonlinear Schrödinger equations and nonlinear groups*, Funkcial. Ekvac. 30 (1987), no. 1, 115–125.

[12] A. Vargas and L. Vega, *Global wellposedness for 1D non-linear Schrödinger equation for data with an infinite L^2 norm*, J. Math. Pures Appl. (9) 80 (2001), no. 10, 1029–1044.

Chapter Seven

Eulerian-Lagrangian Formalism and Vortex Reconnection

P. Constantin

7.1 INTRODUCTION

Incompressible Newtonian fluids are described by the Navier-Stokes equations, the viscous regularization of the friction-free, incompressible Euler equations

$$D_t u + \nabla p = 0, \quad \nabla \cdot u = 0. \tag{7.1}$$

The velocity $u = u(x, t) = (u_1, u_2, u_3)$ is a function of $x \in \mathbf{R}^3$ and $t \in \mathbf{R}$. The material derivative associated to the velocity u is

$$D_t = D_t(u, \nabla) = \partial_t + u \cdot \nabla. \tag{7.2}$$

The Euler equations are conservative, meaning that no dissipation of energy occurs during smooth evolution. The total energy is proportional to the L^2 norm of velocity. The Onsager conjecture ([22], [17]) states that conservation of energy happens if and only if the solutions are smoother than required by the classical Kolmogorov turbulence theory (roughly speaking, Hölder continuous of exponent 1/3). The "if" part was proved ([12]). The "only if" part is far from being proved: there is no known notion of weak solutions dissipating energy except with Hölder continuous velocities ([26], [27]). One of the most difficult problems in modern nonlinear PDE, the problem of the smoothness of solutions of the Euler equation, remains open to date. The trend to blow up is manifested in the rapid growth of line elements, which results in growth of the magnitude of the vorticity, $\omega = \nabla \times u$. There do exist solutions of the Euler equations that blow up ([28], [3], [25], [8]), but these solutions have infinite kinetic energy. Moreover, it is plausible that the blow-up is due to the infinite supply of energy, coming from farfield, and therefore it is unclear whether these solutions can be used to shed light on the physical question of finite energy blow-up. In addition to kinetic energy, smooth solutions of the Euler equations conserve helicity and circulation. In order to describe these, let us recall the Lagrangian description of the fluid. The Lagrangian particle maps are

$$a \mapsto X(a, t), \quad X(a, 0) = a.$$

For fixed a, the trajectories of u obey

$$\frac{dX}{dt} = u(X, t).$$

The incompressibility condition implies

$$det\,(\nabla_a X) = 1.$$

The Euler equations (7.1) are formally equivalent to the requirement that two first-order differential operators commute:

$$[D_t, \Omega] = 0.$$

The first operator $D_t = \partial_t + u \cdot \nabla$ is the material derivative associated to the trajectories of u. The second operator $\Omega = \omega(x, t) \cdot \nabla$ is differentiation along vortex lines, the lines tangent to the vorticity field ω. The commutation means that vortex lines are carried by the flow of u, and is equivalent to the equation

$$D_t\omega = \omega \cdot \nabla u. \qquad (7.3)$$

This is a quadratic equation because ω and u are related, $\omega = \nabla \times u$. If boundary conditions for the divergence-free ω are known (periodic or decay at infinity cases), then one can use the Biot-Savart law

$$u = \mathcal{K}_{3DE} * \omega = \nabla \times (-\Delta)^{-1}\omega \qquad (7.4)$$

coupled with (7.3) as an equivalent formulation of the Euler equations ([4], [2], [20]). A helicity ([21]) density is

$$h = u \cdot \omega + \omega \cdot \nabla\phi.$$

Helicity densities are defined modulo the addition of $\omega \cdot \nabla\phi$ for any smooth ϕ. Helicity integrals

$$\int_T h(x, t)dx = c$$

are constants of motion, for any vortex tube T (a time-evolving region whose boundary is at each point parallel to the vorticity). The constants c have to do with the topological complexity of the flow. The incompressible Euler equations are a Hamiltonian system in infinite dimensions in Clebsch variables ([15], [30]). These are a pair of scalars θ, φ that are constant on particle paths,

$$D_t\varphi = D_t\theta = 0,$$

and also determine the velocity via

$$u^i(x, t) = \theta(x, t)\frac{\partial\varphi(x, t)}{\partial x_i} - \frac{\partial n(x, t)}{\partial x_i}.$$

The helicity constants vanish identically for flows that admit a Clebsch variables representation. This implies that not all flows admit a Clebsch variables representation. But if one uses more variables, then one can represent all flows. This can be done using the Weber formula ([29]). The Weber formula is

$$u^i(x, t) = \left(u^j_{(0)}(A(x, t))\right)\frac{\partial A^j(x, t)}{\partial x_i} - \frac{\partial n(x, t)}{\partial x_i},$$

where

$$A(x, t) = X^{-1}(x, t)$$

is the inverse of the Lagrangian particle map. The Weber formula can be understood as an identity of 1-forms:

$$u(x, t)dx = u_{(0)}(A(x, t))dA(x, t) - dn(x, t).$$

The formula, together with boundary conditions and the divergence-free requirement can be written as

$$u = W[A, v] = \mathbf{P}\left\{(\nabla A)^T v\right\}, \tag{7.5}$$

where $\mathbf{P}$ is the corresponding projector on divergence-free functions and v is the virtual velocity

$$v = u_{(0)} \circ A.$$

In the cases of periodic boundary conditions or whole space,

$$\mathbf{P} = I + R \otimes R$$

holds, with R the Riesz transforms. This procedure ([9]) turns A into an *active scalar system*

$$\begin{cases} D_t A = 0, \\ D_t v = 0, \\ u = W[A, v]. \end{cases} \tag{7.6}$$

Active scalars ([7]) are solutions of passive scalar equations $D_t \theta = 0$ that determine the velocity through a time-independent, possibly nonlocal equation of state $u = U[\theta]$. Knowledge of the values of the active scalars at an instance of time is enough to determine the time derivatives of the active scalar at that instance in time. The Clebsch variables are a pair of active scalars. The Euler equations can be represented with many active scalars. The circulation is the loop integral

$$C_\gamma = \oint_\gamma u \cdot dx,$$

and the conservation of circulation is the statement that

$$\frac{d}{dt} C_{\gamma(t)} = 0$$

for all loops carried by the flow. The Weber formula is equivalent to the conservation of circulation. Differentiating the Weber formula, one obtains

$$\frac{\partial u^i}{\partial x_j} = \mathbf{P}_{ik}\left(Det\left[\frac{\partial A}{\partial x_j}; \frac{\partial A}{\partial x_k}; \omega_{(0)}(A)\right]\right).$$

Here we used the notation

$$\omega_{(0)} = \nabla \times u_{(0)}.$$

Taking the antisymmetric part, one obtains the Cauchy formula:

$$\omega_i = \frac{1}{2}\epsilon_{ijk}\left(Det\left[\frac{\partial A}{\partial x_j}; \frac{\partial A}{\partial x_k}; \omega_{(0)}(A)\right]\right),$$

which we write as

$$\omega = \mathcal{C}[\nabla A, \zeta], \tag{7.7}$$

with ζ the Cauchy invariant

$$\zeta(x, t) = \omega_{(0)} \circ A.$$

Therefore, the active scalar system

$$\begin{cases} D_t A = 0, \\ D_t \zeta = 0, \\ u = \nabla \times (-\Delta)^{-1} \left(\mathcal{C}[\nabla A, \zeta] \right) \end{cases} \tag{7.8}$$

is an equivalent formulation of the Euler equations, in terms of the Cauchy invariant ζ. The PDE formulations of the Euler equations described above are all equivalent formulations, as long as solutions are smooth. Classical local existence results for Euler equations can be proved in either purely Lagrangian formulation ([16]), in Eulerian formulation ([20]), or in Eulerian-Lagrangian formulation ([9]). For instance, one has

THEOREM 7.1 *([9]) Let $\alpha > 0$, and let u_0 be a divergence-free $C^{1,\alpha}$ periodic function of three variables. There exists a time interval $[0, T]$ and a unique $C([0, T];$ $C^{1,\alpha})$ spatially periodic function $\ell(x, t)$ such that*

$$A(x, t) = x + \ell(x, t)$$

solves the active scalar system formulation of the Euler equations,

$$\frac{\partial A}{\partial t} + u \cdot \nabla A = 0,$$

$$u = \mathbf{P} \left\{ (\nabla A(x, t))^T u_0(A(x, t)) \right\}$$

with initial datum $A(x, 0) = x$.

A similar result holds in the whole space, with decay requirements for the vorticity. The blow-up problem can be understood in terms of the growth of vorticity ([1]): on the time interval $[0, T]$ no singularities can arise from smooth initial data if $\int_0^T \|\omega(\cdot, t)\|_{L^\infty(dx)} dt < \infty$. From the Cauchy formula we have immediately that

$$\|\omega(\cdot, t)\|_{L^\infty(dx)} \le C \left(1 + \|\nabla \ell\|^2_{L^\infty(dx)} \right) \|\omega_{(0)}\|_{L^\infty(dx)},$$

and so, blow-up cannot occur without rapid growth of line elements. The Lagrangian representation is done with respect to a reference time (taken to be 0 in the preceding considerations). During a smooth evolution, one may stop at will, relabel, and restart. That means that one may regard the map $X(a, t)$ as a map close to the identity map, or, in other words, one may take small time steps, and keep ℓ small. For a smooth solution, and a short time t, the Eulerian-Lagrangian displacement ℓ can be very well approximated by the displacement obtained by freezing the velocity. Specifically, considering the solution $A_{(0)}(x, t)$ of the equation

$$\left(\partial_t + u_{(0)}(x) \cdot \nabla \right) A_{(0)} = 0$$

with $A_{(0)}(x, 0) = x$, and writing

$$A_{(0)}(x, t) = x + \ell_{(0)}(x, t),$$

we can prove that for smooth $u_{(0)}$ and small t one has

$$\|\ell - \ell_{(0)}\|_{C^{1,\alpha}} = O(t^2),$$

while both ℓ and ℓ_0 are $O(t)$. Knowledge of u at an instance of time gives thus a superlinear approximation of ℓ. On the other hand, knowledge of ℓ gives only a linear approximation for u. Expanding in ℓ in the Weber formula

$$u = \mathbf{P}\left((\mathbf{I} + \nabla\ell)^T u_{(0)}(x + \ell)\right),$$

we obtain

$$u(x, t) = u_{(0)}(x) + \mathbf{P}(\omega_{(0)} \times \ell) + O(\ell^2).$$

Indeed, we Taylor expand: $u_{(0)}(x + \ell) = u_{(0)}(x) + \ell \cdot \nabla u_{(0)}(x) + O(\ell^2)$, which we write as $u_{(0)}(x) + \omega_{(0)} \times \ell + (\nabla u_{(0)})^T \ell + O(\ell^2)$, and the term $(\nabla\ell)^T u_{(0)} + (\nabla u_{(0)})^T \ell$ is canceled by $\mathbf{P}$. Taking the curl of the relation above, we get

$$\omega(x, t) = \omega_{(0)} + \nabla \times \left(\omega_{(0)} \times \ell_{(0)}\right) + O(t^2)$$

for short time. We saw, however, that $\ell_{(0)} = -tu_{(0)} + O(t^2)$, and so we recover

$$\omega(x, t) = \omega_{(0)} + t \left(\nabla \times \left(u_{(0)} \times \omega_{(0)}\right)\right) + O(t^2),$$

which implies the Eulerian equation of evolution of the vorticity (7.3). Thus, the Weber formula, which is almost an algebraic representation of the present-time velocity in terms of the inverse Lagrangian map and a passive, frozen-in reference velocity, embodies the conservation of circulation and contains the dynamics information.

7.2 NAVIER-STOKES EQUATIONS

In view of the above, it is at least natural, if not imperative, to ask: What becomes of the Weber formula in the case of the Navier-Stokes equations?

The Navier-Stokes equations ([14]) can be written as

$$D_\nu u + \nabla p = 0, \tag{7.9}$$

together with the incompressibility condition $\nabla \cdot u = 0$. The operator D_ν

$$D_\nu = D_\nu(u, \nabla) = \partial_t + u \cdot \nabla - \nu\Delta \tag{7.10}$$

describes advection with velocity u and diffusion with kinematic viscosity $\nu > 0$. When $\nu = 0$ we recover formally the Euler equations (7.1), and $D_{\nu|\nu=0} = D_t$. The vorticity $\omega = \nabla \times u$ obeys an equation similar to (7.3):

$$D_\nu\omega = \omega \cdot \nabla u. \tag{7.11}$$

It turns out that Eulerian-Lagrangian equations (7.6) and (7.8) have also viscous counterparts ([10]). The equation corresponding to (7.6) is

$$\begin{cases} D_v A = 0, \\ D_v v = 2\nu C \nabla v, \\ u = W[A, v]. \end{cases} \tag{7.12}$$

The $u = W[A, v]$ is the Weber formula (7.5), exactly the same as in the case of $\nu = 0$. The right-hand side of (7.12) is given in terms of the connection coefficients

$$C_{k;i}^m = \left((\nabla A)^{-1}\right)_{ji} \left(\partial_j \partial_k A^m\right).$$

The detailed form of virtual velocity equation in (7.12) is

$$D_v v_i = 2\nu C_{k;i}^m \partial_k v_m.$$

The connection coefficients are related to the Christoffel coefficients of the flat Riemannian connection in $\mathbf{R}^3$ computed using the change of variables $a = A(x, t)$:

$$C_{k;i}^m(x, t) = -\Gamma_{ji}^m(A(x, t)) \frac{\partial A^j(x, t)}{\partial x_k}.$$

The equation $D_v(u, \nabla) A = 0$ describes advection *and diffusion* of labels.

The diffusion of labels is a consequence of the physically natural idea of adding Brownian motion to the Lagrangian flow. Indeed, if $u(X(a, t), t)$ is known, and if

$$dX(a, t) = u(X(a, t), t)dt + \sqrt{2\nu} dW(t), \quad X(a, 0) = a,$$

with $W(t)$ standard independent Brownian motions in each component, and if

$$Prob\{X(a, t) \in dx\} = \rho(x, t; a)dx,$$

then the expected value of the back to labels map

$$A(x, t) = \int \rho(x, t; a) a da$$

solves

$$D_v(u, \nabla) A = 0.$$

In addition to being well posed, the Eulerian-Lagrangian viscous equations are capable of describing vortex reconnection. We associate to the virtual velocity v the Eulerian-Lagrangian curl of v

$$\zeta = \nabla^A \times v, \tag{7.13}$$

where

$$\nabla_i^A = \left((\nabla A)^{-1}\right)_{ji} \partial_j$$

is the pull-back of the Eulerian gradient. The viscous analog of the Eulerian-Lagrangian Cauchy invariant active scalar system (7.8) is

$$\begin{cases} D_v A = 0, \\ D_v \zeta^q = 2\nu G_p^{qk} \partial_k \zeta^p + \nu T_p^q \zeta^p, \\ u = \nabla \times (-\Delta)^{-1} \left(C[\nabla A, \zeta]\right). \end{cases} \tag{7.14}$$

The Cauchy transformation

$$\mathcal{C}[\nabla A, \zeta] = (det(\nabla A))(\nabla A)^{-1}\zeta$$

is the same as the one used in the Euler equations, (7.7). The specific form of the two terms on the right-hand side of the Cauchy invariant's evolution are

$$G_p^{qk} = \delta_p^q C_{k;m}^m - C_{k;p}^q, \tag{7.15}$$

and

$$T_p^q = \epsilon_{qji}\epsilon_{rmp}C_{k;i}^m C_{k;j}^r. \tag{7.16}$$

The system (7.12) is equivalent to the Navier-Stokes system. When $\nu = 0$ the system reduces to (7.6). The system (7.14) is equivalent to the Navier-Stokes system, and reduces to (7.8) when $\nu = 0$.

The pair (A, v) formed by the diffusive inverse Lagrangian map and the virtual velocity are akin to charts in a manifold. They are a convenient representation of the dynamics of u for some time. When the representation becomes inconvenient, then one has to change the chart. This may (and will) happen if ∇A becomes noninvertible. Likewise, the pair (A, ζ) formed with the "back-to-labels" map A and the diffusive Cauchy invariant ζ are convenient charts. Because the fluid variables u or ω are represented as products of elements in the chart, it is possible for the chart to become singular without the fluid becoming singular. The regularity of the fluid is not equivalent to the regularity of a single chart, but rather to the existence of smooth, compatible charts. In order to quantify this statement let us introduce the terminology of "group expansion" for the procedure of resetting. More precisely, the group expansion for (7.12) is defined as follows. Given a time interval $[0, T]$ we consider resetting times

$$0 = t_0 < t_1 < \cdots < t_n \cdots \leq T.$$

On each interval $[t_i, t_{i+1}]$, $i = 0, \ldots$ we solve the system (7.12):

$$\begin{cases} D_\nu(u, \nabla)A = 0, \\ D_\nu(u, \nabla)v = 2\nu C\nabla v, \\ u = \mathbf{P}\left((\nabla A)^T v\right). \end{cases}$$

with resetting conditions

$$\begin{cases} A(x, t_i) = x, \\ v(x, t_i + 0) = ((\nabla A)^* v)(x, t_i - 0). \end{cases}$$

We require the resetting criterion that $\nabla \ell = (\nabla A) - \mathbf{I}$ must be smaller than a preassigned value ϵ in an analytic norm: $\exists \lambda$ such that for all $i \geq 1$ and all $t \in [t_i, t_{i+1}]$ one has

$$\int e^{\lambda|k|} |\widehat{\ell}(k)| \, dk \leq \epsilon < 1.$$

If there exists N such that $T = \sum_{i=0}^{N}(t_{i+1}-t_i)$ then we say that the group expansion *converges* on $[0, T]$. A group expansion of (7.14) is defined similarly. The resetting conditions are

$$
\begin{cases}
A(x, t_i) = x, \\
\zeta(x, t_i + 0) = C[(\nabla A))(x, t_i - 0), \zeta(x, t_i - 0)].
\end{cases}
$$

The analytic resetting criterion is the same. The first interval of time $[0, t_1)$ is special. The initial value for v is u_0 (the initial datum for the Navier-Stokes solution), and the initial value for ζ is ω_0, the corresponding vorticity. The local time existence is used to guarantee invertibility of the matrix ∇A on $[0, t_1)$ and Gevrey regularity ([18]) to pass from moderately smooth initial data to Gevrey class regular solutions. Note that the resetting conditions are designed precisely so that both u and ω are time continuous.

THEOREM 7.2 ([11]) *Let $u_0 \in H^1(\mathbf{R}^3)$ be divergence-free. Let $T > 0$. Assume that the solution of the Navier-Stokes equations with initial datum u_0 obeys $\sup_{0 \le t \le T} \|\omega(\cdot, t)\|_{L^2(dx)} < \infty$. Then there exists $\lambda > 0$ so that, for any $\epsilon > 0$, there exists $\tau > 0$ such that both group expansions converge on $[0, T]$ and the resetting intervals can be chosen to have any length up to τ, $t_{i+1} - t_i \in [0, \tau]$. The velocity u, solution of the Navier-Stokes equation with initial datum u_0, obeys the Weber formula (7.5). The vorticity $\omega = \nabla \times u$ obeys the Cauchy formula (7.7).*

Conversely, if one group expansion converges, then so does the other, using the same resetting times. The Weber and Cauchy formulas apply and reconstruct the solution of the Navier-Stokes equation. The enstrophy is bounded $\sup_{0 \le t \le T} \|\omega(\cdot, t)\|_{L^2(dx)} < \infty$, and the Navier-Stokes solution is smooth.

The quantity λ can be estimated explicitly in terms of the bound of enstrophy, time T, and kinematic viscosity ν. The bound is algebraic: a negative power of the enstrophy, if all other quanties are fixed. The maximal time step τ is proportional to ϵ, with a coefficient of proportionality that depends algebraically on the bound on enstrophy, time T, and ν. The converse statement, that if the group expansion converges, then the enstrophy is bounded, follows from the fact that there are finitely many resettings. Indeed, the Cauchy formula and the near identity bound on ∇A imply a doubling condition on the enstrophy on each interval. It is well known that the condition regarding the boundedness of the enstrophy implies regularity of the Navier-Stokes solution. Our definition of convergent group expansion is very demanding, and it is justified by the fact that once the enstrophy is bounded, one could mathematically demand analytic norms. But the physical resetting criterion is the invertibility of the matrix ∇A. The Euler equations require no resetting as long as the solution is smooth. The Navier-Stokes equations, at least numerically, require numerous and frequent resettings. There is a deep connection between these resetting times and vortex reconnection. In the Euler equation, as long as the solution is smooth, the Cauchy invariant obeys $\zeta(x, t) = \omega_{(0)}(A(x, t))$, with $\omega_{(0)} = \omega_0$ the initial vorticity. The topology of vortex lines is frozen in time. In the Navier-Stokes system the topology changes. This is the phenomenon of vortex reconnection. There is ample numerical and physical evidence for this phenomenon. In the more complex but similar case of magneto-hydrodynamics, magnetic reconnection occurs and has powerful physical implications. Vortex reconnection is a dynamical dissipative process. The solutions of the Navier-Stokes equations obey a space-time average bound ([6], [11])

$$\int_0^T \int_{\mathbf{R}^3} |\omega(x,t)| \left| \nabla_x \left(\frac{\omega(x,t)}{|\omega(x,t)|} \right) \right|^2 dx dt \leq \frac{1}{2} \nu^{-2} \int_{\mathbf{R}^3} |u_0(x,t)|^2 dx.$$

This bound is consistent with the numerically observed fact that the region of high vorticity is made up of relatively straight vortex filaments (low curvature of vortex lines) separated by distances that vanish with viscosity (Kolmogorov scale). The processes by which these configurations are obtained and sustained are vortex stretching and vortex reconnection. When vortex lines are locally aligned, a geo-metric depletion of nonlinearity occurs, and the local production of enstrophy drops. Indeed, the Navier-Stokes equations have global smooth solutions if the vorticity direction field $\frac{\omega}{|\omega|}$ is Lipschitz continuous ([13]) in regions of high vorticity. Vortex reconnection is a manifestation of a regularizing mechanism. It is difficult to have a precise mathematical definition of vortex reconnection, although the phenomenon can be easily recognized. In numerical simulations, vorticity is placed initially in a region with a certain topology, and the process of change of topology is "watched" (visualized). In numerical work with Ohkitani ([23], [24]), we have proposed a quantitative alternative to "watching," based on the diffusive Eulerian-Lagrangian formalism. We proposed to identify the periods of rapid resetting times with pe-riods of vortex reconnection. The numerical calculations produce periods of rapid resetting; when these are visualized, they coincide with the periods of change of topology. In order to explain the mathematical reason behind this numerical ob-servation, let us recall that the solution of the Navier-Stokes equation is smooth as long as there is an upper bound on the frequency of resetting per unit time. The physical criterion for resetting is the vansihing of $det(\nabla A)$. The equation for the determinant of ∇A is

$$D_\nu \left(\log(det(\nabla A)) \right) = \nu \left\{ C^i_{k;s} C^s_{k;i} \right\}. \tag{7.17}$$

The initial datum vanishes. When $\nu = 0$ we recover conservation of incompress-ibility. In the case $\nu > 0$, the inverse timescale in the right-hand side of this equation is significant for reconnection. Because the equation has a maximum principle it follows that

$$det(\nabla A)(x,t) \geq exp \left\{ -\nu \int_{t_i}^t \sup_x \left\{ C^i_{k;s} C^s_{k;i} \right\} d\sigma \right\}.$$

The $i+1$ resetting time is determined thus by the requirement

$$\int_{t_i}^t \sup_x f_\nu(x,s) ds = M,$$

with $M = \infty$ (in practice, M large), where f_ν is the local resetting frequency

$$f_\nu(x,t) = \nu \left\{ C^i_{k;s} C^s_{k;i} \right\}$$

(because v has units of $(cm^2)(sec)^{-1}$ and C has units of cm^{-1}, f_v has units of $(sec)^{-1}$.)

Because the resetting times are computed using the Christoffel symbols, it is useful to see the form this expression takes using the back-to-labels transformation. Let us recall that using the smooth change of variables $a = A(x, t)$ (at each fixed time t) we compute the Euclidean Riemannian metric by

$$g^{ij}(a, t) = (\partial_k A^i)(\partial_k A^j)(x, t) \tag{7.18}$$

Considering

$$g = det(g_{ij}), \tag{7.19}$$

where g_{ij} is the inverse of g^{ij} and observing that

$$g(A(x, t)) = (det(\nabla A))^{-2},$$

the equation (7.17) becomes

$$\partial_t(\log(\sqrt{g})) = v g^{ij} \partial_i \partial_j \log(\sqrt{g}) - v g^{\alpha\beta} \Gamma^m_{\alpha p} \Gamma^p_{\beta m}. \tag{7.20}$$

The initial datum is zero, the equation is parabolic, has a maximum principle and is driven by the last term. The form (7.20) of the equation (7.17) has the same interpretation: the connection coefficients define an inverse length scale associated to A. The frequency f_v that decides the duration of the time interval of validity of the chart A, and time to reconnection, is

$$f_v(x, t) = v \left\{ C^i_{k;s} C^s_{k;i} \right\} = v \left\{ g^{mn} \Gamma^i_{ms} \Gamma^s_{ni} \right\} \circ A,$$

so its expression in the chart is

$$\phi_v(a, t) = v \left\{ g^{mn} \Gamma^i_{ms} \Gamma^s_{ni} \right\}.$$

The equations for the virtual velocity and for the Cauchy invariant can also be solved by following the path A, i.e., by seeking

$$v(x, t) = v(A(x, t), t),$$
$$\zeta(x, t) = \xi(A(x, t), t). \tag{7.21}$$

The equations for v and ξ become purely diffusive. Using $D_v A = 0$, the operator D_v becomes

$$D_v(f \circ A) = \left((\partial_t - v g^{ij} \partial_i \partial_j) f \right) \circ A. \tag{7.22}$$

The equation for v follows from (7.12):

$$\partial_t v_i = v g^{mn} \partial_m \partial_n v_i - 2v V^{mj}_i \partial_m v_j, \tag{7.23}$$

with

$$V^{mj}_i = g^{mk} \Gamma^j_{ik}.$$

The derivatives are with respect to the Cartesian coordinates a. The equation reduces to $\partial_t v = 0$ when $v = 0$, and in that case we recover $v = u_{(0)}$, the time-independent initial velocity. For $v > 0$, the system is parabolic and wellposed. The equation for ξ follows from (7.14):

$$\partial_t \xi^q = v g^{ij} \partial_i \partial_j \xi^q + 2v W_n^{qk} \partial_k \xi^n + v T_p^q \xi^p, \tag{7.24}$$

with

$$\begin{cases} W_n^{qk} = -\delta_n^q g^{kr} \Gamma_{rp}^p + g^{kp} \Gamma_{pn}^q, \\ T_p^q = \epsilon_{qji} \epsilon_{rmp} \Gamma_{\alpha j}^r \Gamma_{\beta i}^m \cdot g^{\alpha\beta}. \end{cases}$$

Again, when $v = 0$ this reduces to the invariance $\partial_t \xi = 0$. But in the presence of v this is a parabolic system. Both the Cauchy invariant and the virtual velocity equations start out looking like the heat equation, because $g^{mn}(a, 0) = \delta^{mn}$ and $\Gamma_{jk}^i(a, 0) = 0$. The long time behavior depends on the smoothness of the metric.

The inviscid conservation of circulation and of helicity have natural viscous counterparts. Regarding circulation, we note that the Weber formula implies that

$$u dx - v dA = -dn,$$

and therefore

$$\oint_{\gamma \circ A} u dx = \oint_\gamma v da \tag{7.25}$$

holds for any closed loop γ. Regarding helicity, in view of the Cauchy formula one has that a helicity density is $v \cdot \zeta$, the scalar product of the virtual velocity and Cauchy invariant. With the change of variables $a = A(x, t)$ we have

$$v(x) = v(a), \quad \zeta(x) = \xi(a)$$

$$\int_T u \cdot \omega dx = \int_{A(T)} v \cdot \zeta da$$

for any vortex tube T.

The metric g^{ij} determines the connection coefficients. However, the evolution is not purely geometric: the evolution equation of g^{ij} involves ∇u and ∇A. It is remarkable that all the counterparts of the inviscid invariants, virtual velocity, Cauchy invariant, volume element, helicity density, evolve according to equations that do not involve explicitly the velocity u, once one computes in a diffusive Lagrangian frame. This justifies the following terminology: we will say that a function F is *diffusively Lagrangian* under the Navier-Stokes flow if $F = \phi \circ A$ and ϕ obeys a linear, parabolic second-order evolution PDE with coefficients determined locally entirely by the Euclidean Riemannian metric induced by the change of variables A, and which vanish when $v = 0$. More precisely, we require ϕ to obey a linear parabolic PDE

$$\frac{\partial \phi(a, t)}{\partial t} = v \mathcal{L}[g, \partial_a] \phi(a, t),$$

where

$$\mathcal{L}[g, \partial_a]\phi = \nu g^{ij}\partial_i\partial_j\phi + M\nabla\phi + N\phi + P$$

is a linear second-order elliptic differential operator with coefficients M, N, P computed from $g(a, t)$ and finitely many of its a derivatives. The metric itself is not diffusively Lagrangian. Products of diffusive-Lagrangian functions are diffusive Lagrangian. The previous calculations can be summarized thus:

THEOREM 7.3 *The virtual velocity v, the Cauchy invariant ζ the Jacobian determinant $det(\nabla A)$, and the helicity density $v \cdot \zeta$ associated to solutions of the Navier-Stokes equations are diffusively Lagrangian.*

7.3 CONCLUSION

The fundamental physical processes that occur in incompressible, uniform density, viscous fluids are folding, stretching, and reconnection. These processes can be described using a diffusive Lagrangian formalism that uses as basic variables a transformation of space A and a virtual velocity v (or its Lagrangian curl, the Cauchy invariant ζ). In ideal inviscid fluids, folding is represented simply by composition with A. In the presence of viscosity, portions of the fluid that are folded close to one another are mixed by molecular diffusion. Quantities that undergo this folding process are among the diffusively Lagrangian quantities. Diffusive Lagrangian functions obey a certain type of diffusion equation with principal part $\partial_t - \nu g^{ij}\partial_i\partial_j$ and with drift and lower-order coefficients computed solely from the Riemannian metric g^{ij} in the chart A. The quantities that are formally conserved under smooth inviscid evolution become diffusive Lagrangian. The stretching of fluid elements is calculated using the norm of the gradient matrix ∇A. The vanishing of the determinant of the matrix ∇A is the signature of the reconnection process. The frequency of the reconnection process can be defined quantitatively, and is computed using the Riemannian metric and the kinematic viscosity.

REFERENCES

[1] J. T. Beale, T. Kato, and A. Majda, Remarks on the breakdown of smooth solutions for the 3-D Euler equations, Commun. Math. Phys. 94 (1984), 61–66.

[2] A. Chorin, Vorticity and Turbulence, Appl. Math. Sciences 103, Springer-Verlag, 1994.

[3] S. Childress, G. R. Ierley, E. A. Spiegel, W. R. Young, Blow-up of unsteady two-dimensional Euler and Navier-Stokes solutions having stagnation-point form, J. Fluid. Mech. 203 (1989), 1–22.

[4] A. Chorin, Numerical study of slightly viscous flow, J. Fluid. Mech 57 (1973), 785–796.

[5] P. Constantin, Note on loss of regularity for solutions of the 3D incompressible and related equations, Commun. Math. Phys. 106 (1986), 311–325.

[6] P. Constantin, Navier-Stokes equations and area of interfaces, Commun. Math. Phys. 129 (1990), 241–266.

[7] P. Constantin, Geometric and analytic studies in turbulence, in Trends and Perspectives in Appl. Math., L. Sirovich, ed., Appl. Math. Sciences 100, Springer-Verlag (1994).

[8] P. Constantin, The Euler equations and nonlocal conservative Riccati equations, Intern. Math. Res. Notes 9 (2000), 455–465.

[9] P. Constantin, An Eulerian-Lagrangian approach for incompressible fluids: local theory, Journal of the AMS 14 (2001), 263–278.

[10] P. Constantin, An Eulerian-Lagrangian approach to the Navier-Stokes equations, Commun. Math. Phys. 216 (2001), 663–686.

[11] P. Constantin, Near identity transformations for the Navier-Stokes equations, in *Handbook of Mathematical Fluid Dynamics*, vol. 2, S. Friedlander and D. Serre, eds., Elsevier (2003).

[12] P. Constantin, W. E, E. S. Titi, Onsager's conjecture on the energy conservation for solutions of Euler's equations, Commun. Math. Phys. 165 (1994), 207–209.

[13] P. Constantin and C. Fefferman, Direction of vorticity and the problem of global regularity for the Navier-Stokes equations, Indiana Univ. Math. Journal 42 (1993), 775.

[14] P. Constantin and C. Foias, *Navier-Stokes Equations*, University of Chicago Press, Chicago (1988).

[15] B. I . Davydov, Dokl. Akad. Nauk. SSSR 2 (1949), 165.

[16] D. Ebin and J. Marsden, Groups of diffeomorphisms and the motion of an incompressible fluid, Ann. of Math. 92 (1970), 102–163.

[17] G. Eyink, Energy dissipation without viscosity in the ideal hydrodynamics, I. Fourier analysis and local energy transfer, Phys. D 3–4 (1994), 222–240.

[18] C. Foias and R. Temam, Gevrey class regularity for the solutions of the Navier-Stokes equations, J. Funct. Anal. 87 (1989), 359–369.

[19] J. Leray, Essai sur le mouvement d'un liquide visqueux emplissant l'espace, Acta Mathematica 63 (1934), 193–248.

[20] A. Majda and A. Bertozzi, *Vorticity and Incompressible Flow*, Cambridge Texts in Applied Mathematics, Cambridge University Press (2002).

[21] H. K. Moffatt, The degree of knottedness of tangled vortex lines, J. Fluid Mech. 35 (1969), 117–129.

[22] L. Onsager, Statistical hydrodynamics, Nuovo Cimento 6(2) (1949), 279–287.

[23] K. Ohkitani and P. Constantin, Numerical study of the Eulerian-Lagrangian formulation of the Navier-Stokes equations, Phys. Fluids 15–10 (2003), 3251–3254.

[24] K. Ohkitani and P. Constantin, Numerical study of the Eulerian-Lagrangian analysis of the Navier-Stokes turbulence, work in preparation.

[25] K. Ohkitani and J. Gibbon, Numerical study of singularity formation in a class of Euler and Navier-Stokes flows, Phys. Fluids 12 (2000), 3181.

[26] R. Robert, Statistical hydrodynamics (Onsager revisited), in *Handbook of Mathematical Fluid Dynamics*, vol. 2, S. Friedlander and D. Serre, eds., Elsevier (2003), 3–54.

[27] A. Shnirelman, Weak solutions of incompressible Euler equations, in *Handbook of Mathematical Fluid Dynamics*, vol. 2, S. Friedlander and D. Serre, eds., Elsevier (2003), 87–116.

[28] J. T. Stuart, Nonlinear Euler partial differential equations: Singularities in their solution, in *Applied Mathematics, Fluid Mechanics, Astrophysics* Cambridge, MA (1987), World Sci., Singapore (1988), 81–95.

[29] W. Weber, Über eine Transformation der hydrodynamischen Gleichungen, J. Reine Angew. Math. 68 (1868), 286–292.

[30] V. E. Zakharov and E. A. Kuznetsov, Variational principle and canonical variables in magnetohydrodynamics, Doklady Akademii Nauk SSSR 194 (1970), 1288–1289.

Chapter Eight

Long Time Existence for Small Data Semilinear
Klein-Gordon Equations on Spheres

J.-M. Delort and J. Szeftel

8.0 INTRODUCTION

Consider (M, g) a Riemannian manifold and denote by Δ_g (resp. ∇_g) the Laplacian (resp. the gradient) associated to g. Let $m > 0$ be given and consider a local solution u to the Cauchy problem

$$(\partial_t^2 - \Delta_g + m^2)u = f(x, u, \partial_t u, \nabla_g u)$$
$$u|_{t=0} = \epsilon u_0 \tag{8.1}$$
$$\partial_t u|_{t=0} = \epsilon u_1,$$

where f is a real polynomial in $(u, \partial_t u, \nabla_g u)$, vanishing at some order $p \geq 2$ at 0, with C^∞ coefficients in x, where the data (u_0, u_1) are in $C_0^\infty(M)$, real valued, and where $\epsilon > 0$ goes to zero. Denote by $]T_*(\epsilon), T^*(\epsilon)[$ (with $T_*(\epsilon) < 0 < T^*(\epsilon)$) the maximal interval of existence of a smooth solution to (8.1). We aim at giving lower bounds for $T^*(\epsilon), -T_*(\epsilon)$ when ϵ goes to zero.

When $M = \mathbb{R}^d$ is endowed with its standard metric, the problem is well known. If the space dimension d is larger or equal to 3, Klainerman [5] and Shatah [8] proved independently that for small $\epsilon > 0$, one has $T_*(\epsilon) = -\infty$, $T^*(\epsilon) = +\infty$, i.e., the solution is global. The proof of that property relies on the use of dispersive properties of the linear Klein-Gordon equation.

In $d = 2$ space dimensions, the same result was proved by Ozawa, Tsutaya, and Tsutsumi [7] (see also Simon and Taflin [9] and [3]). The difference with higher space dimensions is that dispersion is no longer strong enough to imply directly global existence when $p = 2$. Actually, in contrast with the case $d \geq 3$, the non-linearity is a long-range perturbation of the linear equation, and the proof of global existence relies on a reduction of (8.1), through a method of normal forms, to an equivalent equation with a nonlinearity vanishing at higher order at 0.

We refer to the work of Moriyama, Tonegawa, and Tsutsumi [6] and to [2] for a discussion of equation (8.1) on $\mathbb{R}$.

The problem we are interested in here is the case of a compact manifold M. We thus have no dispersion for the linear equation, and the only general lower bound for the time of existence of smooth solutions with small data is provided by local existence theory. Namely, for a nonlinearity vanishing at order $p \geq 2$ at 0, (8.1) has a solution defined at least on an interval $] - T_\epsilon, T_\epsilon[$ with $T_\epsilon \geq c\epsilon^{-p+1}$. Our aim is to find, when $M = \mathbb{S}^{d-1}$, and under convenient assumptions, a better lower bound when $\epsilon \to 0$. The detailed proofs of the results we shall present are given in [4].

8.1 STATEMENT OF MAIN THEOREM

From now on, we set $M = \mathbb{S}^{d-1}$ and denote by g the standard metric on the sphere. We consider (8.1) and decompose the nonlinearity as

$$f(x, u, \partial_t u, \nabla_g u) = \sum_{q \geq p} f_q(x, u, \partial_t u, \nabla_g u), \qquad (8.2)$$

where f_q is the component homogeneous of degree q in $(u, \partial_t u, \nabla_g u)$ of f. Let us denote by r the largest integer satisfying

$$p \leq r \leq 2p - 1,$$

for any odd k with $p \leq k < r$, f_k depends only on $(u, \partial_t u)$, $\qquad (8.3)$

is even in $\partial_t u$, and independent of x.

Our main theorem is then:

THEOREM 8.1 *There is a zero measure subset $\mathcal{N}$ of $]0, +\infty[$ and for any $m \in]0, +\infty[-\mathcal{N}$, there are $\epsilon_0 > 0, c > 0, s \in \mathbb{N}$ such that for any pair of real valued functions (u_0, u_1) in the unit ball of $H^{s+1}(\mathbb{S}^{d-1}) \times H^s(\mathbb{S}^{d-1})$, any $\epsilon \in]0, \epsilon_0[$, problem (8.1) has a unique solution*

$$u \in C^0(]-T_\epsilon, T_\epsilon[, H^{s+1}(\mathbb{S}^{d-1})) \cap C^1(]-T_\epsilon, T_\epsilon[, H^s(\mathbb{S}^{d-1})), \qquad (8.4)$$

with $T_\epsilon \geq c\epsilon^{-r+1}$.

Example. Assume that f vanishes at some even order $p \geq 2$ at 0. Then condition (8.3) is satisfied taking $r = p + 1$, and Theorem 8.1 gives a lower bound for the time of existence of type $c\epsilon^{-p}$, i.e., we got a gain of one negative power of ϵ in comparison with the estimates given by local existence theory.

Remarks.

- The nonlinearity in (8.1) is a general nonlinearity, depending on u and its first order derivatives, the only restriction being given by (8.3). In particular, we do not assume any Hamiltonian structure for the equation.
- Assumption (8.3) on the components homogeneous of odd degree is certainly not optimal in all cases: we give in [4] examples of cubic or quintic nonlinearities on $\mathbb{S}^1$ that do depend on $\partial_x u$, and for which the conclusion of Theorem 8.1 holds true.
- Remark, anyway, that we cannot expect to remove all assumptions on the odd components of f: it is proved in [1], following an idea of Yordanov [10], that for the equation

$$\Box u + m^2 u = u_t^2 u_x \text{ on } [-T, T] \times \mathbb{S}^1,$$

the solution with Cauchy data of size ϵ blows-up at time of magnitude c/ϵ^2 (for any $m > 0$). In other words, for such an example, the time of existence given by local existence theory is optimal.

- In low space dimensions, namely on $\mathbb{S}^1$ or $\mathbb{S}^2$, it is easy to prove global existence for equations of type $(\partial_t^2 - \Delta_g + m^2)u = f(u)$ for small enough smooth Cauchy data. Actually, such an equation has a conserved energy that controls the H^1 norm for small data, and this is enough to get global solutions for these dimensions.

8.2 STRATEGY OF PROOF

Our strategy will be to eliminate the lowest-order term of the nonlinearity by a method of normal forms. Such an idea was initially introduced in the framework of nonlinear Klein-Gordon equations by Shatah in [8]. If we set $D_t = \frac{1}{i}\frac{\partial}{\partial t}$ and

$$\Lambda_m = \sqrt{-\Delta_g + m^2}$$
$$u_\pm = (D_t \pm \Lambda_m)u,$$
(8.5)

we can write (8.1) as a system, whose first equation would be

$$(D_t - \Lambda_m)u_+ = f(x, \tfrac{1}{2}\Lambda_m^{-1}(u_+ - u_-), \tfrac{i}{2}(u_+ + u_-), \tfrac{1}{2}\nabla_g\Lambda_m^{-1}(u_+ - u_-)), \quad (8.6)$$

the equation for u_- being obtained by conjugation (using that $\bar{u}_+ = -u_-$ since we consider real solutions of (8.1)). We shall explain the strategy of proof on a model equation of the following type:

$$(D_t - \Lambda_m)u = u^\ell \bar{u}^{p-\ell}$$
$$u|_{t=0} = \epsilon u_0,$$
(8.7)

where we wrote u instead of u_+ to simplify notations, where $0 \le \ell \le p$, and where u_0 is a smooth complex valued given function. We shall also assume that p is even, so that according to the example given after Theorem 8.1, our goal is to prove that the solution to (8.7) exists at least over an interval $]-T_\epsilon, T_\epsilon[$ with $T_\epsilon \ge c\epsilon^{-p}$.

We shall look for an operator $u \to B(u, \ldots, u, \bar{u}, \ldots, \bar{u})$, ℓ-linear in u and $p - \ell$-linear in $\bar{u}$, such that the following two properties hold true:

$$(D_t - \Lambda_m)(B(u, \ldots, \bar{u})) = -u^\ell \bar{u}^{p-\ell} + O(u^{p+1})$$
(A)

$$u \to B(u, \ldots, \bar{u}) \text{ is continuous on } H^s(\mathbb{S}^{d-1}) \text{ for } s \text{ large enough.}$$
(B)

Let us remark that (A) and (B) together imply Theorem 8.1. Actually, (A) and (8.7) give

$$(D_t - \Lambda_m)(u + B(u, \ldots, \bar{u})) = O(u^{p+1}),$$
(8.8)

and (B) together with the local inversion theorem ensures that

$$u \to v \stackrel{\text{def}}{=} u + B(u, \ldots, \bar{u})$$
(8.9)

is a diffeomorphism from a neighborhood of 0 in H^s onto another such neighborhood. Consequently, for small enough solutions, (8.8) is equivalent to

$$(D_t - \Lambda_m)v = O(v^{p+1}),\qquad(8.10)$$

which has solutions defined on $]-T_\epsilon, T_\epsilon[$ with $T_\epsilon \geq c\epsilon^{-p}$ by local existence theory, whence Theorem 8.1.

Let us proceed to the construction of B satisfying (A), (B). We shall define B using a decomposition in spherical harmonics. Denote by

$$\lambda_n = \sqrt{n(n+d-2)},\ n \in \mathbb{N}\qquad(8.11)$$

so that the family $(\lambda_n^2)_{n\in\mathbb{N}}$ is the family of eigenvalues of $-\Delta_g$ on $\mathbb{S}^{d-1}$. Denote by E_n the space of spherical harmonics of degree n, i.e., the space of restrictions to $\mathbb{S}^{d-1}$ of harmonic polynomials homogeneous of degree n. Then E_n is the eigenspace associated to λ_n^2, so that if Π_n is the orthogonal projection of $L^2(\mathbb{S}^{d-1})$ onto E_n, one has for $u \in L^2(\mathbb{S}^{d-1})$

$$-\Delta_g \Pi_n u = \lambda_n^2 \Pi_n u,\ \Lambda_m \Pi_n u = \sqrt{m^2 + \lambda_n^2}\Pi_n u.\qquad(8.12)$$

Since we can write

$$u^\ell \bar{u}^{p-\ell} = \sum_{n_1}\cdots\sum_{n_{p+1}}\Pi_{n_{p+1}}[(\Pi_{n_1}u)\cdots(\Pi_{n_\ell}u)(\overline{\Pi_{n_{\ell+1}}u})\cdots(\overline{\Pi_{n_p}u})],$$

it is natural to look for B as given by

$$B(u,\ldots,\bar{u}) = \sum_{n_1}\cdots\sum_{n_{p+1}}a(n_1,\ldots,n_{p+1})\Pi_{n_{p+1}}[(\Pi_{n_1}u)\ldots(\overline{\Pi_{n_p}u})]\quad(8.13)$$

for some coefficients $a(n_1,\ldots,n_{p+1})$ to be determined. Remark that we can write

$$D_t \Pi_{n_j}u = (D_t - \Lambda_m)\Pi_{n_j}u + \Lambda_m \Pi_{n_j}u$$
$$= O(u^p) + \sqrt{m^2 + \lambda_{n_j}^2}\Pi_{n_j}u\qquad(8.14)$$

using the equation and (8.12). If we compute $D_t B$ using these expressions, we see that the first term in the right-hand side of (8.14) will contribute to remainders in the right-hand side of (A), so that the main contribution to $(D_t - \Lambda_m)B(u,\ldots,\bar{u})$ will be given by

$$\sum_{n_1}\cdots\sum_{n_{p+1}}F_m^{p+1,\ell}(\lambda_{n_1},\ldots,\lambda_{n_{p+1}})a(n_1,\ldots,n_{p+1})\Pi_{n_{p+1}}[(\Pi_{n_1}u)\cdots(\overline{\Pi_{n_p}u})],$$
$$(8.15)$$

where in general, for $0 \leq \ell \leq q$,

$$F_m^{q,\ell}(\xi_1,\ldots,\xi_q) = \sum_{j=1}^{\ell}\sqrt{m^2+\xi_j^2} - \sum_{j=\ell+1}^{q}\sqrt{m^2+\xi_j^2}.\qquad(8.16)$$

We thus see that we shall get (A) if we can choose

$$a(n_1, \ldots, n_{p+1}) = -(F_m^{p+1,\ell}(\lambda_{n_1}, \ldots, \lambda_{n_{p+1}}))^{-1}. \tag{8.17}$$

This is possible only if $F_m^{p+1,\ell}$ does not vanish at $(\lambda_{n_1}, \ldots, \lambda_{n_{p+1}})$. We shall see that this can be achieved if m stays outside a set of zero measure. Actually, we shall be able to obtain good enough upper bounds on the absolute value of (8.17) so that (B) holds true. This is the content of the following proposition.

PROPOSITION 8.2 *For any family of integers $n_1, \ldots, n_p$ denote*

$$\mu(n_1, \ldots, n_p) = 1 + \text{ second largest among } n_1, \ldots, n_p. \tag{8.18}$$

There is a zero measure subset $\mathcal{N}$ of $]0, +\infty[$ such that for any $m \in]0, +\infty[-\mathcal{N}$, there are $c > 0$, $N_1 \in \mathbb{N}$ and for any $(n_1, \ldots, n_{p+1})$ with

$$\Pi_{n_{p+1}}[(\Pi_{n_1}u) \cdots (\overline{\Pi_{n_p}u})] \not\equiv 0 \tag{8.19}$$

we have

$$|F_m^{p+1,\ell}(\lambda_{n_1}, \ldots, \lambda_{n_{p+1}})| \geq c\mu(n_1, \ldots, n_p)^{-N_1}. \tag{8.20}$$

Remark. Condition (8.19) originates from the fact that we want to estimate a given by (8.17) only when its coefficient in (8.13) is not identically zero. We shall actually use only some inequalities between $n_1, \ldots, n_{p+1}$ implied by (8.19).

Let us show that Proposition 8.2 implies that condition (B) holds true for m outside $\mathcal{N}$ and if s is large enough relatively to N_1. Let us estimate the L^2 norm of the general term of the sum in (8.13) for a multi-index $(n_1, \ldots, n_{p+1})$ for which for instance $n_p = \max(n_1, \ldots, n_p)$. We then have

$$\mu(n_1, \ldots, n_p) \sim 1 + n_1 + \cdots + n_{p-1}, \tag{8.21}$$

and we write

$$\|a\Pi_{n_{p+1}}(\Pi_{n_1}u \cdots \overline{\Pi_{n_p}u})\|_{L^2} \leq |a| \prod_{j=1}^{p-1} \|\Pi_{n_j}u\|_{L^\infty} \|\Pi_{n_p}u\|_{L^2}. \tag{8.22}$$

Using (8.17), (8.20), (8.21), and Sobolev injection, we get an upper bound in terms of

$$(1 + n_1 + \cdots + n_{p-1})^{N_1} \prod_{j=1}^{p-1} (1 + \lambda_{n_j})^{\frac{d-1}{2}+\delta-s} \|\Pi_{n_p}u\|_{L^2} \tag{8.23}$$

if we assume $u \in H^s$ and if $\delta > 0$ is small. Since $\lambda_{n_j} \sim n_j$, for s large enough relatively to N_1, we bound (8.23) by $C\|\Pi_{n_p}u\|_{L^2}$, which essentially shows that B has also H^s smoothness. Remark that it was essential for this proof to work that we had in (8.20) a lower bound in terms of $\mu(n_1, \ldots, n_p)$ and not $\max(n_1, \ldots, n_p)$ in order to lose in (8.23) derivatives only on low frequencies.

The proof of Proposition 8.2 will be a consequence of Diophantine-like estimates.

8.3 PROOF OF LOWER BOUNDS

Set $q = p + 1$, so that q is an odd integer in the case we are treating, remember that we defined $F_m^{q,\ell}(\xi_1, \ldots, \xi_q)$ in (8.16), and set

$$G_m^{q,\ell}(\xi_1, \ldots, \xi_{q+1}) = F_m^{q,\ell}(\xi_1, \ldots, \xi_q) + \xi_{q+1}. \tag{8.24}$$

The first step in the proof of Proposition 8.2 is the following geometrical result:

THEOREM 8.3 *There is a zero measure subset $\mathcal{N}$ of $]0, +\infty[$ and for any $m \in]0, +\infty[- \mathcal{N}$ there are $c > 0$, $N_0 \in \mathbb{N}$ such that*

$$
\begin{aligned}
|F_m^{q,\ell}(\xi_1, \ldots, \xi_q)| &\geq c(1 + |\xi_1| + \cdots + |\xi_q|)^{-N_0} \\
|G_m^{q,\ell}(\xi_1, \ldots, \xi_{q+1})| &\geq c(1 + |\xi_1| + \cdots + |\xi_{q+1}|)^{-N_0}
\end{aligned}
\tag{8.25}
$$

for any $\xi_1, \ldots, \xi_q \in \mathcal{H} \overset{\text{def}}{=} \{\sqrt{n(n + d - 2)}, n \in \mathbb{N}\}, \xi_{q+1} \in \mathbb{Z}$.

Let us make a few comments on this theorem: in general, functions $F_m^{q,\ell}, G_m^{q,\ell}$ have a nonempty zero set on $\mathbb{R}^q, \mathbb{R}^{q+1}$. The theorem asserts that, when q is odd, we can choose m outside a set of null measure, in such a way that these zero sets do not contain points in the discrete subset at which we want to estimate our functions. Moreover, there is enough distance between these two sets so that (8.25) holds true.

We refer to [4] for a proof of Theorem 8.3. By elementary measure theory Theorem 8.3 follows from estimates for the volume of tubes given by subanalytic functions depending on parameters. Subanalytic geometry allows one to deduce these volume estimates from Łojaciewiecz inequalities.

Let us show how Theorem 8.3 implies Proposition 8.2. Note first that we have for products of spherical harmonics the inclusion

$$E_p \cdot E_q \subset \bigoplus_{|p-q| \leq n \leq p+q} E_n. \tag{8.26}$$

Applying this property several times, one sees that if the inequalities

$$
\begin{aligned}
n_{p+1} &\leq n_1 + n_2 + \cdots + n_p \\
n_1 &\leq n_2 + \cdots + n_p + n_{p+1} \\
&\;\;\vdots \\
n_p &\leq n_{p+1} + n_1 + \cdots + n_{p-1}
\end{aligned}
\tag{8.27}
$$

do not hold true, then for any function $u \in L^2$, $\Pi_{n_{p+1}}(\Pi_{n_1} u \cdots \overline{\Pi_{n_p} u}) \equiv 0$. The right way to interpret condition (8.19) in the statement of Proposition 8.2 is thus to say that (8.20) should hold true under conditions (8.27).

By (8.25) and the first relation (8.27) we have, using $\lambda_{n_j} \sim n_j$,

$$
\begin{aligned}
|F_m^{p+1,\ell}(\lambda_{n_1}, \ldots, \lambda_{n_{p+1}})| &\geq c(1 + n_1 + \cdots + n_{p+1})^{-N_0} \\
&\geq c(1 + n_1 + \cdots + n_p)^{-N_0},
\end{aligned}
\tag{8.28}
$$

and to get (8.20) we need to replace the right-hand side by $c\mu(n_1, \ldots, n_p)^{-N_1}$.

CASE 8.1 *Assume that for some* $\delta > 0$, $\mu(n_1, \ldots, n_p) \geq \delta(1 + n_1 + \cdots + n_p)^\delta$. *Then* (8.20) *follows from* (8.28) *with* $N_1 = N_0/\delta$.

CASE 8.2 *Assume that*

$$\mu(n_1, \ldots, n_p) < \delta(1 + n_1 + \cdots + n_p)^\delta, \tag{8.29}$$

and split the argument considering the cases when $\max(n_1, \ldots, n_p)$ *is reached at* n_j *with either* $j \leq \ell$ *or* $j > \ell$.

- If $\max(n_1, \ldots, n_p) = n_p$, then $\mu(n_1, \ldots, n_p) \sim 1 + n_1 + \cdots + n_{p-1}$ and we write

$$F_m^{p+1,\ell} = \left[\sum_1^\ell \sqrt{m^2 + \lambda_{n_j}^2} - \sum_{\ell+1}^{p-1} \sqrt{m^2 + \lambda_{n_j}^2} \right] - \left(\sqrt{m^2 + \lambda_{n_p}^2} + \sqrt{m^2 + \lambda_{n_{p+1}}^2} \right).$$

Since the term between brackets is $O(\mu) = O(\delta n_p^\delta)$ by (8.29), we get for δ small enough $|F_m^{p+1,\ell}| \geq c n_p \to +\infty$ if $n_p \to +\infty$, which is much better than the wanted estimate (8.20).

- If $\max(n_1, \ldots, n_p) = n_1$, then $\mu(n_1, \ldots, n_p) \sim 1 + n_2 + \cdots + n_p$ and we write

$$F_m^{p+1,\ell} = \left[\sqrt{m^2 + \lambda_{n_1}^2} - \sqrt{m^2 + \lambda_{n_{p+1}}^2} \right] + F_m^{p-1,\ell-1}(\lambda_{n_2}, \ldots, \lambda_{n_p}). \tag{8.30}$$

Using the explicit value of λ_{n_j} we expand the term between brackets as

$$(n_1 - n_{p+1}) + O(1/n_1) + O(1/n_{p+1}), \quad n_1, n_{p+1} \to +\infty \tag{8.31}$$

so that plugging into (8.30) and using the notation defined by (8.24)

$$F_m^{p+1,\ell}(\lambda_{n_1}, \ldots, \lambda_{n_{p+1}}) = G_m^{p-1,\ell-1}(\lambda_{n_2}, \ldots, \lambda_{n_p}, n_1 - n_{p+1}) \tag{8.32}$$
$$+ O(1/n_1) + O(1/n_{p+1}).$$

Since $\lambda_{n_j} \in \mathcal{H}$, $n_1 - n_{p+1} \in \mathbb{Z}$, theorem 8.3 gives for $|G_m^{p-1,\ell-1}|$ a lower bound of type $C(1 + n_2 + \cdots + n_p + |n_1 - n_{p+1}|)^{-N_0}$. But (8.27) implies that $|n_1 - n_{p+1}| \leq n_2 + \cdots + n_p$ whence

$$|G_m^{p-1,\ell-1}| \geq c(1 + n_2 + \cdots + n_p)^{-N_0} \sim \mu(n_1, \ldots, n_p)^{-N_0}. \tag{8.33}$$

If we remember (8.29), we see that $\frac{1}{n_1} = O(c_\delta \mu^{-1/\delta})$ with $c_\delta \to 0$ if $\delta \to 0$, so for δ small enough the remainders in the right-hand side of (8.32) are harmless, and we get for $|F_m^{p+1,\ell}|$ a lower bound as in (8.33). This concludes the proof of Proposition 8.2.

8.4 THE CASE OF ODD NONLINEARITIES, AND FINAL COMMENTS

In the preceding section, the assumption that p was even (and so $q = p + 1$ odd) was essential to get Theorem 8.3. If p is odd, $q = p + 1$ and $\ell = \frac{q}{2}$, the function

$F_m^{q,\ell}(\xi_1, \ldots, \xi_q)$ of (8.25) vanishes for any value of m on a subset of $\mathcal{H} \times \cdots \times \mathcal{H}$: for instance $F_m^{4,2}(\xi_1, \xi_2, \xi_3, \xi_4) \equiv 0$ if $(\xi_1^2 = \xi_3^2, \xi_2^2 = \xi_4^2)$ or $(\xi_1^2 = \xi_4^2, \xi_2^2 = \xi_3^2)$. In other words, when p is odd, $\ell = \frac{p+1}{2}$, we have resonances preventing us from eliminating the whole homogeneous part of degree p in the right-hand side of (8.7). If we look for an operator B of type (8.13), the best we can do is to reduce (8.7) to an equation of type

$$(D_t - \Lambda_m)v = M'(v, \ldots, \bar{v}) + O(v^{p+1})$$
$$v|_{t=0} = v_0,$$
(8.34)

where M' is the contribution to the homogeneous part of degree p that cannot be eliminated. Remark, nevertheless, that if M' satisfies an L^2 antisymmetry property, of type

$$\mathrm{Im} \int_{\mathbb{S}^{d-1}} M'(v, \ldots, \bar{v}) \bar{v} \, dx = 0,$$
(8.35)

the M' contribution will disappear in an L^2-energy inequality. In the same way, an H^s-antisymmetry property makes M' disappear in any H^s-energy inequality, whence solutions to (8.34) on an interval of length c/ϵ^p by the energy method. It turns out that condition (8.3) on f_k (k odd) implies that the reduced system (8.6) satisfies such an antisymmetry property. This allows one to treat such nonlinearities as we did for p even in Section 8.3.

Let us remark that there are severe restrictions to extending our strategy of Sections 8.2 and 8.3 to more general manifolds. Actually, we used two very specific properties of the sphere: on the one hand, when we use (8.31) we exploit the fact that differences of square roots of eigenvalues stay in a discrete subset—namely, $\mathbb{Z}$—up to some small remainders. Moreover, when we make use of (8.27) we take into account a very special property of eigenfunctions of the sphere. If we wanted to generalize our result, for instance, to Zoll manifolds, we would still know very precisely the location of eigenvalues, but we would lack information on products of eigenfunctions. On the other hand, if we were trying to treat the case of the torus $\mathbb{T}^d$ ($d \geq 2$), we would have very good properties for the product of eigenfunctions, but the differences of square roots of eigenvalues, $|n| - |m|, n, m \in \mathbb{Z}^d$, describe a dense subset in $\mathbb{R}$, which prevents us from proving an analog of Theorem 8.3.

There is nevertheless a special case that we can treat on $\mathbb{T}^d$: this is the case of quadratic nonlinearities. Actually, when $p = 2$ in (8.25), $F_m^{3,\ell}$ has no real zeros when restricted to $\xi_3 = \xi_1 + \xi_2$, and (8.20), with $N_1 = 1$, follows from an explicit calculation. Because of that, and due to the fact that we have a Fourier analysis on $\mathbb{T}^d$, we prove in [4] existence of solutions of a class of quasi-linear Klein-Gordon equations over an interval of time of size c/ϵ^2. In the semilinear case, a (more complicated) proof of this fact has already been published in [1].

REFERENCES

[1] J.-M. Delort, *Temps d'existence pour l'équation de Klein-Gordon semi-linéaire à données petites périodiques*, Amer. J. Math. 120 (1998), no. 3, 663–689.

[2] J.-M. Delort, *Existence globale et comportement asymptotique pour l'équation de Klein-Gordon quasi linéaire à données petites en dimension 1*, Ann. Sci. École Norm. Sup. (4) 34 (2001), no. 1, 1–61.

[3] J.-M. Delort, D. Fang, and R. Xue, *Global existence of small solutions for quadratic quasi-linear Klein-Gordon systems in two space dimensions*, J. Funct. Anal. 211 (2004), no. 2, 288–323.

[4] J.-M. Delort and J. Szeftel, *Long time existence for small data nonlinear Klein-Gordon equations on tori and spheres*, Int. Math. Res. Not. 2004, no. 37, 1897–1966.

[5] S. Klainerman, *Global existence of small amplitude solutions to nonlinear Klein-Gordon equations in four space-time dimensions*, Comm. Pure Appl. Math. 38 (1985), 631–641.

[6] K. Moriyama, S. Tonegawa, and Y. Tsutsumi, *Almost global existence of solutions for the quadratic semi-linear Klein-Gordon equation in one space dimension*, Funkcialaj Ekvacioj 40, no. 2 (1997), 313–333.

[7] T. Ozawa, K. Tsutaya, and Y. Tsutsumi, *Global existence and asymptotic behavior of solutions for the Klein-Gordon equations with quadratic nonlinearity in two space dimensions*, Math. Z. 222 (1996), 341–362.

[8] J. Shatah, *Normal forms and quadratic nonlinear Klein-Gordon equations*, Comm. Pure Appl. Math. 38 (1985), 685–696.

[9] J.C.H. Simon and E. Taflin, *The Cauchy problem for nonlinear Klein-Gordon equations*, Commun. Math. Phys. 152 (1993), 433–478.

[10] B. Yordanov, *Blow-up for the one-dimensional Klein-Gordon equation with a cubic nonlinearity*, preprint (1996).

Added on proof: An extension of Theorem 8.1 to Zoll manifolds has been published by the authors in: *Long-time existence for semi-linear Klein-Gordon equations with small Cauchy data on Zoll manifolds*, Amer. J. Math. 128 (2006), 1187–1218.

Chapter Nine

Local and Global Wellposedness
of Periodic KP-I Equations

A. D. Ionescu and C. E. Kenig

9.1 INTRODUCTION

Let $\mathbb{T} = \mathbb{R}/(2\pi\mathbb{Z})$. In this paper we consider the Kadomstev-Petviashvili initial value problems

$$\begin{cases} \partial_t u + P(\partial_x)u - \partial_x^{-1}\partial_y^2 u + u\partial_x u = 0; \\ u(0) = \phi, \end{cases} \tag{9.1.1}$$

on $\mathbb{T} \times \mathbb{T}$ and $\mathbb{R} \times \mathbb{T}$, where $P(\partial_x) = \partial_x^3$ (the third-order KP-I) or $P(\partial_x) = -\partial_x^5$ (the fifth-order KP-I). Our goal is to prove local and global wellposedness theorems for the third-order KP-I initial value problem on $\mathbb{T} \times \mathbb{T}$ and $\mathbb{R} \times \mathbb{T}$, and for the fifth-order KP-I initial value problem on $\mathbb{R} \times \mathbb{T}$.

KP-I equations, as well as KP-II equations in which the sign of the term $\partial_x^{-1}\partial_y^2 u$ in (9.1.1) is $+$ instead of $-$ arise naturally in physical contexts as models for the propagation of dispersive long waves, with weak transverse effects. The KP-II initial value problems are much better understood from the point of view of wellposedness, due mainly to the X_b^s method of J. Bourgain [2]. For instance, the third-order KP-II initial value problem is globally wellposed in L^2, on both $\mathbb{R} \times \mathbb{R}$ and $\mathbb{T} \times \mathbb{T}$ (see J. Bourgain [2]), as well as in some spaces larger than L^2 (see H. Takaoka and N. Tzvetkov [18] and the references therein). The fifth-order KP-II initial value problem is also globally wellposed in L^2, on both $\mathbb{R} \times \mathbb{R}$ and $\mathbb{T} \times \mathbb{T}$ (see J.-C. Saut and N. Tzvetkov [15]).

On the other hand, it has been shown in [10] that certain KP-I initial value problems are badly behaved with respect to Picard iterative methods in the standard Sobolev spaces, since the flow map fails to be C^2 at the origin in these spaces. Due to this fact, the wellposedness theory of these equations is more limited. For example, global wellposedness of the third-order KP-I initial value problem in the natural energy space $Z_{(3)}^1(\mathbb{R} \times \mathbb{R})$ (see the definition below) remains an open problem. It is known, however, that the third-order KP-I initial value problem on $\mathbb{R} \times \mathbb{R}$ is globally wellposed in the "second" energy space $Z_{(3)}^2(\mathbb{R} \times \mathbb{R})$, as well as locally wellposed in a larger space (using conservation laws, Strichartz estimates, and an interpolation

The first author was supported in part by an NSF grant, an Alfred P. Sloan research fellowship, and a David and Lucile Packard fellowship. The second author was supported in part by an NSF grant.

argument as in [9] and [3]) (see C. E. Kenig [7] and the references therein). On $\mathbb{T} \times \mathbb{T}$, the third-order KP-I initial value problem is known to be globally wellposed in the "third" energy space (see J. Colliander [4]), as well as locally wellposed in a larger space (see R. J. Iorio and W.V.L. Nunes [5]) (using conservation laws and energy estimates). The fifth-order KP-I initial value problem is known to be globally wellposed in the energy spaces $Z_{(5)}^1(\mathbb{R} \times \mathbb{R})$ and $Z_{(5)}^1(\mathbb{T} \times \mathbb{R})$, as well as locally wellposed in larger spaces (using Picard iterative methods) (see J.-C. Saut and N. Tzvetkov [15] and [16]).

To motivate the definition of our Banach spaces, we recall several KP-I conservation laws: for the third-order KP-I equation (on both $\mathbb{T} \times \mathbb{T}$ and $\mathbb{R} \times \mathbb{T}$), the quantities

$$E_{(3)}^0(g) = \int g^2 \, dxdy, \tag{9.1.2}$$

$$E_{(3)}^1(g) = \frac{1}{2}\int (\partial_x g)^2 \, dxdy + \frac{1}{2}\int (\partial_x^{-1}\partial_y g)^2 \, dxdy - \frac{1}{6}\int g^3 \, dxdy, \tag{9.1.3}$$

and

$$\begin{aligned}
E_{(3)}^2(g) = {} & \frac{3}{2}\int (\partial_x^2 g)^2 \, dxdy + \frac{5}{6}\int (\partial_x^{-2}\partial_y^2 g)^2 \, dxdy + 5\int (\partial_y g)^2 \, dxdy \\
& - \frac{5}{6}\int g^2(\partial_x^{-2}\partial_y^2 g) \, dxdy - \frac{5}{6}\int g(\partial_x^{-1}\partial_y g)^2 \, dxdy \\
& + \frac{5}{4}\int g^2\partial_x^2 g \, dxdy + \frac{5}{24}\int g^4 \, dxdy
\end{aligned} \tag{9.1.4}$$

are formally conserved by the flow, where the integration is over $\mathbb{T} \times \mathbb{T}$ or $\mathbb{R} \times \mathbb{T}$. For the fifth-order KP-I equation on $\mathbb{R} \times \mathbb{T}$, the quantities

$$E_{(5)}^0(g) = \int g^2 \, dxdy \tag{9.1.5}$$

and

$$E_{(5)}^1(g) = \frac{1}{2}\int (\partial_x^2 g)^2 \, dxdy + \frac{1}{2}\int (\partial_x^{-1}\partial_y g)^2 \, dxdy - \frac{1}{6}\int g^3 \, dxdy \tag{9.1.6}$$

are formally conserved.

For $g \in L^2(\mathbb{T} \times \mathbb{T})$ let $\widehat{g}(m, n)$, $m, n \in \mathbb{Z}$, denote its Fourier transform; similarly, for $g \in L^2(\mathbb{R} \times \mathbb{T})$ let $\widehat{g}(\xi, n)$, $\xi \in \mathbb{R}$, $n \in \mathbb{Z}$, denote its Fourier transform. Related to the conservation laws above, we define the energy spaces $Z_{(3)}^s(\mathbb{T} \times \mathbb{T})$, $s = 0, 1, 2$, $Z_{(3)}^s(\mathbb{R} \times \mathbb{T})$, $s = 0, 1, 2$, and $Z_{(5)}^s(\mathbb{R} \times \mathbb{T})$, $s = 0, 1$:

$$\begin{aligned}
Z_{(3)}^s(\mathbb{T} \times \mathbb{T}) = {} & \{g : \mathbb{T} \times \mathbb{T} \to \mathbb{R} : \widehat{g}(0, n) = 0 \text{ for any } n \in \mathbb{Z} \setminus \{0\} \text{ and} \\
& \|g\|_{Z_{(3)}^s} = \|\widehat{g}(m, n)[1 + |m|^s + |n/m|^s]\|_{L^2(\mathbb{Z}\times\mathbb{Z})} < \infty\},
\end{aligned} \tag{9.1.7}$$

$$Z_{(3)}^s(\mathbb{R} \times \mathbb{T}) = \{g : \mathbb{R} \times \mathbb{T} \to \mathbb{R} :$$
$$\|g\|_{Z_{(3)}^s} = \|\widehat{g}(\xi, n)[1 + |\xi|^s + |n/\xi|^s]\|_{L^2(\mathbb{R} \times \mathbb{Z})} < \infty\}, \qquad (9.1.8)$$

and

$$Z_{(5)}^s(\mathbb{R} \times \mathbb{T}) = \{g : \mathbb{R} \times \mathbb{T} \to \mathbb{R} :$$
$$\|g\|_{Z_{(5)}^s} = \|\widehat{g}(\xi, n)[1 + \xi^{2s} + |n/\xi|^s]\|_{L^2(\mathbb{R} \times \mathbb{Z})} < \infty\}. \qquad (9.1.9)$$

Our main theorems concern global well-posedness of the third-order KP-I initial value problem in $Z_{(3)}^2(\mathbb{T} \times \mathbb{T})$ and $Z_{(3)}^2(\mathbb{R} \times \mathbb{T})$, and global wellposedness of the fifth-order KP-I initial value problem in $Z_{(5)}^1(\mathbb{R} \times \mathbb{T})$. We do not know if global wellposedness holds for the fifth-order KP-I initial value problem on $\mathbb{T} \times \mathbb{T}$ (for any type of initial data). For $s \in \mathbb{R}$ let $H^s(\mathbb{T} \times \mathbb{T})$ and $H^s(\mathbb{R} \times \mathbb{T})$, denote the standard Sobolev spaces on $\mathbb{T} \times \mathbb{T}$ and $\mathbb{R} \times \mathbb{T}$.

THEOREM 9.1.1 *Assume that $\phi \in Z_{(3)}^2(\mathbb{T} \times \mathbb{T})$. Then the initial value problem*

$$\begin{cases} \partial_t u + \partial_x^3 u - \partial_x^{-1} \partial_y^2 u + u \partial_x u = 0 \text{ on } \mathbb{T} \times \mathbb{T} \times \mathbb{R}; \\ u(0) = \phi, \end{cases} \qquad (9.1.10)$$

admits a unique solution $u \in C(\mathbb{R} : Z_{(3)}^2(\mathbb{T} \times \mathbb{T})) \cap C^1(\mathbb{R} : H^{-1}(\mathbb{T} \times \mathbb{T}))$. In addition, $u \in L^\infty(\mathbb{R} : Z_{(3)}^2(\mathbb{T} \times \mathbb{T}))$, $\partial_x u \in L^1_{\text{loc}}(\mathbb{R} : L^\infty(\mathbb{T} \times \mathbb{T}))$,

$$E_{(3)}^s(u(t)) = E_{(3)}^s(\phi), \quad s = 0, 1, 2, \quad t \in \mathbb{R}, \qquad (9.1.11)$$

and the mapping $\phi \to u$ is continuous from $Z_{(3)}^2(\mathbb{T} \times \mathbb{T})$ to $C([-T, T] : Z_{(3)}^2(\mathbb{T} \times \mathbb{T}))$ for any $T \in [0, \infty)$.

THEOREM 9.1.2 *Assume that $\phi \in Z_{(3)}^2(\mathbb{R} \times \mathbb{T})$. Then the initial value problem*

$$\begin{cases} \partial_t u + \partial_x^3 u - \partial_x^{-1} \partial_y^2 u + u \partial_x u = 0 \text{ on } \mathbb{R} \times \mathbb{T} \times \mathbb{R}; \\ u(0) = \phi, \end{cases} \qquad (9.1.12)$$

admits a unique solution $u \in C(\mathbb{R} : Z_{(3)}^2(\mathbb{R} \times \mathbb{T})) \cap C^1(\mathbb{R} : H^{-1}(\mathbb{R} \times \mathbb{T}))$. In addition, $u \in L^\infty(\mathbb{R} : Z_{(3)}^2(\mathbb{R} \times \mathbb{T}))$, $\partial_x u \in L^1_{\text{loc}}(\mathbb{R} : L^\infty(\mathbb{R} \times \mathbb{T}))$,

$$E_{(3)}^s(u(t)) = E_{(3)}^s(\phi), \quad s = 0, 1, 2, \quad t \in \mathbb{R}, \qquad (9.1.13)$$

and the mapping $\phi \to u$ is continuous from $Z_{(3)}^2(\mathbb{R} \times \mathbb{T})$ to $C([-T, T] : Z_{(3)}^2(\mathbb{R} \times \mathbb{T}))$ for any $T \in [0, \infty)$.

THEOREM 9.1.3 *Assume that $\phi \in Z_{(5)}^1(\mathbb{R} \times \mathbb{T})$. Then the initial value problem*

$$\begin{cases} \partial_t u - \partial_x^5 u - \partial_x^{-1} \partial_y^2 u + u \partial_x u = 0 \text{ on } \mathbb{R} \times \mathbb{T} \times \mathbb{R}; \\ u(0) = \phi, \end{cases} \qquad (9.1.14)$$

*admits a unique solution $u \in C(\mathbb{R} : Z_{(5)}^1(\mathbb{R} \times \mathbb{T})) \cap C^1(\mathbb{R} : H^{-3}(\mathbb{R} \times \mathbb{T}))$. In
addition, $u \in L^\infty(\mathbb{R} : Z_{(5)}^1(\mathbb{R} \times \mathbb{T}))$, $u, \partial_x u \in L_{loc}^1(\mathbb{R} : L^\infty(\mathbb{R} \times \mathbb{T}))$*

$$E_{(5)}^s(u(t)) = E_{(5)}^s(\phi), \quad s = 0, 1, \ t \in \mathbb{R}, \tag{9.1.15}$$

*and the mapping $\phi \to u$ is continuous from $Z_{(5)}^1(\mathbb{R} \times \mathbb{T})$ to $C([-T, T] : Z_{(5)}^1(\mathbb{R} \times \mathbb{T}))$
for any $T \in [0, \infty$.*

In addition, we prove that sufficiently high Sobolev regularity is globally preserved by the flow (see Theorems 9.5.1, 9.5.2, and 9.5.3). We also prove local
wellposedness theorems below the energy spaces $Z_{(3)}^2(\mathbb{T} \times \mathbb{T})$, $Z_{(3)}^2(\mathbb{R} \times \mathbb{T})$, and
$Z_{(5)}^1(\mathbb{R} \times \mathbb{T})$, respectively (see Theorems 9.6.1, 9.6.2, and 9.6.3).

As in [7], our proofs are based on controlling $\|\partial_x u\|_{L_{loc}^1 L^\infty}$, where u is a solution
of one of the equations (9.1.10), (9.1.12), or (9.1.14). The main difficulty is that
the Strichartz estimates of J.-C. Saut [14] for the free KP-I flow on $\mathbb{R} \times \mathbb{R}$, which
are used in [7], fail in periodic settings. We replace these Strichartz estimates with
certain time-frequency localized Strichartz estimates (see Theorems 9.3.1, 9.3.2, and
9.3.3), which are still sufficient for our purpose. We mention that Picard iterative
methods in the energy spaces do not work to prove Theorems 9.1.1, 9.1.2, and 9.1.3
(N. Tzvetkov, personal communication).

The rest of the paper is organized as follows: in Section 9.2 we summarize our
notation. In Section 9.3 we prove three localized Strichartz estimates for the homogeneous KP-I flow, as well as three corollaries. In Section 9.4 we prove Theorems
9.1.1, 9.1.2, and 9.1.3. In Section 9.5 we prove that sufficiently high Sobolev
regularity is globally preserved by the flow. In Section 9.6 we prove three local
wellposedness theorems. In Appendix A, we review the Kato-Ponce commutator
estimate and a Leibniz rule for fractional derivatives, on both $\mathbb{R}$ and $\mathbb{T}$.

9.2 NOTATION

In this section we summarize some of our notation.

1. The anisotropic Sobolev spaces $H^{s_1, s_2}(\mathbb{T} \times \mathbb{T})$, $s_1, s_2 \geq 0$, are defined by

$$H^{s_1, s_2}(\mathbb{T} \times \mathbb{T}) = \{g \in L^2(\mathbb{T} \times \mathbb{T}) :$$

$$\|g\|_{H^{s_1, s_2}} = \|\widehat{g}(m, n)[(1 + m^2)^{s_1/2} + (1 + n^2)^{s_2/2}]\|_{L^2(\mathbb{Z} \times \mathbb{Z})} < \infty\}. \tag{9.2.1}$$

We also define the standard Sobolev spaces $H^s(\mathbb{T} \times \mathbb{T})$, $s \in \mathbb{R}$, and
$H^\infty(\mathbb{T} \times \mathbb{T})$:

$$H^s(\mathbb{T} \times \mathbb{T}) = \{g \in \mathcal{S}'(\mathbb{T} \times \mathbb{T}) :$$

$$\|g\|_{H^s} = \|\widehat{g}(m, n)[(1 + m^2 + n^2)^{s/2}]\|_{L^2(\mathbb{Z} \times \mathbb{Z})} < \infty\},$$

and

$$H^\infty(\mathbb{T} \times \mathbb{T}) = \cap_{k=0}^\infty H^k(\mathbb{T} \times \mathbb{T}).$$

The Sobolev spaces $H^{s_1, s_2}(\mathbb{R} \times \mathbb{T})$, $s_1, s_2 \geq 0$, $H^s(\mathbb{R} \times \mathbb{T})$, $s \in \mathbb{R}$, and $H^\infty(\mathbb{R} \times \mathbb{T})$ are defined in a similar way.

2. We defined the energy spaces $Z^s_{(3)}(\mathbb{T} \times \mathbb{T})$, $s = 0, 1, 2$, $Z^s_{(3)}(\mathbb{R} \times \mathbb{T})$, $s = 0, 1, 2$, and $Z^s_{(5)}(\mathbb{R} \times \mathbb{T})$, $s = 0, 1$, in (9.1.7), (9.1.8), and (9.1.9).

3. For $s_1, s_2 \geq 0$ and $s \in \mathbb{R}$ we define the Banach spaces

$$Y^{s_1, s_2, s}(\mathbb{T} \times \mathbb{T}) = \{g \in Z^0_{(3)}(\mathbb{T} \times \mathbb{T}) : \|g\|_{Y^{s_1, s_2, s}} = \|\widehat{g}(m, n)$$
$$[(1 + m^2)^{s_1/2} + (1 + n^2)^{s_2/2} + (1 + m^2 + n^2)^{s/2}|n/m|]\|_{L^2(\mathbb{Z} \times \mathbb{Z})} < \infty\},$$
$$\text{(9.2.2)}$$

and

$$Y^{s_1, s_2, s}(\mathbb{R} \times \mathbb{T}) = \{g \in Z^0_{(3)}(\mathbb{R} \times \mathbb{T}) : \|g\|_{Y^{s_1, s_2, s}} = \|\widehat{g}(\xi, n)$$
$$[(1 + \xi^2)^{s_1/2} + (1 + n^2)^{s_2/2} + (1 + \xi^2 + n^2)^{s/2}|n/\xi|]\|_{L^2(\mathbb{R} \times \mathbb{Z})} < \infty\}.$$
$$\text{(9.2.3)}$$

We also define $Y^s(\mathbb{T} \times \mathbb{T}) = Y^{s,s,s}(\mathbb{T} \times \mathbb{T})$, $Y^s(\mathbb{R} \times \mathbb{T}) = Y^{s,s,s}(\mathbb{R} \times \mathbb{T})$, $Y^\infty(\mathbb{T} \times \mathbb{T}) = \cap_{k=0}^\infty Y^k(\mathbb{T} \times \mathbb{T})$, and $Y^\infty(\mathbb{R} \times \mathbb{T}) = \cap_{k=0}^\infty Y^k(\mathbb{R} \times \mathbb{T})$.

4. For $s \in \mathbb{R}$ we define the operators J^s_x, J^s_y, and J^s by

$$\widehat{J^s_x g}(m, n) = (1 + m^2)^{s/2} \widehat{g}(m, n);$$
$$\widehat{J^s_y g}(m, n) = (1 + n^2)^{s/2} \widehat{g}(m, n); \qquad \text{(9.2.4)}$$
$$\widehat{J^s g}(m, n) = (1 + m^2 + n^2)^{s/2} \widehat{g}(m, n)$$

on $S'(\mathbb{T} \times \mathbb{T})$. By a slight abuse of notation, we define the operators J^s_x, J^s_y, and J in the same way on $S'(\mathbb{R} \times \mathbb{T})$.

5. For any set A let $\mathbf{1}_A$ denote its the characteristic function. Given a Banach space X, a measurable function $u : \mathbb{R} \to X$, and an exponent $p \in [1, \infty]$, we define

$$\|u\|_{L^p X} = \left[\int_{\mathbb{R}} (\|u(t)\|_X)^p \, dt \right]^{1/p} \quad \text{if } p \in [1, \infty) \text{ and } \|u\|_{L^\infty X}$$
$$= \text{esssup}_{t \in \mathbb{R}} \|u(t)\|_X.$$

Also, if $I \subseteq \mathbb{R}$ is a measurable set, and $u : I \to X$ is a measurable function, we define

$$\|u\|_{L^p_I X} = \|\mathbf{1}_I(t) u\|_{L^p X}.$$

For $T \geq 0$, we define $\|u\|_{L^p_T X} = \|u\|_{L^p_{[-T, T]} X}$.

9.3 LOCALIZED STRICHARTZ ESTIMATES

In this section we prove three Strichartz estimates, localized in both frequency and time. For $t \in \mathbb{R}$ let $W_{(3)}(t)$ denote the operator on $Y^\infty(\mathbb{R} \times \mathbb{T})$ defined by the Fourier multiplier $(\xi, n) \to e^{i(\xi^3 + n^2/\xi)t}$ and $W_{(5)}(t)$ denote the operator on $Y^\infty(\mathbb{R} \times \mathbb{T})$

defined by the Fourier multiplier $(\xi, n) \rightarrow e^{i(\xi^5 + n^2/\xi)t}$. By a slight abuse of notation we define the operator $W_{(3)}(t)$ in the same way on $Y^\infty(\mathbb{T} \times \mathbb{T})$.

For integers $k = 0, 1, \dots$ we define the operators Q_x^k, Q_y^k, $\tilde{Q}_x^k$, and $\tilde{Q}_y^k$ on $H^\infty(\mathbb{T} \times \mathbb{T})$ by

$$\widehat{Q_x^k g}(m, n) = \mathbf{1}_{[2^{k-1}, 2^k)}(|m|)\widehat{g}(m, n) \text{ if } k \geq 1 \text{ and } \widehat{Q_x^0 g}(m, n)$$
$$= \mathbf{1}_{[0,1)}(|m|)\widehat{g}(m, n),$$

$$\widehat{Q_y^k g}(m, n) = \mathbf{1}_{[2^{k-1}, 2^k)}(|n|)\widehat{g}(m, n) \text{ if } k \geq 1 \text{ and } \widehat{Q_y^0 g}(m, n)$$
$$= \mathbf{1}_{[0,1)}(|n|)\widehat{g}(m, n),$$

$$\tilde{Q}_x^k = \sum_{k'=0}^{k} Q_x^{k'}, \quad \tilde{Q}_y^k = \sum_{k'=0}^{k} Q_y^{k'}, \ k = 0, 1, \dots.$$

By a slight abuse of notation, we define the operators Q_x^k, Q_y^k, $\tilde{Q}_x^k$, and $\tilde{Q}_y^k$ in the same way on $H^\infty(\mathbb{R} \times \mathbb{T})$.

THEOREM 9.3.1 *Assume $\phi \in Y^\infty(\mathbb{T} \times \mathbb{T})$. Then, for any $\epsilon > 0$,*

$$||W_{(3)}(t)\tilde{Q}_y^{2j} Q_x^j \phi||_{L^2_{2^{-j}} L^\infty} \leq C_\epsilon 2^{(3/8+\epsilon)j} ||\tilde{Q}_y^{2j} Q_x^j \phi||_{L^2}, \tag{9.3.1}$$

and

$$||W_{(3)}(t)Q_y^{2j+k} Q_x^j \phi||_{L^2_{2^{-j-k}} L^\infty} \leq C_\epsilon 2^{(3/8+\epsilon)j} ||Q_y^{2j+k} Q_x^j \phi||_{L^2}, \tag{9.3.2}$$

for any integers $j \geq 0$ and $k \geq 1$.

THEOREM 9.3.2 *Assume $\phi \in Y^\infty(\mathbb{R} \times \mathbb{T})$. Then, for any $\epsilon > 0$,*

$$||W_{(3)}(t)\tilde{Q}_y^{2j} Q_x^j \phi||_{L^2_{2^{-j}} L^\infty} \leq C_\epsilon 2^{\epsilon j} ||\tilde{Q}_y^{2j} Q_x^j \phi||_{L^2}, \tag{9.3.3}$$

and

$$||W_{(3)}(t)Q_y^{2j+k} Q_x^j \phi||_{L^2_{2^{-j-k}} L^\infty} \leq C_\epsilon 2^{\epsilon j} ||Q_y^{2j+k} Q_x^j \phi||_{L^2}, \tag{9.3.4}$$

for any integers $j \geq 0$ and $k \geq 1$.

THEOREM 9.3.3 *Assume $\phi \in Y^\infty(\mathbb{R} \times \mathbb{T})$. Then, for any $\epsilon > 0$,*

$$||W_{(5)}(t)\tilde{Q}_y^{3j} Q_x^j \phi||_{L^2_{2^{-2j}} L^\infty} \leq C_\epsilon 2^{(-1/2+\epsilon)j} ||\tilde{Q}_y^{3j} Q_x^j \phi||_{L^2}, \tag{9.3.5}$$

and

$$||W_{(5)}(t)Q_y^{3j+k} Q_x^j \phi||_{L^2_{2^{-2j-k}} L^\infty} \leq C_\epsilon 2^{(-1/2+\epsilon)j} ||Q_y^{3j+k} Q_x^j \phi||_{L^2}. \tag{9.3.6}$$

for any integers $j \geq 0$ and $k \geq 1$.

Proof of Theorem 9.3.1. We first prove (9.3.1). Let $a(m, n) = (\widetilde{Q}_y^{2j} Q_x^j \phi)\widehat{\;}(m, n)$, so $a(0, n) = 0$. Let $\psi_0 : \mathbb{R} \to [0, 1]$ denote a smooth even function supported in the interval $[-2, 2]$ and equal to 1 in the interval $[-1, 1]$ and $\psi_1 : \mathbb{R} \to [0, 1]$ a smooth even function supported in the set $\{|r| \in [1/4, 4]\}$ and equal to 1 in the set $\{|r| \in [1/2, 2]\}$. Then,

$$W_{(3)}(t)\widetilde{Q}_y^{2j} Q_x^j \phi(x, y) = C \sum_{m,n\in\mathbb{Z}} a(m, n)\psi_1(m/2^j)\psi_0(n/2^{2j})e^{i(mx+ny+F(m,n)t)},$$

where

$$F(m, n) = n^2/m + m^3. \tag{9.3.7}$$

Thus, for (9.3.1), it suffices to prove that

$$\left\|\mathbf{1}_{[0,2^{-j}]}(|t|) \sum_{m,n\in\mathbb{Z}} a(m, n)\psi_1(m/2^j)\psi_0(n/2^{2j})e^{i(mx(t)+ny(t)+F(m,n)t)}\right\|_{L_t^2}$$

$$\leq C_\epsilon 2^{(3/8+\epsilon)j}\|a\|_{L^2(\mathbb{Z}\times\mathbb{Z})}$$

for any measurable functions $x, y : [-2^{-j}, 2^{-j}] \to \mathbb{T}$. By duality, this is equivalent to proving that

$$\left\|\int_{\mathbb{R}} g(t)\mathbf{1}_{[0,2^{-j}]}(|t|)\psi_1(m/2^j)\psi_0(n/2^{2j})e^{i(mx(t)+ny(t)+F(m,n)t)}\,dt\right\|_{L^2(\mathbb{Z}\times\mathbb{Z})}$$

$$\leq C_\epsilon 2^{(3/8+\epsilon)j}\|g\|_{L_t^2},$$

for any $g \in L^2(\mathbb{R})$. By expanding the L^2 norm in the left-hand side, it suffices to prove that

$$\left|\int_{\mathbb{R}}\int_{\mathbb{R}} g(t)g(t')K_j(t, t')\,dt\,dt'\right| \leq C_\epsilon 2^{(3/4+2\epsilon)j}\|g\|_{L_t^2}^2, \tag{9.3.8}$$

where

$$K_j(t, t') = \mathbf{1}_{[0,2^{-j}]}(|t|)\mathbf{1}_{[0,2^{-j}]}(|t'|) \sum_{m,n\in\mathbb{Z}} \psi_j^2(m)\psi_0^2(n/2^{2j})$$

$$\times e^{i[m(x(t)-x(t'))+n(y(t)-y(t'))+F(m,n)(t-t')]}.$$

For integers $l \geq j$ let

$$K_j^l(t, t') = \mathbf{1}_{[2^{-l},2\cdot2^{-l}]}(|t - t'|)K_j(t, t').$$

Clearly, for (9.3.8), it suffices to prove that

$$|K_j^l(t, t')| \leq C_\epsilon 2^l 2^{(3/4+2\epsilon)j}2^{(j-l)/10}$$

for any $l \geq j$. To summarize, it suffices to prove that

$$\left|\sum_{m,n\in\mathbb{Z}} \psi_1^2(m/2^j)\psi_0^2(n/2^{2j})e^{i(mx+ny+F(m,n)t)}\right| \leq C_\epsilon 2^l 2^{(3/4+2\epsilon)j}2^{(j-l)/10} \tag{9.3.9}$$

for any $x, y \in [0, 2\pi)$, $|t| \in [2^{-l}, 2\cdot2^{-l}], l \geq j$.

To prove (9.3.9), we use the formula (9.3.7) and estimate the summation in n first. Let

$$A_j(m, t, y) = \sum_{n \in \mathbb{Z}} \psi_0^2(n/2^{2j}) e^{i(ny + n^2 t/m)}.$$

We use the Poisson summation formula

$$\sum_{n \in \mathbb{Z}} F(n) = \sum_{\nu \in \mathbb{Z}} \widehat{F}(2\pi \nu) \tag{9.3.10}$$

for any $F \in \mathcal{S}(\mathbb{R})$. Thus,

$$A_j(m, t, y) = \sum_{\nu \in \mathbb{Z}} \int_{\mathbb{R}} \psi_0^2(\eta/2^{2j}) e^{i(\eta(y + 2\pi \nu) + \eta^2 t/m)} \, d\eta.$$

Since $y \in [0, 2\pi)$, $\eta \le 2^{2j+1}$, $t \le 2 \cdot 2^{-j}$, and $|m| \ge 2^{j-1}$, it follows by integration by parts that

$$\left| \int_{\mathbb{R}} \psi_0^2(\eta/2^{2j}) e^{i(\eta(y + 2\pi \nu) + \eta^2 t/m)} \, d\eta \right| \le C/\nu^2$$

if $|\nu| \ge 100$. Thus,

$$A_j(m, t, y) = \sum_{|\nu| \le 100} \int_{\mathbb{R}} \psi_0^2(\eta/2^{2j}) e^{i(\eta(y + 2\pi \nu) + \eta^2 t/m)} \, d\eta + O(1). \tag{9.3.11}$$

The term $O(1)$ in the right-hand side of (9.3.11) contributes $O(2^j)$ after taking the summation in m, which is dominated by the right-hand side of (9.3.9). Thus, for (9.3.9) it suffices to prove that

$$\left| \sum_{m \in \mathbb{Z}} \psi_1^2(m/2^j) e^{i(mx + m^3 t)} \int_{\mathbb{R}} \psi_0^2(\eta/2^{2j}) e^{i(\eta y' + \eta^2 t/m)} \, d\eta \right| \tag{9.3.12}$$

$$\le C_\epsilon 2^l 2^{(3/4 + 2\epsilon)j} 2^{(j-l)/10},$$

for any $l \ge j$, $x, y' \in \mathbb{R}$, $|t| \in [2^{-l}, 2 \cdot 2^{-l}]$. We write

$$\int_{\mathbb{R}} \psi_0^2(\eta/2^{2j}) e^{i(\eta y' + \eta^2 t/m)} \, d\eta$$

$$= C|m/t|^{1/2} \int_{\mathbb{R}} \widehat{\psi_0^2}(\xi) e^{-i(y' + \xi/2^{2j})^2 |m|/(4|t|)} \, d\xi. \tag{9.3.13}$$

By substituting into (9.3.12) and using the fact that $\|\widehat{\psi_0^2}\|_{L^1(\mathbb{R})} \le C$, it suffices to prove that

$$\left| \sum_{m \ge 0} (m/2^j)^{1/2} \psi_1^2(m/2^j) e^{i(mx' + m^3 t)} \right| \le C_\epsilon 2^{(3/4 + 2\epsilon)j} 2^{2(l-j)/5} \tag{9.3.14}$$

for any $l \ge j$, $x' \in \mathbb{R}$, $|t| \in [2^{-l}, 2 \cdot 2^{-l}]$.

To prove (9.3.14), we note first that we may assume j large and $l \in [j, 7j/4]$. We use a basic lemma of H. Weyl (see, for example, [13, Chapter 4]):

Lemma 9.3.4 *If* $h(x) = \alpha_d x^d + \cdots + \alpha_1 x + \alpha_0$ *is a polynomial with real coefficients, and* $|\alpha_d - a/q| \le 1/q^2$ *for some* $a \in \mathbb{Z}$, $q \in \mathbb{Z} \setminus \{0\}$ *with* $(a, q) = 1$, *then, for any* $\delta > 0$,

$$\left| \sum_{m=1}^{N} e^{2\pi i h(m)} \right| \le C_\delta N^{1+\delta} [q^{-1} + N^{-1} + q N^{-d}]^{1/2^{d-1}},$$

uniformly in N *and* q.

We also use the following standard consequence of Dirichlet's principle:

Lemma 9.3.5 *For any integer* $\Lambda \ge 1$ *and any* $r \in \mathbb{R}$, *there are integers* $q \in \{1, 2, \ldots, \Lambda\}$ *and* $a \in \mathbb{Z}$, $(a, q) = 1$, *such that*

$$|r - a/q| \le 1/(\Lambda q).$$

To prove (9.3.14) we use the summation by parts formula,

$$\sum_{m=1}^{\infty} a_m b_m = \sum_{N=0}^{\infty} \left(\sum_{m=0}^{N} a_m \right) (b_N - b_{N+1}),$$

for any compactly supported sequences a_m and b_m. Since the function $m \to (m/2^j)^{1/2} \psi_1^2(m/2^j)$ has bounded variation on $\mathbb{Z} \cap [0, \infty)$, for (9.3.14) it suffices to prove that

$$\left| \sum_{m=0}^{N} e^{i(mx' + m^3 t)} \right| \le C_\epsilon 2^{(3/4+2\epsilon)j} \tag{9.3.15}$$

for any integer $N \in [2^{j-1}, 2^{j+1}]$, any $l \in [j, 7j/4]$, $x' \in \mathbb{R}$, and $|t| \in [2^{-l}, 2 \cdot 2^{-l}]$. We fix $\Lambda = 2^{2j}$ and apply Lemma 9.3.5 to $r = t/(2\pi)$. Then,

$$|t/(2\pi) - a/q| \le 1/(2^{2j}q) \tag{9.3.16}$$

for some integers $q \in \{1, 2, \ldots, 2^{2j}\}$ and $a \in \mathbb{Z}$, $(a, q) = 1$. Since $|t|/2\pi \approx 2^{-l}$, $l \in [j, 7j/4]$, and j is large, the denominator q in (9.3.16) is in the interval $[c2^j, 2^{2j}]$ (the restriction $l \le 7j/4$ guarantees that $a/q \ne 0/1$). The bound (9.3.15) then follows from Lemma 9.3.4, by taking $\delta = 2\epsilon$. This completes the proof of (9.3.1).

The proof of (9.3.2) is similar. The bound (9.3.2) is a consequence of the uniform bound,

$$\left| \sum_{m,n} \psi_1^2(m/2^j) \psi_1^2(n/2^{2j+k}) e^{i(mx+ny+F(m,n)t)} \right| \le C_\epsilon 2^l 2^{(3/4+2\epsilon)j} 2^{(j-l)/10},$$

$$\tag{9.3.17}$$

for any $l \ge j+k$, $|t| \in [2^{-l}, 2 \cdot 2^{-l}]$, $x, y \in [0, 2\pi)$, which is the analog of (9.3.9). As before, we estimate the sum in n first. We use the Poisson summation formula and the fact that $n \approx 2^{2j+k}$, $|t/m| \le C2^{-2j-k}$ to replace the sum in n with a sum of C integrals, modulo acceptable errors, as in (9.3.11). Finally, we use an identity similar to (9.3.13) to reduce the proof of (9.3.17) to (9.3.14) for $l \geq j + k$. This completes the proof of Theorem 9.3.1. □

Proof of Theorem 9.3.2. This is similar to the proof of Theorem 9.3.1. We first prove (9.3.3). We may assume $j \geq 1$. Let $a(\xi, n) = (\tilde{Q}_y^{2j} Q_x^j \phi)\hat{}(\xi, n)$. Then,

$$W_{(3)}(t)\tilde{Q}_y^{2j} Q_x^j \phi(x, y) = C \int_{\mathbb{R}} \sum_{n \in \mathbb{Z}} a(\xi, n)\psi_1(\xi/2^j)\psi_0(n/2^{2j})e^{i(\xi x + ny + F(\xi, n)t)} \, d\xi,$$

where

$$F(\xi, n) = n^2/\xi + \xi^3. \tag{9.3.18}$$

The same argument as in the proof of Theorem 9.3.1 shows that, for (9.3.3), it suffices to prove that

$$\left| \int_{\mathbb{R}} \sum_{n \in \mathbb{Z}} \psi_1^2(\xi/2^j)\psi_0^2(n/2^{2j})e^{i(\xi x + ny + F(\xi, n)t)} \, d\xi \right| \leq C2^l \tag{9.3.19}$$

for any $x \in \mathbb{R}$, $y \in [0, 2\pi)$, $|t| \in [2^{-l}, 2 \cdot 2^{-l}]$, $l \in [j, 3j]$. We use the formula (9.3.18), and estimate the summation in n first, using the Poisson summation formula. The same argument as before (see (9.3.11) and (9.3.13)) shows that, for (9.3.19), it suffices to prove that

$$\left| \int_{\mathbb{R}_+} (\xi/2^j)^{1/2}\psi_1^2(\xi/2^j)e^{i(\xi x' + \xi^3 t)} \, d\xi \right| \leq C2^{(l-j)/2} \tag{9.3.20}$$

for any $l \geq j$, $x' \in \mathbb{R}$, $|t| \in [2^{-l}, 2 \cdot 2^{-l}]$. This follows from the van der Corput lemma (see, for instance, [17, Chapter 8]).

The proof of (9.3.4) is similar. The same argument as before shows that the bound (9.3.4) is a consequence of the uniform bound,

$$\left| \int_{\mathbb{R}} \sum_{n} \psi_1^2(\xi/2^j)\psi_1^2(n/2^{2j+k})e^{i(\xi x + ny + F(\xi, n)t)} \, d\xi \right| \leq C2^l, \tag{9.3.21}$$

for any $l \in [j + k, 3j + k]$, $|t| \in [2^{-l}, 2 \cdot 2^{-l}]$, $x \in \mathbb{R}$, $y \in [0, 2\pi)$, which is the analog of (9.3.19). The case $j = 0$ is trivial. If $j \geq 1$, we estimate the sum in n first, using the Poisson summation formula. Since $n \approx 2^{2j+k}$, $|t/\xi| \leq C2^{-2j-k}$, we replace the sum in n with a sum of C integrals, modulo acceptable errors. Finally, we use an identity similar to (9.3.13) to reduce the proof of (9.3.21) to (9.3.20) for $l \geq j + k$. This completes the proof of Theorem 9.3.2. □

Proof of Theorem 9.3.3. This is also similar to the proof of Theorem 9.3.1. We first prove (9.3.5). We may assume $j \geq 1$. Let $a(\xi, n) = (\tilde{Q}_y^{3j} Q_x^j \phi)\hat{}(\xi, n)$. Then

$$W_{(5)}(t)\tilde{Q}_y^{3j} Q_x^j \phi(x, y) = C \int_{\mathbb{R}} \sum_{n \in \mathbb{Z}} a(\xi, n)\psi_1(\xi/2^j)\psi_0(n/2^{3j})e^{i(\xi x + ny + F(\xi, n)t)} \, d\xi,$$

where

$$F(\xi, n) = n^2/\xi + \xi^5. \tag{9.3.22}$$

The same argument as in the proof of Theorem 9.3.1 shows that, for (9.3.5), it suffices to prove that

$$\left| \int_{\mathbb{R}} \sum_{n \in \mathbb{Z}} \psi_1^2(\xi/2^j) \psi_0^2(n/2^{3j}) e^{i(\xi x + ny + F(\xi,n)t)} \, d\xi \right| \leq C2^{l-j} \tag{9.3.23}$$

for any $x \in \mathbb{R}$, $y \in [0, 2\pi)$, $|t| \in [2^{-l}, 2 \cdot 2^{-l}]$, $l \in [2j, 5j]$. We use the formula (9.3.22), and estimate the summation in n first, using the Poisson summation formula. The same argument as before (see (9.3.11) and (9.3.13)) shows that, for (9.3.23), it suffices to prove that

$$\left| \int_{\mathbb{R}_+} (\xi/2^j)^{1/2} \psi_1^2(\xi/2^j) e^{i(\xi x' + \xi^5 t)} \, d\xi \right| \leq C2^{(l-3j)/2} \tag{9.3.24}$$

for any $l \geq 2j$, $x' \in \mathbb{R}$, $|t| \in [2^{-l}, 2 \cdot 2^{-l}]$. This follows from the van der Corput lemma (see, for instance, [17, Chapter 8]).

The proof of (9.3.6) is similar. The same argument as before shows that the bound (9.3.6) is a consequence of the uniform bound,

$$\left| \int_{\mathbb{R}} \sum_n \psi_1^2(\xi/2^j) \psi_1^2(n/2^{3j+k}) e^{i(\xi x + ny + F(\xi,n)t)} \, d\xi \right| \leq C2^{l-j}, \tag{9.3.25}$$

for any $l \in [2j + k, 5j + k]$, $|t| \in [2^{-l}, 2 \cdot 2^{-l}]$, $x \in \mathbb{R}$, $y \in [0, 2\pi)$, which is the analog of (9.3.23). The case $j = 0$ is trivial. If $j \geq 1$, we estimate the sum in n first, using the Poisson summation formula. Since $n \approx 2^{3j+k}$, $|t/\xi| \leq C2^{-3j-k}$ we replace the sum in n with a sum of C integrals, modulo acceptable errors. Finally, we use an identity similar to (9.3.13) to reduce the proof of (9.3.25) to (9.3.24) for $l \geq 2j + k$. This completes the proof of Theorem 9.3.3. □

As in [7] we have the following corollaries of Theorems 9.3.1, 9.3.2, and 9.3.3:

COROLLARY 9.3.6 *Assume $N \geq 4$, $u \in C([-T, T] : Y^{0,0,-N}(\mathbb{T} \times \mathbb{T})) \cap C^1([-T, T] : H^{-N-1}(\mathbb{T} \times \mathbb{T}))$, $f \in C([-T, T] : H^{-N}(\mathbb{T} \times \mathbb{T}))$, $T \in [0, 1/2]$, and*

$$[\partial_t + \partial_x^3 - \partial_x^{-1} \partial_y^2] u = \partial_x f \text{ on } \mathbb{T} \times \mathbb{T} \times [-T, T].$$

Then, for any $s_1 > 7/8$ and $s_2 > 1/2$,

$$\|u\|_{L_T^1 L^\infty} \leq C_{s_1, s_2} T^{1/2} [\|J_x^{s_1} u\|_{L_T^\infty L^2} + \|J_x^{s_1-1} J_y^{s_2} u\|_{L_T^\infty L^2} + \|J_x^{s_1} f\|_{L_T^1 L^2}]. \tag{9.3.26}$$

COROLLARY 9.3.7 *Assume $N \geq 4$, $u \in C([-T, T] : Y^{0,0,-N}(\mathbb{R} \times \mathbb{T})) \cap C^1([-T, T] : H^{-N-1}(\mathbb{R} \times \mathbb{T}))$, $f \in C([-T, T] : H^{-N}(\mathbb{R} \times \mathbb{T}))$, $T \in [0, 1/2]$, and*

$$[\partial_t + \partial_x^3 - \partial_x^{-1} \partial_y^2] u = \partial_x f \text{ on } \mathbb{R} \times \mathbb{T} \times [-T, T].$$

Then, for any $s_1 > 1/2$ and $s_2 > 1/2$,

$$\|u\|_{L_T^1 L^\infty} \leq C_{s_1,s_2} T^{1/2}[\|J_x^{s_1} u\|_{L_T^\infty L^2} + \|J_x^{s_1-1} J_y^{s_2} u\|_{L_T^\infty L^2} + \|J_x^{s_1} f\|_{L_T^1 L^2}]. \tag{9.3.27}$$

COROLLARY 9.3.8 *Assume $N \geq 4$, $u \in C([-T, T] : Y^{0,0,-N}(\mathbb{R} \times \mathbb{T})) \cap C^1([-T, T] : H^{-N-1}(\mathbb{R} \times \mathbb{T}))$, $f \in C([-T, T] : H^{-N}(\mathbb{R} \times \mathbb{T}))$, $T \in [0, 1/2]$, and*

$$[\partial_t - \partial_x^5 - \partial_x^{-1} \partial_y^2] u = \partial_x f \text{ on } \mathbb{R} \times \mathbb{T} \times [-T, T].$$

Then, for any $s_1 > 1/2$ and $s_2 > 1/2$,

$$\|u\|_{L_T^1 L^\infty} \leq C_{s_1,s_2} T^{1/2}[\|J_x^{s_1} u\|_{L_T^\infty L^2} + \|J_x^{s_1-3/2} J_y^{s_2} u\|_{L_T^\infty L^2} + \|J_x^{s_1} f\|_{L_T^1 L^2}]. \tag{9.3.28}$$

The assumption $N \geq 4$ in Corollaries 9.3.6, 9.3.7, and 9.3.8 (and also in Theorems 9.6.1, 9.6.2, and 9.6.3 in Section 9.6) is related to $\partial_t u \in C([-T, T] : H^{-N-1})$.

Proof of Corollary 9.3.6. Without loss of generality, we may assume that $u \in C([-T, T] : Y^\infty(\mathbb{T} \times \mathbb{T})) \cap C^1([-T, T] : H^\infty(\mathbb{T} \times \mathbb{T}))$, and $f \in C([-T, T] : H^\infty(\mathbb{T} \times \mathbb{T}))$. It suffices to prove that if $s_1 = 7/8 + \epsilon$ and $s_2 = 1/2 + \epsilon$, $\epsilon \in (0, 10^{-3}]$, then

$$\|\tilde{Q}_y^{2j} Q_x^j u\|_{L_T^1 L^\infty} \leq C_\epsilon 2^{-\epsilon j/2} T^{1/2}[\|J_x^{s_1} u\|_{L_T^\infty L^2} + \|J_x^{s_1} f\|_{L_T^1 L^2}], \tag{9.3.29}$$

and

$$\|Q_y^{2j+k} Q_x^j u\|_{L_T^1 L^\infty} \leq C_\epsilon 2^{-\epsilon(j+k)/2} T^{1/2}[\|J_x^{s_1-1} J_y^{s_2} u\|_{L_T^\infty L^2} + \|J_x^{s_1} f\|_{L_T^1 L^2}], \tag{9.3.30}$$

for any integers $j \geq 0$ and $k \geq 1$.

For (9.3.29), we partition the interval $[-T, T]$ into 2^j equal intervals of length $2T2^{-j}$, denoted by $[a_{j,l}, a_{j,l+1})$, $l = 1, \ldots, 2^j$. The term in the left-hand side of (9.3.29) is dominated by

$$\sum_{l=1}^{2^j} \|\mathbf{1}_{[a_{j,l},a_{j,l+1})}(t) \tilde{Q}_y^{2j} Q_x^j u\|_{L^1 L^\infty}$$

$$\leq C 2^{-j/2} T^{1/2} \sum_{l=1}^{2^j} \|\mathbf{1}_{[a_{j,l},a_{j,l+1})}(t) \tilde{Q}_y^{2j} Q_x^j u\|_{L^2 L^\infty}. \tag{9.3.31}$$

By Duhamel's formula, for $t \in [a_{j,l}, a_{j,l+1}]$,

$$u(t) = W_{(3)}(t - a_{j,l})[u(a_{j,l})] + \int_{a_{j,l}}^t W_{(3)}(t - s)[\partial_x f(s)] \, ds.$$

It follows from (9.3.1) that

$$\|\mathbf{1}_{[a_{j,l},a_{j,l+1})}(t) \tilde{Q}_y^{2j} Q_x^j u\|_{L^2 L^\infty} \leq C_\epsilon 2^{(3/8+\epsilon/2)j} \|\tilde{Q}_y^{2j} Q_x^j u(a_{j,l})\|_{L^2}$$

$$+ C_\epsilon 2^{(3/8+\epsilon/2)j} 2^j \|\mathbf{1}_{[a_{j,l},a_{j,l+1})}(t) \tilde{Q}_y^{2j} Q_x^j f\|_{L^1 L^2}. \tag{9.3.32}$$

Combining (9.3.31) and (9.3.32), the left-hand side of (9.3.29) is dominated by

$$C_\epsilon 2^{-\epsilon j/2} T^{1/2} \left[2^{-j} \sum_{l=1}^{2^j} ||J_x^{s_1} u(a_{j,l})||_{L^2} + ||J_x^{s_1} f||_{L^1 L^2} \right].$$

This leads to (9.3.29).

The proof of (9.3.30) is similar: the only difference is that we partition the interval $[-T, T]$ into 2^{j+k} equal intervals of length $2T2^{-j-k}$, denoted by $[b_{j,l}, b_{j,l+1})$, $l = 1, \dots, 2^{j+k}$ and use (9.3.2) instead of (9.3.1). The term in the left-hand side of (9.3.30) is dominated by

$$C2^{-(j+k)/2} T^{1/2} \sum_{l=1}^{2^{j+k}} ||\mathbf{1}_{[b_{j,l}, b_{j,l+1})}(t) Q_y^{2j+k} Q_x^j u||_{L^2 L^\infty}. \tag{9.3.33}$$

It follows from Duhamel's formula and (9.3.2) that

$$||\mathbf{1}_{[b_{j,l}, b_{j,l+1})}(t) Q_y^{2j+k} Q_x^j u||_{L^2 L^\infty} \le C_\epsilon 2^{(3/8+\epsilon/2)j} ||Q_y^{2j+k} Q_x^j u(b_{j,l})||_{L^2}$$

$$+ C_\epsilon 2^{(3/8+\epsilon/2)j} 2^j ||\mathbf{1}_{[b_{j,l}, b_{j,l+1})}(t) Q_y^{2j+k} Q_x^j f||_{L^1 L^2}. \tag{9.3.34}$$

Combining (9.3.33) and (9.3.34), the left-hand side of (9.3.30) is dominated by

$$C_\epsilon 2^{-\epsilon(j+k)/2} T^{1/2} \left[2^{-j-k} \sum_{l=1}^{2^{j+k}} ||J_x^{s_1-1} J_y^{s_2} u(b_{j,l})||_{L^2} + ||J_x^{s_1} f||_{L^1 L^2} \right].$$

This leads to (9.3.30). □

Proof of Corollary 9.3.7. This is similar to the proof of Corollary 9.3.6. To control $||\tilde{Q}_y^{2j} Q_x^j u||_{L_T^1 L^\infty}$, we partition the interval $[-T, T]$ into 2^j equal subintervals, and use Duhamel's formula and the bound (9.3.3). To control $||Q_y^{2j+k} Q_x^j u||_{L_T^1 L^\infty}$, we partition the interval $[-T, T]$ into 2^{j+k} equal subintervals, and use Duhamel's formula and the bound (9.3.4). □

Proof of Corollary 9.3.8. This is also similar to the proof of Corollary 9.3.6. To control $||\tilde{Q}_y^{3j} Q_x^j u||_{L_T^1 L^\infty}$, we partition the interval $[-T, T]$ into 2^{2j} equal subintervals, and use Duhamel's formula and the bound (9.3.5). To control $||Q_y^{3j+k} Q_x^j u||_{L_T^1 L^\infty}$, we partition the interval $[-T, T]$ into 2^{2j+k} equal subintervals, and use Duhamel's formula and the bound (9.3.6). □

9.4 PROOF OF THEOREMS 9.1.1, 9.1.2, AND 9.1.3

In this section we prove Theorems 9.1.1, 9.1.2, and 9.1.3. For Theorems 9.1.1 and 9.1.2 we rely on the bounds (9.3.26) and (9.3.27) with $s_1 + 2s_2 = 3$. For Theorem 9.1.3 we rely on the bound (9.3.6) with $s_1 + 3s_2 = 5/2$. We write in detail only the proof of Theorem 9.1.2 and indicate the minor changes needed for Theorems 9.1.1 and 9.1.3. For simplicity of notation, in this section we write H^∞, H^s, H^{s_1,s_2}, Y^∞, Y^s, $Y^{s_1,s_2,s}$, and $Z^s_{(3)}$ instead of $H^\infty(\mathbb{R} \times \mathbb{T})$, $H^s(\mathbb{R} \times \mathbb{T})$, $H^{s_1,s_2}(\mathbb{R} \times \mathbb{T})$, $Y^\infty(\mathbb{R} \times \mathbb{T})$, $Y^s(\mathbb{R} \times \mathbb{T})$, $Y^{s_1,s_2,s}(\mathbb{R} \times \mathbb{T})$, and $Z^s_{(3)}(\mathbb{R} \times \mathbb{T})$, respectively.

For $\epsilon \in (0, 1/2]$, we define the operators R^ϵ on $H^\infty(\mathbb{R} \times \mathbb{T})$ by

$$\widehat{R^\epsilon g}(\xi, n) = [\psi_0(\epsilon \xi) - \psi_0(\xi/\epsilon)]\psi_0(\epsilon n)\widehat{g}(\xi, n), \qquad (9.4.1)$$

where ψ_0 is defined in the proof of Theorem 9.3.1. Clearly, R^ϵ extends to a bounded operator on $L^p(\mathbb{R} \times \mathbb{T})$, $p \in [1, \infty]$, uniformly in ϵ.

We start with a local wellposedness result:

LEMMA 9.4.1 *Assume $\phi \in Y^\infty$. Then there is $T = T(\|\phi\|_{Y^3}) > 0$ and a solution $u \in C([-T, T] : Y^\infty) \cap C^1([-T, T] : H^\infty)$ of the initial value problem*

$$\begin{cases} \partial_t u + \partial_x^3 u - \partial_x^{-1}\partial_y^2 u + u\partial_x u = 0 \text{ in } C([-T, T] : H^\infty); \\ u(0) = \phi. \end{cases} \qquad (9.4.2)$$

Proof of Lemma 9.4.1. This is known. See, for instance, [5, Section 4] for the proof in the case of functions in $Y^\infty(\mathbb{R} \times \mathbb{R})$; the same argument applies in the case of functions in $Y^\infty(\mathbb{R} \times \mathbb{T})$. □

We assume now that $\phi \in Y^\infty \cap Z^2_{(3)}$.

LEMMA 9.4.2 *Assume $u \in C([-T, T] : Y^\infty) \cap C^1([-T, T] : H^\infty)$ is a solution of (9.4.2) and $\phi \in Z^2_{(3)}$. Then,*

$$\|u\|_{L^\infty_T Z^2_{(3)}} \leq \overline{C}(\|\phi\|_{Z^2_{(3)}}) \qquad (9.4.3)$$

for some constant $\overline{C}(\|\phi\|_{Z^2_{(3)}})$ independent of T.

Remark. For Theorem 9.1.3 the bound (9.4.3) is replaced with

$$\|u\|_{L^\infty_T Z^1_{(5)}} \leq \overline{C}(\|\phi\|_{Z^1_{(5)}}).$$

Also, the space $Z^2_{(3)}$ is replaced with $Z^1_{(5)}$ throughout.

Proof of Lemma 9.4.2. This is also known. The main ingredient is the conservation of the energies $E^s_{(3)}$, $s = 0, 1, 2$, defined in (9.1.2), (9.1.3), and (9.1.4),

$$E^s_{(3)}(u(t)) = E^s_{(3)}(\phi), \quad s = 0, 1, 2, \qquad (9.4.4)$$

for any $t \in [-T, T]$. The identities (9.4.4) are justified in [11, section 3] (on $\mathbb{R} \times \mathbb{R}$), using suitable approximations of the solution u and estimating the commutators. The proof in [11, section 3] depends only on the Sobolev imbedding theorem and clearly works in the case of functions defined on $\mathbb{R} \times \mathbb{T}$. Given (9.4.4), the proof of (9.4.3) is standard (see, for instance, [11, Proposition 5]). □

Next, we prove a global bound for $\|u\|_{L^1_T L^\infty} + \|\partial_x u\|_{L^1_T L^\infty}$.

LEMMA 9.4.3 *Assume $u \in C([-T, T] : Y^\infty) \cap C^1([-T, T] : H^\infty)$ is a solution of (9.4.2) and $\phi \in Z_{(3)}^2$. Then there is a constant $\bar{c}(||\phi||_{Z_{(3)}^2})$ such that*

$$\int_{t_1}^{t_2} ||u(t)||_{L^\infty} + ||\partial_x u(t)||_{L^\infty} \, dt \leq 1, \tag{9.4.5}$$

for any $t_1, t_2 \in [-T, T]$ with

$$t_2 - t_1 \leq \bar{c}(||\phi||_{Z_{(3)}^2}). \tag{9.4.6}$$

In particular,

$$||u||_{L_T^1 L^\infty} + ||\partial_x u||_{L_T^1 L^\infty} \leq C(T, ||\phi||_{Z_{(3)}^2}). \tag{9.4.7}$$

Proof of Lemma 9.4.3. First, we observe that

$$Z_{(3)}^2 \hookrightarrow H^{2,1}.$$

We apply Corollary 9.3.7 to both u and $\partial_x u$, with $s_1 = s_2 = 2/3$. We note also that $||J_x^2 g||_{L^2} + ||J_x^{2/3} J_y^{2/3} g||_{L^2} \leq C||g||_{H^{2,1}}$, for any $g \in H^\infty$. It follows from (9.3.27) that

$$\int_{t_1}^{t_2} ||u(t)||_{L^\infty} + ||\partial_x u(t)||_{L^\infty} \, dt$$

$$\leq C(t_2 - t_1)^{1/2} \left[||u||_{L_T^\infty H^{2,1}} + \int_{t_1}^{t_2} ||J_x^2(u^2)(t)||_{L^2} \, dt \right] \leq C(t_2 - t_1)^{1/2} A,$$

where

$$A = ||u||_{L_T^\infty H^{2,1}} + \int_{t_1}^{t_2} ||u^2(t)||_{L^2} + ||(u \partial_x^2 u)(t)||_{L^2} + ||(\partial_x u)^2(t)||_{L^2} \, dt$$

$$\leq C||u||_{L_T^\infty Z_{(3)}^2} \left[1 + \int_{t_1}^{t_2} ||u(t)||_{L^\infty} + ||\partial_x u(t)||_{L^\infty} \, dt \right].$$

The inequality (9.4.5) follows from (9.4.3) and the restriction (9.4.6). □

We can now control $||u||_{L_T^\infty Y^s}$, $s > 7$.

LEMMA 9.4.4 *Assume $u \in C([-T, T] : Y^\infty) \cap C^1([-T, T] : H^\infty)$ is a solution of (9.4.2) and $\phi \in Z_{(3)}^2$. Then, for $s > 7$,*

$$||u||_{L_T^\infty Y^s} \leq C(s, T, ||\phi||_{Z_{(3)}^2 \cap Y^s}). \tag{9.4.8}$$

Proof of Lemma 9.4.4. We show first that for any $s \geq 1$,

$$||J_x^s u||_{L_T^\infty L^2} \leq C(s, T, ||\phi||_{Z_{(3)}^2}) ||J_x^s \phi||_{L^2}. \tag{9.4.9}$$

We start from the identity (9.4.2), apply J_x^s, multiply by $2J_x^s u$, and integrate in x and y. The result is

$$\partial_t \int_{\mathbb{R} \times \mathbb{T}} (J_x^s u(t))^2 \, dx dy = - \int_{\mathbb{R} \times \mathbb{T}} J_x^s(u(t)\partial_x u(t)) \cdot J_x^s u(t) \, dx dy.$$

We apply the Kato-Ponce commutator estimate (Lemma 9.A.1) with $f = u(t)$ and $g = \partial_x u(t)$. Integration by parts then shows that

$$\partial_t \|J_x^s u(t)\|_{L^2}^2 \le C_s \|J_x^s u(t)\|_{L^2}^2 (\|u(t)\|_{L^\infty} + \|\partial_x u(t)\|_{L^\infty}). \tag{9.4.10}$$

The bound (9.4.9) then follows from (9.4.7).

A similar argument, using the operator ∂_y instead of J_x^s, shows that

$$\|J_y^1 u\|_{L_T^\infty L^2} \le C(T, \|\phi\|_{Z_{(3)}^2}) \|J_y^1 \phi\|_{L^2}. \tag{9.4.11}$$

Next, we show that for any $\epsilon > 0$,

$$\|J_x^{3+\epsilon} u\|_{L_T^\infty L^\infty} \le C(\epsilon, T, \|\phi\|_{Z_{(3)}^2 \cap H^{7+3\epsilon,1}}). \tag{9.4.12}$$

In view of (9.4.9) and (9.4.11), it suffices to prove that

$$\|J_x^{3+\epsilon} g\|_{L^\infty} \le C_\epsilon \|g\|_{H^{7+3\epsilon,1}}, \ \epsilon > 0 \tag{9.4.13}$$

for any $g \in Y^\infty$. This is elementary:

$$\|J_x^{3+\epsilon} g\|_{L^\infty} \le C \sum_{j,k=0}^{\infty} 2^{(3+\epsilon)j} \|\widehat{Q_x^j Q_y^k g}\|_{L^1} \le C \sum_{j,k=0}^{\infty} 2^{(3+\epsilon)j} 2^{(j+k)/2} \|\widehat{Q_x^j Q_y^k g}\|_{L^2}$$

$$\le C\|g\|_{H^{7+3\epsilon,1}} \sum_{j,k=0}^{\infty} 2^{(7j+2\epsilon j+k)/2} (2^{(7+3\epsilon)j} + 2^k)^{-1},$$

which gives (9.4.13).

Next, we show that for $s > 7$,

$$\|\partial_x \partial_y u\|_{L_T^\infty L^2} \le C(s, T, \|\phi\|_{Z_{(3)}^2 \cap H^s}). \tag{9.4.14}$$

We start from the identity (9.4.2), apply $\partial_x \partial_y$, multiply by $2\partial_x \partial_y u$, and integrate in x and y. After integration by parts, the result is

$$\partial_t \|\partial_x \partial_y u(t)\|_{L^2}^2 \le C \left| \int_{\mathbb{R} \times \mathbb{T}} \partial_x u(t) \cdot (\partial_x \partial_y u(t))^2 \, dx dy \right|$$

$$+ C \left| \int_{\mathbb{R} \times \mathbb{T}} \partial_x^2 u(t) \cdot \partial_y u \cdot \partial_x \partial_y u(t) \, dx dy \right|$$

$$\le C \|\partial_x u(t)\|_{L^\infty} \|\partial_x \partial_y u(t)\|_{L^2}^2 + C \|\partial_x^2 u(t)\|_{L^\infty} \|\partial_y u(t)\|_{L^2} \|\partial_x \partial_y u(t)\|_{L^2}.$$

The bound (9.4.14) follows from (9.4.11) and (9.4.12).

A similar argument shows that for $s > 7$,

$$||\partial_x^2 \partial_y u||_{L_T^\infty L^2} \leq C(s, T, ||\phi||_{Z_{(3)}^2 \cap H^s}).   \tag{9.4.15}$$

We also note that, for any $t \in [-T, T]$,

$$||\partial_y u(t)||_{L^\infty} \leq C(||\partial_y^2 u(t)||_{L^2} + ||\partial_x^2 \partial_y u(t)||_{L^2})   \tag{9.4.16}$$

(see the proof of (9.4.13)). We apply ∂_y^2 to the identity (9.4.2), multiply by $2\partial_y^2 u$, and integrate in x and y. After integration by parts, the result is

$$\partial_t ||\partial_y^2 u(t)||_{L^2}^2 \leq C||\partial_x u(t)||_{L^\infty} ||\partial_y^2 u(t)||_{L^2}^2$$
$$+ C||\partial_y u(t)||_{L^\infty} ||\partial_x \partial_y u(t)||_{L^2} ||\partial_y^2 u(t)||_{L^2}$$
$$\leq C(s, T, ||\phi||_{Z_{(3)}^2 \cap H^s})(||\partial_y^2 u(t)||_{L^2}^2 + 1),$$

by using (9.4.12), (9.4.14), (9.4.15), and (9.4.16). It follows that

$$||\partial_y^2 u||_{L_T^\infty L^2} \leq C(s, T, ||\phi||_{Z_{(3)}^2 \cap H^s}),$$

for $s > 7$. Thus, using (9.4.15) and (9.4.16),

$$||\partial_y u||_{L_T^\infty L^\infty} \leq C(s, T, ||\phi||_{Z_{(3)}^2 \cap H^s}), \quad s > 7.   \tag{9.4.17}$$

An argument as in the proof of (9.4.9) shows that for any $s \geq 1$ and $t \in [0, T]$,

$$||J_y^s u(t)||_{L^2}^2 \leq C_s \int_0^t ||u(t')||_{Y^s}^2 (||u(t')||_{L^\infty}$$
$$+ ||\partial_x u(t')||_{L^\infty} + ||\partial_y u(t')||_{L^\infty}) \, dt'.   \tag{9.4.18}$$

We show now that for any $s \geq 1$ and $t \in [0, T]$,

$$||J_x^s (\partial_x^{-1} \partial_y) u(t)||_{L^2}^2$$
$$\leq C_s \int_0^t ||u(t')||_{Y^s}^2 (||u(t')||_{L^\infty} + ||\partial_x u(t')||_{L^\infty} + ||\partial_y u(t')||_{L^\infty}) \, dt'.   \tag{9.4.19}$$

We start from the identity (9.4.2), apply $J_x^s (\partial_x^{-1} \partial_y) R^\epsilon$, multiply by $2J_x^s (\partial_x^{-1} \partial_y) R^\epsilon u$, and integrate in x and y. The result is

$$\partial_t ||J_x^s (\partial_x^{-1} \partial_y) R^\epsilon u(t)||_{L^2}^2 \leq C \left| \int_{\mathbb{R} \times \mathbb{T}} J_x^s R^\epsilon (u(t) \partial_y u(t)) \cdot J_x^s (\partial_x^{-1} \partial_y) R^\epsilon u(t) \, dx dy \right|$$

$$\leq C \left| \int_{\mathbb{R} \times \mathbb{T}} J_x^s [u(t) R^\epsilon \partial_y u(t)] \cdot J_x^s (\partial_x^{-1} \partial_y) R^\epsilon u(t) \, dx dy \right|$$
$$+ C||J_x^s (\partial_x^{-1} \partial_y) R^\epsilon u(t)||_{L^2} ||J_x^s [u(t) R^\epsilon \partial_y u(t) - R^\epsilon (u(t) \partial_y u(t))]||_{L^2}.$$

Using the Kato-Ponce inequality (Lemma 9.A.1) and the Sobolev imbedding theorem, it follows easily that

$$\partial_t ||J_x^s (\partial_x^{-1} \partial_y) R^\epsilon u(t)||_{L^2}^2 \leq C_s ||u(t)||_{Y^s}^2 (||u(t)||_{L^\infty} + ||\partial_x u(t)||_{L^\infty} + ||\partial_y u(t)||_{L^\infty})$$
$$+ C_s ||u(t)||_{Y^s} [||(R^\epsilon - I) J^{s+3} u(t)||_{L^2} ||J^{s+3} u(t)||_{L^2}$$
$$+ ||(R^\epsilon - I) J^{s+1} (u^2(t))||_{L^2}].$$

We integrate this inequality from 0 to t and let $\epsilon \to 0$. The bound (9.4.19) follows from the Lebesgue dominated convergence theorem.

A similar argument shows that for any $s \geq 1$ and $t \in [0, T]$,

$$\|J_y^s(\partial_x^{-1}\partial_y)u(t)\|_{L^2}^2$$

$$\leq C_s \int_0^t \|u(t')\|_{Y^s}^2 (\|u(t')\|_{L^\infty} + \|\partial_x u(t')\|_{L^\infty} + \|\partial_y u(t')\|_{L^\infty})\, dt'.$$
(9.4.20)

The bound (9.4.8) now follows from (9.4.9), (9.4.17), (9.4.18), (9.4.19), and (9.4.20). ☐

The following corollary follows from Lemma 9.4.1 and Lemma 9.4.4 :

COROLLARY 9.4.5 *Assume $\phi \in Y^\infty \cap Z_{(3)}^2$. Then there is a global solution $u \in C(\mathbb{R} : Y^\infty) \cap C^1(\mathbb{R} : H^\infty)$ of the initial value problem (9.4.2) with*

$$\|u\|_{L^\infty Z_{(3)}^2} \leq \overline{C}(\|\phi\|_{Z_{(3)}^2}).$$

We turn now to the proof of Theorem 9.1.2. For later use, we prove a stronger uniqueness property.

LEMMA 9.4.6 *Assume $N \geq 4$ and $u_1, u_2 \in C([a, b] : Y^{2s,s,-N}) \cap C^1([a, b] : H^{-N-1})$, $s > 3/4$, are solutions of the equation*

$$\partial_t u + \partial_x^3 u - \partial_x^{-1}\partial_y^2 u + u\partial_x u = 0 \text{ on } \mathbb{R} \times \mathbb{T} \times [a, b],$$
(9.4.21)

and $u_1(t_0) = u_2(t_0)$ for some $t_0 \in [a, b]$. Then $u_1 \equiv u_2$.

Remark. For Theorems 9.1.1 and 9.1.3, the space $C([a, b] : Y^{3/2+,3/4+,-N}) \cap C^1([a, b] : H^{-N-1})$ has to be replaced with $C([a, b] : Y^{15/8+,15/16+,-N}) \cap C^1([a, b] : H^{-N-1})$ and $C([a, b] : Y^{3/2+,1/2+,-N}) \cap C^1([a, b] : H^{-N-1})$, respectively. Compare with the local wellposedness theorems 9.6.1, 9.6.2, and 9.6.3.

Proof of Lemma 9.4.6. Clearly,

$$\|u_1\|_{L^\infty_{[a,b]}Y^{2s,s,-N}} + \|u_2\|_{L^\infty_{[a,b]}Y^{2s,s,-N}} \leq C_0.$$
(9.4.22)

Thus,

$$\|u_1\|_{L^\infty_{[a,b]}L^p} + \|u_2\|_{L^\infty_{[a,b]}L^p} \leq C \text{ for } p \in [2, 4].$$
(9.4.23)

Using Corollary 9.3.7 for u_j and $\partial_x u_j$, $j = 1, 2$, we have

$$\|u_j\|_{L^1_{[a',b']}L^\infty} + \|\partial_x u_j\|_{L^1_{[a',b']}L^\infty}$$

$$\leq C_s(b' - a')^{1/2}[\|J_x^{2s}u_j\|_{L^\infty_{[a',b']}L^2} + \|J_y^s u_j\|_{L^\infty_{[a',b']}L^2} + \|J_x^{2s}u_j^2\|_{L^1_{[a',b']}L^2}]$$

for any $[a', b'] \subseteq [a, b]$, with $b' - a' \leq 1$. We use the Kato-Ponce commutator estimate (Lemma 9.A.1) on the last term and the bound (9.4.22). As in Lemma 9.4.3, it follows that

$$||u_j||_{L^1_{[a',b']}L^\infty} + ||\partial_x u_j||_{L^1_{[a',b']}L^\infty} \leq 1/10, \quad j = 1, 2, \tag{9.4.24}$$

whenever $b' - a' \leq c_0$. It suffices to prove that $u_1(t) = u_2(t)$ for $|t - t_0| \leq c_0$. Let $u = u_1 - u_2$ and $v = (u_1 + u_2)/2$. It follows from (9.4.21) that

$$\begin{cases} \partial_t u + \partial_x^3 u - \partial_x^{-1}\partial_y^2 u + \partial_x(uv) = 0 \text{ on } \mathbb{R} \times \mathbb{T} \times [a, b]; \\ u(t_0) = 0. \end{cases} \tag{9.4.25}$$

We apply the operator R^ϵ (defined in (9.4.1)) to (9.4.25), multiply by $2R^\epsilon u$, and integrate in x and y. The result is

$$\begin{aligned} \partial_t ||R^\epsilon u(t)||_{L^2}^2 &= 2\int_{\mathbb{R}\times\mathbb{T}} R^\epsilon(u(t)v(t))\partial_x R^\epsilon u(t)\,dxdy \\ &\leq 2||\partial_x v(t)||_{L^\infty}||R^\epsilon u(t)||_{L^2}^2 + ||\partial_x R^\epsilon u(t)||_{L^2}||R^\epsilon(u(t)v(t)) \\ &\quad - R^\epsilon(u(t))v(t)||_{L^2}. \end{aligned}$$

For $|t - t_0| \leq c_0$, we integrate from t_0 to t, use (9.4.24), and absorb the first term of the right-hand side into the left-hand side. The result is

$$||R^\epsilon u||_{L^\infty_{\{|t-t_0|\leq c_0\}}L^2}^2 \leq C\int_{|t-t_0|\leq c_0} ||R^\epsilon(u(t)v(t)) - R^\epsilon(u(t))v(t)||_{L^2}\,dt.$$

We let $\epsilon \to 0$. By (9.4.23) and the Lebesgue dominated convergence theorem, it follows that $u(t) = 0$ if $|t - t_0| \leq c_0$, as desired. $\square$

We construct now the solution u.

LEMMA 9.4.7 *Assume $\phi \in Z_{(3)}^2$. Then there is a solution $u \in C(\mathbb{R} : Y^{2,1,0}) \cap C^1(\mathbb{R} : H^{-1})$ of the initial value problem (9.1.12). In addition,*

$$\sup_{t\in\mathbb{R}} ||u(t)||_{Z_{(3)}^2} \leq \overline{C}(||\phi||_{Z_{(3)}^2}), \tag{9.4.26}$$

and

$$||u||_{L^1_T L^\infty} + ||\partial_x u||_{L^1_T L^\infty} \leq C(T, ||\phi||_{Z_{(3)}^2}), \tag{9.4.27}$$

for any $T \geq 0$.

Remark. In Theorem 9.1.3, the solution u constructed at this stage is in $C(\mathbb{R} : Y^{2,2/3,-1}) \cap C^1(\mathbb{R} : H^{-3})$.

Proof of Lemma 9.4.7. We adapt the Bona-Smith approximations [1] to our anisotropic setting. Assume $\phi \in Z_{(3)}^2$ and let $\phi_k = P_{(3)}^k\phi$, $k = 1, 2, \ldots$, where

$$\widehat{P_{(3)}^k g}(\xi, n) = \widehat{g}(\xi, n)\mathbf{1}_{[0,k]}(|\xi|)\mathbf{1}_{[0,k^2]}(|n|). \tag{9.4.28}$$

For $k \geq 1$ let

$$h_\phi(k) = \left[\sum_{n \in \mathbb{Z}} \int_{\xi^2 + |n| \geq k^2} |\widehat{\phi}(\xi, n)|^2 (1 + \xi^2 + n^2/\xi^2)^2 \, d\xi \right]^{1/2}. \qquad (9.4.29)$$

Clearly, h_ϕ is nonincreasing in k and

$$\lim_{k \to \infty} h_\phi(k) = 0. \qquad (9.4.30)$$

Using Plancherel theorem, it is easy to see that

$$\|\phi - \phi_k\|_{L^2} \leq C(\|\phi\|_{Z_{(3)}^2}) k^{-2} h_\phi(k), \; k \geq 1, \qquad (9.4.31)$$

and

$$\|J_x^s \phi_k\|_{L^2} \leq C(\|\phi\|_{Z_{(3)}^2}) k^{s-2}, \; k \geq 1, s = 3, 4. \qquad (9.4.32)$$

Let $u_k \in C(\mathbb{R} : Y^\infty)$ denote global solutions of (9.4.2), with $u_k(0) = \phi_k$. Using (9.4.3) and (9.4.7)

$$\|u_k\|_{L^\infty Z_{(3)}^2} \leq \overline{C}(\|\phi\|_{Z_{(3)}^2}), \qquad (9.4.33)$$

and

$$\|u_k\|_{L_T^1 L^\infty} + \|\partial_x u_k\|_{L_T^1 L^\infty} \leq C(T, \|\phi\|_{Z_{(3)}^2}), \; T \geq 0, \qquad (9.4.34)$$

uniformly in $k = 1, 2, \dots$.

For $1 \leq k \leq k'$ let $v_{k,k'} = u_k - u_{k'}$. For any $T \geq 0$ we show that

$$\|v_{k,k'}\|_{L_T^\infty Y^{2,1,0}} \leq C(T, \|\phi\|_{Z_{(3)}^2}) h_\phi(k)^{1/8}, \; 1 \leq k \leq k'. \qquad (9.4.35)$$

The bound (9.4.31) and the uniqueness argument show that

$$\|v_{k,k'}\|_{L_T^\infty L^2} \leq C(T, \|\phi\|_{Z_{(3)}^2}) k^{-2} h_\phi(k), \; 1 \leq k \leq k'. \qquad (9.4.36)$$

By interpolating with (9.4.33),

$$\|\partial_x^{-1} \partial_y v_{k,k'}\|_{L_T^\infty L^2} \leq C(T, \|\phi\|_{Z_{(3)}^2}) k^{-1} [h_\phi(k)]^{1/2}, \; 1 \leq k \leq k', \qquad (9.4.37)$$

and

$$\|J_x^1 v_{k,k'}\|_{L_T^\infty L^2} \leq C(T, \|\phi\|_{Z_{(3)}^2}) k^{-1} [h_\phi(k)]^{1/2}, \; 1 \leq k \leq k'. \qquad (9.4.38)$$

Also, using (9.4.9) and (9.4.32),

$$\|J_x^s u_k\|_{L_T^\infty L^2} \leq C(T, \|\phi\|_{Z_{(3)}^2}) k^{s-2}, \; k \geq 1, s = 3, 4. \qquad (9.4.39)$$

We control now $||J_x^2 v_{k,k'}||_{L_T^\infty L^2}$. Note that $u_k^2 - u_{k'}^2 = 2u_k v_{k,k'} - v_{k,k'}^2$ and

$$\begin{cases} \partial_t v_{k,k'} + \partial_x^3 v_{k,k'} - \partial_x^{-1} \partial_y^2 v_{k,k'} + \partial_x (u_k v_{k,k'} - v_{k,k'}^2/2) = 0 \text{ in } C(\mathbb{R} : H^\infty); \\ v_{k,k'}(0) = \phi_{k,k'}. \end{cases}$$

(9.4.40)

We apply ∂_x^2 to this identity, multiply by $2\partial_x^2 v_{k,k'}$, and integrate in x and y. After integration by parts, the result is

$$\partial_t ||\partial_x^2 v_{k,k'}(t)||_{L^2}^2 \leq C \left| \int_{\mathbb{R} \times \mathbb{T}} \partial_x v_{k,k'}(t) [\partial_x^2 v_{k,k'}(t)]^2 \, dxdy \right|$$

$$+ C \left| \int_{\mathbb{R} \times \mathbb{T}} \partial_x u_k(t) [\partial_x^2 v_{k,k'}(t)]^2 \, dxdy \right|$$

$$+ C \left| \int_{\mathbb{R} \times \mathbb{T}} \partial_x^3 u_k(t) [\partial_x v_{k,k'}(t)]^2 \, dxdy \right|$$

$$+ C \left| \int_{\mathbb{R} \times \mathbb{T}} \partial_x^4 u_k(t) v_{k,k'}(t) \partial_x v_{k,k'}(t) \, dxdy \right|.$$

Using (9.4.36), (9.4.38), and (9.4.39), it follows that

$$\partial_t ||\partial_x^2 v_{k,k'}(t)||_{L^2}^2 \leq C(T, ||\phi||_{Z_{(3)}^2})(||\partial_x u_k(t)||_{L^\infty} + ||\partial_x u_{k'}(t)||_{L^\infty})$$
$$(h_\phi(k)^{1/2} + ||\partial_x^2 v_{k,k'}(t)||_{L^2}^2).$$

The bound (9.4.34) shows that $||\partial_x^2 v_{k,k'}||_{L_T^\infty L^2} \leq C(T, ||\phi||_{Z_{(3)}^2}) h_\phi(k)^{1/4}$, thus

$$||J_x^2 v_{k,k'}||_{L_T^\infty L^2} \leq C(T, ||\phi||_{Z_{(3)}^2}) h_\phi(k)^{1/4}.$$ (9.4.41)

By interpolating with (9.4.33),

$$||J_y^1 v_{k,k'}||_{L_T^\infty L^2} \leq C(T, ||\phi||_{Z_{(3)}^2}) h_\phi(k)^{1/8}.$$ (9.4.42)

The bound (9.4.35) follows from (9.4.30), (9.4.37), (9.4.41), and (9.4.42).

Remark. For Theorem 9.1.3 we also need the bound

$$||J^{-1}(\partial_x^{-1} \partial_y) v_{k,k'}||_{L_T^\infty L^2} \leq C(T, ||\phi||_{Z_{(5)}^1}) k^{-1}.$$ (9.4.43)

To prove (9.4.43) we start from the identity

$$\partial_t v_{k,k'} - \partial_x^5 v_{k,k'} - \partial_x^{-1} \partial_y^2 v_{k,k'} = -(1/2)\partial_x [v_{k,k'}(u_k + u_{k'})]$$
$$\text{in } C([-T, T] : H^\infty),$$

apply the operator $R^\epsilon J^{-1}(\partial_x^{-1}\partial_y)$, multiply by $2R^\epsilon J^{-1}(\partial_x^{-1}\partial_y)v_{k,k'}$, and integrate in x and y. The result is

$$\partial_t \| R^\epsilon J^{-1}(\partial_x^{-1}\partial_y)v_{k,k'}(t) \|_{L^2}^2$$

$$\leq C \left| \int_{\mathbb{R}\times\mathbb{T}} R^\epsilon J^{-1}\partial_y(v_{k,k'}(t)(u_k(t)+u_{k'}(t))) \cdot R^\epsilon J^{-1}(\partial_x^{-1}\partial_y)v_{k,k'}(t)\,dxdy \right|$$

$$\leq C \|v_{k,k'}(t)\|_{L^2}(\|u_k(t)\|_{L^\infty}+\|u_{k'}(t)\|_{L^\infty})\|R^\epsilon J^{-1}(\partial_x^{-1}\partial_y)v_{k,k'}(t)\|_{L^2}.$$

The bound (9.4.43) follows from (9.4.34) and (9.4.36).

It follows from (9.4.35) that $\{u_k\}_{k=1}^\infty$ is a Cauchy sequence in $C([-T,T] : Y^{2,1,0})$, for any $T \geq 0$. Let $u \in C(\mathbb{R} : Y^{2,1,0})$ denote its limit. Since $\partial_x^3 u_k - \partial_x^{-1}\partial_y^2 u_k + u_k\partial_x u_k \to \partial_x^3 u - \partial_x^{-1}\partial_y^2 u + u\partial_x u$ in $C([-T,T] : H^{-1})$, for any $T \geq 0$, and

$$u_k(t) = \phi_k - \int_0^t \partial_x^3 u_k(t') - \partial_x^{-1}\partial_y^2 u_k(t') + u_k(t')\partial_x u_k(t')\,dt', \qquad (9.4.44)$$

we have $\partial_t u \in C(\mathbb{R} : H^{-1})$ and u is a solution of (9.1.12). The bounds (9.4.26) and (9.4.27) follow from the uniform bounds (9.4.33) and (9.4.34). $\qquad\square$

We prove now the identities (9.1.13).

LEMMA 9.4.8 *Let u denote the solution constructed in Lemma 9.4.7. Then*

$$E^s_{(3)}(u(t)) = E^s_{(3)}(\phi)$$

for $s = 0, 1, 2$ and $t \in \mathbb{R}$.

Remark. For Theorem 9.1.3, the corresponding identities are $E^s_{(5)}(u(t)) = E^s_{(5)}(\phi)$ for $s = 0, 1$ and $t \in \mathbb{R}$. The imbedding $Y^{2,1,0} \hookrightarrow L^\infty$, which is used in the proofs of Lemmas 9.4.8 and Lemma 9.4.9 below, has to be replaced with the imbedding $Y^{2,2/3,-1} \hookrightarrow L^3$, which is easy to prove.

Proof of Lemma 9.4.8. For $s = 0, 1$, the identities follow directly from the corresponding identities for the functions u_k (see (9.4.4)) and the fact that $u_k \to u$ in $C([-T,T] : Y^{2,1,0})$, $T \in [0,\infty)$. For $s = 2$, the uniqueness argument shows that it suffices to prove that

$$E^2_{(3)}(u(t)) \leq E^2_{(3)}(\phi) \qquad (9.4.45)$$

for any $t \in \mathbb{R}$.

From the definition of the functions ϕ_k,

$$\lim_{k\to\infty} \phi_k = \phi \text{ in } Z^2_{(3)}.$$

In particular, $\phi_k \to \phi \in L^2 \cap L^\infty$. Thus, each one of the seven terms in the definition (9.1.4) of $E^2_{(3)}(\phi_k)$ converges to the corresponding term in $E^2_{(3)}(\phi)$. Thus,

$$\lim_{k\to\infty} E^2_{(3)}(\phi_k) = E^2_{(3)}(\phi). \qquad (9.4.46)$$

We show now that

$$E^2_{(3)}(u(t)) \leq \liminf_{k \to \infty} E^2_{(3)}(u_k(t)) \qquad (9.4.47)$$

for any $t \in \mathbb{R}$. We combine three of the terms in the definition of $E^2_{(3)}$ and rewrite

$$E^2_{(3)}(g) = \frac{5}{6} \int (\partial_x^{-2} \partial_y^2 g - g^2/2)^2 \, dx dy + \frac{3}{2} \int (\partial_x^2 g)^2 \, dx dy + 5 \int (\partial_y g)^2 \, dx dy$$
$$- \frac{5}{6} \int g(\partial_x^{-1} \partial_y g)^2 \, dx dy + \frac{5}{4} \int g^2 \partial_x^2 g \, dx dy, \qquad (9.4.48)$$

for $g \in Z^2_{(3)}$. For any fixed t, $u_k(t) \to u(t)$ in $Y^{2,1,0} \hookrightarrow L^\infty$. Thus, all the terms in the definition (9.4.48) of $E^2_{(3)}(u_k(t))$ converge to the corresponding terms in $E^2_{(3)}(u(t))$, with the possible exception of the first term. For the first term, we note that

$$||\partial_x^{-2} \partial_y^2 u_k(t) - u_k(t)^2/2||_{L^2} \leq C.$$

Thus, there is a subsequence k_l such that

$$\partial_x^{-2} \partial_y^2 u_{k_l}(t) - u_{k_l}(t)^2/2 \rightharpoonup v \text{ in } L^2.$$

Since $u_{k_l}(t) \to u(t)$ in $L^2 \cap L^\infty$,

$$\partial_x^{-2} \partial_y^2 u_{k_l}(t) \rightharpoonup v + u(t)^2/2 \text{ in } L^2.$$

Since $u_{k_l}(t) \to u(t)$ in L^2, we have $v + u(t)^2/2 = \partial_x^{-2} \partial_y^2 u(t)$, and (9.4.47) follows. Since $E^2_{(3)}(u_k(t)) = E^2_{(3)}(\phi_k)$ (see (9.4.4)), the inequality (9.4.45) follows. $\square$

Next, we show that $u \in C(\mathbb{R} : Z^2_{(3)})$.

LEMMA 9.4.9 *Let u denote the solution constructed in Lemma 9.4.7. Then $u \in C(\mathbb{R} : Z^2_{(3)})$.*

Proof of Lemma 9.4.9. Assume $t_k \in \mathbb{R}$ is a sequence, $t_k \to t_0$. It suffices to prove that

$$\lim_{l \to \infty} \partial_x^{-2} \partial_y^2 u(t_{k_l}) = \partial_x^{-2} \partial_y^2 u(t_0) \text{ in } L^2 \qquad (9.4.49)$$

for some subsequence t_{k_l} of t_k. We examine the formula (9.4.48). Since $u \in C(\mathbb{R} : Y^{2,1,0})$ and $Y^{2,1,0} \hookrightarrow L^\infty$, all the terms in the definition of $E^2_{(3)}(u(t))$ are continuous in t, with the possible exception of the first term. However, since $E^2_{(3)}(u(t))$ is constant, it follows that

$$t \to ||\partial_x^{-2} \partial_y^2 u(t) - u(t)^2/2||_{L^2} \text{ is continuous on } \mathbb{R}. \qquad (9.4.50)$$

Thus, for some subsequence t_{k_l} of t_k,

$$\partial_x^{-2} \partial_y^2 u(t_{k_l}) - u(t_{k_l})^2/2 \rightharpoonup v \text{ in } L^2.$$

Since $u(t_{k_l}) \to u(t_0)$ in $L^2 \cap L^\infty$,

$$\partial_x^{-2}\partial_y^2 u(t_{k_l}) \rightharpoonup v + u(t_0)^2/2 \text{ in } L^2.$$

Since $u(t_{k_l}) \to u(t_0)$ in L^2, we have $v + u(t_0)^2/2 = \partial_x^{-2}\partial_y^2 u(t_0)$. Thus

$$\partial_x^{-2}\partial_y^2 u(t_{k_l}) - u(t_{k_l})^2/2 \rightharpoonup \partial_x^{-2}\partial_y^2 u(t_0) - u(t_0)^2/2 \text{ in } L^2.$$

Since $||\partial_x^{-2}\partial_y^2 u(t_{k_l}) - u(t_{k_l})^2/2||_{L^2} \to ||\partial_x^{-2}\partial_y^2 u(t_0) - u(t_0)^2/2||_{L^2}$ (due to (9.4.50)), it follows that

$$\partial_x^{-2}\partial_y^2 u(t_{k_l}) - u(t_{k_l})^2/2 \to \partial_x^{-2}\partial_y^2 u(t_0) - u(t_0)^2/2 \text{ in } L^2,$$

which gives (9.4.49).   □

Finally, we prove the continuity of the flow map.

LEMMA 9.4.10 *Assume $T \in [0, \infty)$ and $\phi^l \to \phi$ in $Z_{(3)}^2$ as $l \to \infty$. Then $u^l \to u$ in $C([-T, T] : Z_{(3)}^2)$ as $l \to \infty$, where u^l and u are the solutions to the initial value problem (9.1.12) corresponding to initial data ϕ^l and ϕ.*

Proof of Lemma 9.4.10. We show first that

$$u^l \to u \text{ in } C([-T, T] : Y^{2,1,0}). \tag{9.4.51}$$

For $k \geq 1$ let, as before, $\phi_k^l = P_{(3)}^k \phi^l$ and $u_k^l \in C(\mathbb{R}_+ : Y^\infty)$ the corresponding solutions. The formula (9.4.29) shows easily that

$$h_{\phi^l}(k) \leq C(||\phi||_{Z_{(3)}^2})[h_\phi(k) + ||\phi - \phi^l||_{Z_{(3)}^2}]. \tag{9.4.52}$$

It follows, using (9.4.35), that

$$\begin{cases} ||u_k - u||_{L_T^\infty Y^{2,1,0}} \leq C(T, ||\phi||_{Z_{(3)}^2})h_\phi(k)^{1/8}; \\ ||u_k^l - u^l||_{L_T^\infty Y^{2,1,0}} \leq C(T, ||\phi||_{Z_{(3)}^2})(h_\phi(k) + ||\phi - \phi^l||_{Z_{(3)}^2})^{1/8}. \end{cases}$$

An argument similar to the proof of (9.4.35), using the bounds

$$||\phi_k^l - \phi_k||_{L^2} \leq C(||\phi||_{Z_{(3)}^2})||\phi - \phi^l||_{Z_{(3)}^2}, \quad k, l \geq 1$$

and

$$||J_x^4 \phi_k^l||_{L^2} + ||J_x^4 \phi_k||_{L^2} \leq C(||\phi||_{Z_{(3)}^2})k^2, \quad k, l \geq 1,$$

instead of (9.4.31) and (9.4.32), shows that

$$||u_k^l - u_k||_{L_T^\infty Y^{2,1,0}} \leq C(T, ||\phi||_{Z_{(3)}^2})k||\phi - \phi^l||_{Z_{(3)}^2}^{1/4}.$$

The limit (9.4.51) follows.

Finally, we show that

$$\lim_{l \to \infty} \partial_x^{-2} \partial_y^2 u^l = \partial_x^{-2} \partial_y^2 u \text{ in } C([-T, T] : L^2). \tag{9.4.53}$$

Assume, by contradiction, that (9.4.53) was not true. Then there is $\varepsilon_0 > 0$, a subsequence l_j, and a sequence $t_j \in [-T, T]$, $j = 1, 2, \ldots$, with the property that

$$\|\partial_x^{-2} \partial_y^2 u^{l_j}(t_j) - \partial_x^{-2} \partial_y^2 u(t_j)\| \geq \varepsilon_0.$$

We may assume that $t_j \to t_0$ as $j \to \infty$. Using Lemma 9.4.9, it follows that

$$\|\partial_x^{-2} \partial_y^2 u^{l_j}(t_j) - \partial_x^{-2} \partial_y^2 u(t_0)\| \geq \varepsilon_0/2, \tag{9.4.54}$$

for j large enough. On the other hand, $E^2_{(3)}(u^{l_j}(t_j)) = E^2_{(3)}(\phi^{l_j})$, $E^2_{(3)}(u(t_0)) = E^2_{(3)}(\phi)$, using Lemma 9.4.8, and $E^2_{(3)}(\phi^{l_j}) \to E^2_{(3)}(\phi)$ as $j \to \infty$ (since $\phi^{l_j} \to \phi$ in $Z^2_{(3)}$). Thus,

$$\lim_{j \to \infty} E^2_{(3)}(u^{l_j}(t_j)) = E^2_{(3)}(u(t_0)).$$

However, it follows from Lemma 9.4.7 and (9.4.51) that

$$\lim_{j \to \infty} u^{l_j}(t_j) = u(t_0) \text{ in } Y^{2,1,0}.$$

It follows from the last two limits and the formula (9.4.48) that

$$\lim_{j \to \infty} \|\partial_x^{-2} \partial_y^2 u^{l_j}(t_j) - u^{l_j}(t_j)^2/2\|_{L^2} = \|\partial_x^{-2} \partial_y^2 u(t_0) - u(t_0)^2/2\|_{L^2}.$$

The same argument as in the proof of Lemma 9.4.9, using a weakly convergent subsequence, gives a contradiction with (9.4.54). $\qquad\qquad\square$

9.5 GLOBAL EXISTENCE OF SMOOTH SOLUTIONS

In this section we prove that high Sobolev regularity is globally preserved by the flow.

THEOREM 9.5.1 *Assume that $s > 7$ and $\phi \in H^s(\mathbb{T} \times \mathbb{T}) \cap Z^2_{(3)}(\mathbb{T} \times \mathbb{T})$. Then the initial value problem (9.1.10) admits a unique solution $u \in C(\mathbb{R} : H^s(\mathbb{T} \times \mathbb{T}) \cap Z^2_{(3)}(\mathbb{T} \times \mathbb{T})) \cap C^1(\mathbb{R} : H^{-1}(\mathbb{T} \times \mathbb{T}))$. In addition, the mapping $\phi \to u$ is continuous from $H^s(\mathbb{T} \times \mathbb{T}) \cap Z^2_{(3)}(\mathbb{T} \times \mathbb{T})$ to $C([-T, T] : H^s(\mathbb{T} \times \mathbb{T}) \cap Z^2_{(3)}(\mathbb{T} \times \mathbb{T}))$ for any $T \in [0, \infty)$.*

THEOREM 9.5.2 *Assume that $s > 7$ and $\phi \in H^s(\mathbb{R} \times \mathbb{T}) \cap Z^2_{(3)}(\mathbb{R} \times \mathbb{T})$. Then the initial value problem (9.1.12) admits a unique solution $u \in C(\mathbb{R} : H^s(\mathbb{R} \times \mathbb{T}) \cap Z^2_{(3)}(\mathbb{R} \times \mathbb{T})) \cap C^1(\mathbb{R} : H^{-1}(\mathbb{R} \times \mathbb{T}))$. In addition, the mapping $\phi \to u$ is continuous from $H^s(\mathbb{R} \times \mathbb{T}) \cap Z^2_{(3)}(\mathbb{R} \times \mathbb{T})$ to $C([-T, T] : H^s(\mathbb{R} \times \mathbb{T}) \cap Z^2_{(3)}(\mathbb{R} \times \mathbb{T}))$ for any $T \in [0, \infty)$.*

THEOREM 9.5.3 *Assume that $s > 7$ and $\phi \in H^s(\mathbb{R} \times \mathbb{T}) \cap Z^1_{(5)}(\mathbb{R} \times \mathbb{T})$. Then the initial value problem (9.1.14) admits a unique solution $u \in C(\mathbb{R} : H^s(\mathbb{R} \times \mathbb{T}) \cap Z^1_{(5)}(\mathbb{R} \times \mathbb{T})) \cap C^1(\mathbb{R} : H^{-1}(\mathbb{R} \times \mathbb{T}))$. In addition, the mapping $\phi \rightarrow u$ is continuous from $H^s(\mathbb{R} \times \mathbb{T}) \cap Z^1_{(5)}(\mathbb{T} \times \mathbb{T})$ to $C([-T, T] : H^s(\mathbb{R} \times \mathbb{T}) \cap Z^1_{(5)}(\mathbb{R} \times \mathbb{T}))$ for any $T \in [0, \infty)$.*

The restriction $s > 7$ is related to Lemma 9.4.4. While this restriction may be easily relaxed, we do not know if Theorems 9.5.1, 9.5.2, and 9.5.3 hold for $s = 2$. Variants of Theorems 9.5.1, 9.5.2, and 9.5.3, using the anisotropic Sobolev spaces H^{s_1, s_2} or the interpolation spaces $Y^{s_1, s_2, s}$, are also possible.

As in Section 9.4, we only indicate the proof of Theorem 9.5.2. We keep the convention of notation explained at the beginning of Section 9.4.

Proof of Theorem 9.5.2. In view of Theorem 9.1.2, we only need to show that $u \in C(\mathbb{R} : H^s)$ and the mapping $\phi \rightarrow u$ is continuous from $H^s \cap Z^2_{(3)}$ to $C([-T, T] : H^s)$. With the notation in the proof of Lemma 9.4.7, we have the uniform bounds

$$\sup_{t \in \mathbb{R}} \|u_k(t)\|_{Z^2_{(3)}} \leq \overline{C}(\|\phi\|_{Z^2_{(3)}}), \tag{9.5.1}$$

and, using Lemma 9.4.4,

$$\|u_k\|_{L^\infty_T H^s} \leq C(s, T, \|\phi\|_{Z^2_{(3)} \cap H^s}), \quad T \geq 0, \tag{9.5.2}$$

uniformly in $k = 1, 2, \ldots$. An argument similar to the proof of (9.4.35), using the Kato-Ponce commutator estimate (Lemma 9.A.1), shows that

$$\|v_{k,k'}\|_{L^\infty_T Y^{s,s,0}} \leq C(s, T, \|\phi\|_{Z^2_{(3)} \cap H^s})[\|\phi - \phi_k\|_{H^s} + k^{-1/10}], \quad 1 \leq k \leq k'. \tag{9.5.3}$$

Thus $u \in C(\mathbb{R} : H^s)$, as desired.

Assume now that $\phi^l \rightarrow \phi$ in $H^s \cap Z^2_{(3)}$ as $l \rightarrow \infty$. The same argument as in the proof of (9.4.51), using the bound (9.5.3), shows that $u^l \rightarrow u$ in $C([-T, T] : H^s)$, which completes the proof of Theorem 9.5.2. $\square$

9.6 LOCAL WELLPOSEDNESS THEOREMS

In this section we prove local well-posedness theorems in the spaces $Y^{s_1, s_2, -N}$, for suitable s_1 and s_2. For any Banach space X and $r > 0$ let $B(X, r) = \{x \in X : \|x\|_X \leq r\}$.

THEOREM 9.6.1 *Assume that $N \geq 4$, $s \in (15/16, 1)$, and $r > 0$. Then there is $T = T(s, r) > 0$ with the property that for any $\phi \in B(Y^{2s, s, -N}(\mathbb{T} \times \mathbb{T}), r)$ there is a unique solution $u \in C([-T, T] : B(Y^{2s, s, -N}(\mathbb{T} \times \mathbb{T}), Cr)) \cap C^1(\mathbb{R} : H^{-N-1}(\mathbb{T} \times \mathbb{T}))$ of the initial value problem (9.1.10).*

THEOREM 9.6.2 *Assume that $N \geq 4$, $s \in (3/4, 1)$, and $r > 0$. Then there is $T = T(s, r) > 0$ with the property that for any $\phi \in B(Y^{2s, s, -N}(\mathbb{R} \times \mathbb{T}), r)$ there is a unique solution $u \in C([-T, T] : B(Y^{2s, s, -N}(\mathbb{R} \times \mathbb{T}), Cr)) \cap C^1(\mathbb{R} : H^{-N-1}(\mathbb{R} \times \mathbb{T}))$ of the initial value problem (9.1.12).*

THEOREM 9.6.3 *Assume that* $N \geq 4$, $s \in (1/2, 1)$ *and* $r > 0$. *Then there is* $T = T(s, r) > 0$ *with the property that for any* $\phi \in B(Y^{3s,s,-N}(\mathbb{R} \times \mathbb{T}), r)$ *there is a unique solution* $u \in C([-T, T] : B(Y^{3s,s,-N}(\mathbb{R} \times \mathbb{T}), Cr)) \cap C^1(\mathbb{R} : H^{-N-1}(\mathbb{R} \times \mathbb{T}))$ *of the initial value problem* (9.1.14).

As in Sections 9.4 and 9.5, we only indicate the proof of Theorem 9.6.2. We keep the convention of notation explained at the beginning of section 9.4.

Proof of Theorem 9.6.2. The uniqueness of the solution follows from Lemma 9.4.6. To construct the solution, assume $\phi \in B(Y^{2s,s,-N}, r)$ and let $\phi_k = R^{1/2k}\phi$, $k = 1, 2, \ldots$, where $R^{1/2k}$ is defined in (9.4.1). Clearly, $\phi_k \in Z^2_{(3)} \cap Y^\infty$, $\|\phi_k\|_{Y^{2s,s,-N}} \leq r$ and $\lim_{k\to\infty} \phi_k = \phi$ in $Y^{2s,s,-N}$. Let $u_k \in C(\mathbb{R} : Y^\infty)$, $k = 1, 2, \ldots$, denote the global solutions of (9.1.12) corresponding to the initial data ϕ_k (these global solutions were constructed in Corollary 9.4.5). The bound (9.4.10) shows that

$$\|J_x^{2s} u_k\|_{L_T^\infty L^2} \leq \|J_x^{2s}\phi_k\|_{L^2} \exp\left(C \int_{-T}^{T} \|u_k(t)\|_{L^\infty} + \|\partial_x u_k(t)\|_{L^\infty}\, dt\right), \quad (9.6.1)$$

for any $T \geq 0$. A similar estimate, using the Leibniz rule in Lemma 9.A.2 instead of the Kato-Ponce commutator estimate, shows that

$$\|J_y^{s} u_k\|_{L_T^\infty L^2} \leq \|J_y^{s}\phi_k\|_{L^2} \exp\left(C_s \int_{-T}^{T} \|u_k(t)\|_{L^\infty} + \|\partial_x u_k(t)\|_{L^\infty}\, dt\right). \quad (9.6.2)$$

As in Lemma 9.4.6, it follows from Corollary 9.3.7 and the Kato–Ponce commutator estimate (Lemma 9.A.1) that

$$\int_{-T}^{T} \|u_k(t)\|_{L^\infty} + \|\partial_x u_k(t)\|_{L^\infty}\, dt$$

$$\leq CT^{1/2}(\|J_x^{2s} u_k\|_{L_T^\infty L^2} + \|J_y^{s} u_k\|_{L_T^\infty L^2} + \|J_x^{2s}(u_k^2)\|_{L_T^1 L^2})$$

$$\leq CT^{1/2}(\|J_x^{2s} u_k\|_{L_T^\infty L^2} + \|J_y^{s} u_k\|_{L_T^\infty L^2})\left(1 + \int_{-T}^{T} \|u_k(t)\|_{L^\infty} + \|\partial_x u_k(t)\|_{L^\infty}\, dt\right).$$

$$(9.6.3)$$

We substitute (9.6.1) and (9.6.2) into (9.6.3) to show that

$$\int_{-T}^{T} \|u_k(t)\|_{L^\infty} + \|\partial_x u_k(t)\|_{L^\infty}\, dt$$

$$\leq CrT^{1/2} \exp\left(C_s \int_{-T}^{T} \|u_k(t)\|_{L^\infty} + \|\partial_x u_k(t)\|_{L^\infty}\, dt\right).$$

Thus, if $T \leq T_0(s, r)$ is sufficiently small,

$$\int_{-T}^{T} \|u_k(t)\|_{L^\infty} + \|\partial_x u_k(t)\|_{L^\infty}\, dt \leq C(s, r)T^{1/2}, \quad (9.6.4)$$

uniformly in k. For such a T, it follows from (9.6.1) and (9.6.2) that

$$\|u_k\|_{L_T^\infty H^{2s,s}} \leq Cr$$

uniformly in k. An argument similar to the proof of (9.4.43) shows that

$$||J^{-N}(\partial_x^{-1}\partial_y)u_k||_{L_T^\infty L^2} \leq Cr.$$

Thus,

$$||u_k||_{L_T^\infty Y^{2s,s,-N}} \leq Cr, \qquad (9.6.5)$$

uniformly in k.

We now let $k \to \infty$. Let $v_{k,k'} = u_k - u_{k'}$, $1 \leq k \leq k'$. The uniqueness argument in Lemma 9.4.6 and the bound (9.6.4) show that

$$||v_{k,k'}||_{L_T^\infty L^2} \leq C||\phi_{k,k'}||_{Y^{2s,s,-N}}.$$

An argument similar to the proof of (9.4.43) shows that

$$||J^{-N}(\partial_x^{-1}\partial_y)v_{k,k'}||_{L_T^\infty L^2} \leq C||\phi_{k,k'}||_{Y^{2s,s,-N}}.$$

Therefore, u_k is a Cauchy sequence in $C([-T, T] : Y^{0,0,-N})$. Let $u \in C([-T, T] : Y^{0,0,-N})$ denote its limit. As in (9.4.44), u is a solution of the initial value problem (9.1.12) and $u \in C^1([-T, T] : H^{-N-1})$. The uniform bound (9.6.5) shows that

$$||u||_{L_T^\infty Y^{2s,s,-N}} \leq Cr. \qquad (9.6.6)$$

Finally, we show that $u \in C([-T, T] : H^{2s,s})$. The same argument as in the proof of Lemma 9.4.9, using weakly convergent subsequences, shows that it suffices to prove that the maps $t \to ||J_x^{2s}u(t)||_{L^2}$ and $t \to ||J_y^s u(t)||_{L^2}$ are (uniformly) continuous on $[-T, T]$. Assume, by contradiction, that the map $t \to ||J_x^{2s}u(t)||_{L^2}$ is not uniformly continuous on $[-T, T]$, so there is $\varepsilon_0 > 0$ and sequences $t_l, t_l' \subset [-T, T]$, $l = 1, 2, \ldots$, $|t_l - t_l'| \to 0$ as $l \to \infty$, with the property that

$$||J_x^{2s}u(t_l')||_{L^2} - ||J_x^{2s}u(t_l)||_{L^2} \geq \varepsilon_0. \qquad (9.6.7)$$

We fix l large enough such that $|t_l - t_l'|$ is sufficiently small (compared to ε_0). Without loss of generality (due to the uniqueness of solutions), we may assume $t_l = 0$, so $u_{t_l} = \phi$. We repeat the approximation argument at the beginning of the proof and use (9.6.1) and (9.6.4) with $T = T(\varepsilon_0, s, r)$ sufficiently small. It follows that

$$||J_x^{2s}u_k(t_l')||_{L^2} \leq ||J_x^{2s}\phi||_{L^2} + \varepsilon_0/2,$$

provided that t_l' is sufficiently small, which is in contradiction with (9.6.7). The proof of the continuity of the map $t \to ||J_y^s u(t)||_{L^2}$ is similar, using (9.6.2) instead of (9.6.1). $\qquad \square$

APPENDIX 9A. TWO LEMMAS

For $s \geq 0$ let $J_{\mathbb{R}}^s$ denote the operator on $\mathcal{S}'(\mathbb{R})$ defined by the Fourier multiplier $\xi \to (1 + \xi^2)^{s/2}$ and $J_{\mathbb{T}}^s$ denote the operator on $\mathcal{S}'(\mathbb{T})$ defined by the Fourier multiplier $m \to (1 + m^2)^{s/2}$.

LEMMA 9.A.1 (a) *If $s \geq 1$ and $f, g \in H^{3/2+}(\mathbb{R})$, then*

$$||J_{\mathbb{R}}^s(fg) - fJ_{\mathbb{R}}^s g||_{L^2} \leq C_s[||J_{\mathbb{R}}^s f||_{L^2}||g||_{L^\infty}$$
$$+ (||f||_{L^\infty} + ||\partial f||_{L^\infty})||J_{\mathbb{R}}^{s-1} g||_{L^2}]. \quad (9.A.1)$$

(b) *If $s \geq 1$ and $f, g \in H^{3/2+}(\mathbb{T})$, then*

$$||J_{\mathbb{T}}^s(fg) - fJ_{\mathbb{T}}^s g||_{L^2} \leq C_s[||J_{\mathbb{T}}^s f||_{L^2}||g||_{L^\infty}$$
$$+ (||f||_{L^\infty} + ||\partial f||_{L^\infty})||J_{\mathbb{T}}^{s-1} g||_{L^2}]. \quad (9.A.2)$$

Proof of Lemma 9.A.1. Part (a) follows from the Kato-Ponce commutator estimate [6, Lemma X1]. We use part (a) to prove part (b). We may assume $f, g \in C^\infty(\mathbb{T})$ and fix $1 = \psi_1 + \psi_2$ a partition of unity, with $\psi_1, \psi_2 : \mathbb{R} \to [0, 1]$, ψ_1, ψ_2 smooth and periodic with period 2π, ψ_1 supported in $\bigcup_{m \in \mathbb{Z}}[2\pi m + \pi/4, 2\pi m + 7\pi/4]$, and ψ_2 supported in $\bigcup_{m \in \mathbb{Z}}[2\pi m - 3\pi/4, 2\pi m + 3\pi/4]$. Given $f, g \in C^\infty(\mathbb{T})$, let $\tilde{f} \in C^\infty(\mathbb{R})$ denote the periodic extension of f, and

$$\tilde{g}_1(x) = g(x)\psi_1(x)\mathbf{1}_{[0,2\pi]}(x), \quad \tilde{g}_2(x) = g(x)\psi_2(x)\mathbf{1}_{[-\pi,\pi]}(x).$$

Clearly, $\tilde{g}_1, \tilde{g}_2 \in C_0^\infty(\mathbb{R})$. Also,

$$\mathcal{F}_{\mathbb{T}}(fg)(m) = \sum_{l \in \{1,2\}} \mathcal{F}_{\mathbb{R}}(\tilde{f}\tilde{g}_l)(m).$$

Using the Poisson summation formula (9.3.10),

$$J_{\mathbb{T}}^s(fg)(x) = (2\pi)^{-1} \sum_{m \in \mathbb{Z}} \mathcal{F}_{\mathbb{T}}(fg)(m)(1 + m^2)^{s/2} e^{imx}$$

$$= (2\pi)^{-1} \sum_{l \in \{1,2\}} \sum_{m \in \mathbb{Z}} \mathcal{F}_{\mathbb{R}}(\tilde{f}\tilde{g}_l)(m)(1 + m^2)^{s/2} e^{imx}$$

$$= \sum_{l \in \{1,2\}} \sum_{\mu \in \mathbb{Z}} J_{\mathbb{R}}^s(\tilde{f}\tilde{g}_l)(x + 2\pi\mu),$$

for any $x \in \mathbb{T}$. A similar computation shows that

$$f(x)J_{\mathbb{T}}^s g(x) = \sum_{l \in \{1,2\}} \sum_{\mu \in \mathbb{Z}} f(x)J_{\mathbb{R}}^s \tilde{g}_l(x + 2\pi\mu),$$

for any $x \in \mathbb{T}$. Thus,

$$||J_{\mathbb{T}}^s(fg) - fJ_{\mathbb{T}}^s g||_{L^2(\mathbb{T})} \leq C \sum_{l \in \{1,2\}} ||J_{\mathbb{R}}^s(\tilde{f}\tilde{g}_l) - \tilde{f}J_{\mathbb{R}}^s \tilde{g}_l||_{L^2(\mathbb{R})}. \quad (9.A.3)$$

We fix two smooth functions $\tilde{\psi}_1, \tilde{\psi}_2 : \mathbb{R} \to [0, 1]$, $\tilde{\psi}_1$ supported in $[\pi/10, 19\pi/10]$ and equal to 1 in $[\pi/5, 9\pi/5]$, $\tilde{\psi}_2$ supported in $[-9\pi/10, 9\pi/10]$ and equal to 1 in $[-4\pi/5, 4\pi/5]$. Then $\tilde{f}\tilde{g}_l = (\tilde{f}\tilde{\psi}_l)\tilde{g}_l$, $l = 1, 2$. Thus, the right-hand side of (9.A.3) is dominated by

$$C \sum_{l \in \{1,2\}} \left[||J_{\mathbb{R}}^s((\tilde{f}\tilde{\psi}_l)\tilde{g}_l) - (\tilde{f}\tilde{\psi}_l)J_{\mathbb{R}}^s \tilde{g}_l||_{L^2(\mathbb{R})} + ||\tilde{f}(1 - \tilde{\psi}_l)J_{\mathbb{R}}^s \tilde{g}_l||_{L^2(\mathbb{R})} \right]. \quad (9.A.4)$$

Using (9.A.1) and the simple observations $||J_\mathbb{R}^s(\widetilde{f}\widetilde{\psi}_l)||_{L^2(\mathbb{R})} \leq C_s||J_\mathbb{T}^s f||_{L^2(\mathbb{T})}$ and $||J_\mathbb{R}^{s-1}\widetilde{g}_l||_{L^2(\mathbb{R})} \leq C_s||J_\mathbb{T}^{s-1}g||_{L^2(\mathbb{T})}$, $l = 0, 1$, the first term in (9.A.4) is dominated by the right-hand side of (9.A.2). For the second term in (9.A.4), we notice that

$$|J_\mathbb{R}^s\widetilde{g}_l(x)| \leq C_s||g||_{L^\infty}(1+|x|)^{-2},$$

for x in the support of $(1 - \widetilde{\psi}_l)$. Thus, the second term in (9.A.4) is dominated by $C||f||_{L^2}||g||_{L^\infty}$, which completes the proof of (9.A.2). □

LEMMA 9.A.2 (a) *If $s \in (0, 1)$ and $f, g \in H^{1/2+}(\mathbb{R})$ then*

$$||J_\mathbb{R}^s(fg) - fJ_\mathbb{R}^s g||_{L^2} \leq C_s||J_\mathbb{R}^s f||_{L^2}||g||_{L^\infty}. \tag{9.A.5}$$

(b) *If $s \in (0, 1)$ and $f, g \in H^{1/2+}(\mathbb{T})$ then*

$$||J_\mathbb{T}^s(fg) - fJ_\mathbb{T}^s g||_{L^2} \leq C_s||J_\mathbb{T}^s f||_{L^2}||g||_{L^\infty}. \tag{9.A.6}$$

Proof of Lemma 9.A.2. Part (a) follows from [8, Theorem A.12]. Part (b) then follows from part (a) using the same argument as in Lemma 9.A.1. □

REFERENCES

[1] J. L. Bona and R. Smith, The initial-value problem for the Korteweg-de Vries equation, Philos. Trans. Roy. Soc. London Ser. A 278 (1975), 555–601.

[2] J. Bourgain, On the Cauchy problem for the Kadomstev-Petviashvili Equation, Geom. Funct. Anal. 3 (1993), 315–341.

[3] N. Burq, P. Gérard, and N. Tzvetkov, Strichartz inequalities and the nonlinear Schrödinger equation on compact manifolds, Amer. J. Math. 126 (2004), 569–605.

[4] J. Colliander, Globalizing estimates for the periodic KPI equation, Illinois J. Math. 40 (1996), 692–698.

[5] R. J. Iorio and W. V. L. Nunes, On equations of KP-type, Proc. Roy. Soc. Edinburgh Sect. A 128 (1998), 725–743.

[6] T. Kato and G. Ponce, Commutator estimates and the Euler and Navier–Stokes equations, Comm. Pure Appl. Math. 41 (1988), 891–907.

[7] C. E. Kenig, On the local and global well-posedness theory for the KP-I equation, Ann. Inst. H. Poincaré Anal. Non Linéaire 21 (2004), 827–838.

[8] C. E. Kenig, G. Ponce, and L. Vega, Well-posedness and scattering results for the generalized Korteweg–de Vries equation via the contraction principle, Comm. Pure Appl. Math. 46 (1993), 527–620.

[9] H. Koch and N. Tzvetkov, Local well-posedness of the Benjamin—One equation in $H^s(\mathbb{R})$, Int. Math. Res. Not. 2003 (2003), 1449–1464.

[10] L. Molinet, J.-C. Saut, and N. Tzvetkov, Well-posedness and ill-posedness results for the Kadomtsev-Petviashvili-I equation, Duke Math. J. 115 (2002), 353–384.

[11] L. Molinet, J.-C. Saut, and N. Tzvetkov, Global well-posedness for the KP-I equation, Math. Ann. 324 (2002), 255–275.

[12] L. Molinet, J.-C. Saut, and N. Tzvetkov, Correction: Global well-posedness for the KP-I equation, Math. Ann. 328 (2004), 707–710.

[13] M. B. Nathanson, *Additive Number Theory: The Classical Bases,* Graduate Texts in Mathematics 164, Springer-Verlag, New York (1996).

[14] J.-C. Saut, Remarks on the generalized Kadomstev-Petviashvili equations, Indiana Univ. Math. J. 42 (1993), 1011–1026.

[15] J.-C. Saut and N. Tzvetkov, The Cauchy problem for the fifth order KP equations, J. Math. Pures Appl. 79 (2000), 307–338.

[16] J.-C. Saut and N. Tzvetkov, On periodic KP-I type equations, Comm. Math. Phys. 221 (2001), 451–476.

[17] E. M. Stein, *Harmonic Analysis: Real-Variable Methods, Orthogonality, and Oscillatory Integrals,* Princeton Mathematical Series 43, Princeton University Press, Princeton, NJ (1993).

[18] H. Takaoka and N. Tzvetkov, On the local regularity of the Kadomtsev-Petviashvili-II equation, Int. Math. Res. Not. 2001 (2001), 77–114.

Chapter Ten

The Cauchy Problem for the Navier-Stokes Equations with Spatially Almost Periodic Initial Data

Y. Giga, A. Mahalov, and B. Nicolaenko

10.1 INTRODUCTION

We consider the Cauchy problem for the Navier-Stokes equations ($n \geq 2$) :

$$\partial_t u - \nu \Delta u + (u, \nabla)u + \nabla p = 0, \quad \operatorname{div} u = 0 \quad \text{in} \quad \mathbf{R}^n \times (0, T), \quad (10.1.1)$$

$$u|_{t=0} = u_0 \quad (\operatorname{div} u_0 = 0) \quad \text{in} \quad \mathbf{R}^n, \quad (10.1.2)$$

when the initial data u_0 is spatially nondecreasing, in particular almost periodic. We use a standard convention of notation; $u(x, t) = (u^1(x, t), \ldots, u^n(x, t))$ represents the unknown velocity field while $p(x, t)$ represents the unknown pressure field; $\nu > 0$ denotes the kinematic viscosity and $\partial_t u = \partial u / \partial t$, $(u, \nabla) = \sum_{i=1}^{n} u^i \partial_{x_i}$, $\partial_{x_i} = \partial / \partial x_i$, $\nabla p = (\partial_{x_1} p, \ldots, \partial_{x_n} p)$ with $x = (x_1, \ldots, x_n)$.

It is by now well known ([CK], [Ca], [CaM], [GIM], [KT]) that the problem (10.1.1)–(10.1.2) admits a local in time-classical solution for any bounded initial data. It is unique under an extra assumption for pressure. It is also well known [GMS] that the solution is global in time if the space dimension $n = 2$. By the translation invariance in space variables for (10.1.1)–(10.1.2) and the uniqueness of the solution it is clear that if the initial data u_0 is spatially periodic, so is the solution.

Our goal in this note is to show that if u_0 is spatially almost periodic in the sense of Bohr [AG], [Co], so is the solution. This fact follows from continuity of the solution with respect to initial data in uniform topology. Fortunately, such a result follows from analysis developed in [GIM]. However, this persistency of almost periodicity is not noted elsewhere, so we shall give its statement as well as its proof. We note that considerations of solutions not decaying at infinity such as corresponding to almost periodic initial data are essential in the development of mathematical theory for homogeneous statistical solutions [FMRT], [VF].

It turns out that this persistency property also holds for the three-dimensional Navier-Stokes equation in a rotating frame with almost periodic initial data (10.1.2):

$$u_t - \nu \Delta u + (u, \nabla)u + \Omega e_3 \times u + \nabla p = 0, \quad \operatorname{div} u = 0 \text{ in} \quad \mathbf{R}^3 \times (0, T), \quad (10.1.3)$$

with $\Omega \in \mathbf{R}$, where $e_3 = (0, 0, 1)$ is the direction of the axis of rotation. In [GIMM] we have proved its local solvability for some class of bounded initial data not necessarily decaying at space infinity. Since we are forced to use a homogeneous

Besov space, a more specific argument is necessary to prove the persistence of almost periodicity. We shall prove this result in Section 10.3.

An almost periodic function always has its mean value [AG], [Co]. However, the converse does not hold. Nevertheless, we also prove that existence of mean value is preserved for the Navier-Stokes flow even if the initial velocity is not necessarily almost periodic.

10.2 PERSISTENCE OF ALMOST PERIODICITY

We first recall a well-known existence result. Let $L_\sigma^\infty(\mathbf{R}^n)$ be the space of all divergence-free essentially bounded vector fields on $\mathbf{R}^n$ equipped with essential supremum norm $\|\cdot\|_\infty$. Let $BUC(\mathbf{R}^n)$ be the space of all bounded uniformly continuous functions on $\mathbf{R}^n$. The space $BUC_\sigma(\mathbf{R}^n)(= BUC(\mathbf{R}^n) \cap L_\sigma^\infty)$ is evidently a closed subspace of $L_\sigma^\infty(\mathbf{R}^n)$. The space $C(I, X)$ denotes the space of all continuous functions from I to X, where X is a Banach space. We do not distinguish the space of vector-valued and scalar functions.

PROPOSITION 10.2.1 ([GIM]). (Existence and uniqueness). *Assume that $u_0 \in BUC_\sigma(\mathbf{R}^n)$. There exists $T_0 > 0$ and a unique classical solution $(u, \nabla p)$ of (10.1.1)–(10.1.2) such that*

(i) *$u \in C([0, T_0], BUC_\sigma(\mathbf{R}^n))$ and u is smooth in $\mathbf{R}^n \times (0, T_0)$;*
(ii) *$\sup_{0<t<T_0} \|u\|_\infty(t) < \infty$;*
(iii) *$\nabla p = \nabla \sum_{i,j=1}^n R_i R_j u^i u^i$, where $R_i = \partial_{x_i}(-\Delta)^{-1/2}$.*

Moreover, $t^{1/2}\nabla u \in C([0, T_0], L_\sigma^\infty(\mathbf{R}^n))$ and $T_0\|u_0\|_\infty^2 \geq C_0$ with a constant depending only on n.

Remark 10.2.2

(i) One is able to take T_0 arbitrary large number for $n = 2$. In particular, there is a global in time unique smooth solution u for any $u_0 \in BUC_\sigma(\mathbf{R}^n)$ (and more generally for $u_0 \in L_\sigma^\infty(\mathbf{R}^n)$) [GMS].
(ii) As observed in [GIM], we need some restriction of the form of ∇p to have the uniqueness.

LEMMA 10.2.3 (Continuity with respect to initial data). *Let u_k be the bounded (smooth) solution $\mathbf{R}^n \times (0, T)$ of the Navier-Stokes equations for initial data $u_{0k} \in BUC_\sigma(\mathbf{R}^n)$ for $k = 1, 2, \ldots$. Assume that $\sup_{0 \leq t \leq T} \|u_k\|_\infty(t) \leq M$. Then the following two properties hold.*

(i) *$\|u_1(\cdot, t) - u_2(\cdot, t)\|_\infty \leq C\|u_{01} - u_{02}\|_\infty, t \in [0, T]$ with C depending only on n, T, and M.*
(ii) *If u_{0k} converges to u_0 in $L_\sigma^\infty(\mathbf{R}^n)$, then the solution u with initial data u_0 exists for $\mathbf{R}^n \times (0, T)$, and it is the uniform limit of u_k in $\mathbf{R}^n \times (0, T)$.*

Proof. This is easy to prove by applying arguments in [GIM]. The solution in [GIM] is the mild solution of an integral equation so u_k satisfies

$$u_k(t) = e^{t\Delta}u_{0k} - \int_0^t \text{div } e^{(t-s)\Delta}\mathbf{P}(u_k \otimes u_k)ds, \qquad (10.2.1)$$

where $\mathbf{P} = (\delta_{ij} + R_i R_j)$ and $e^{t\Delta}f$ is the solution of the heat equation with initial data f, i.e., $(e^{t\Delta}f)(x) = (G_t * f)(x)$ with $G_t(x) = (4\pi t)^{-n/2} \exp(-|x|^2/4t)$. By (10.2.1) it is clear that the difference $u_1 - u_2$ fulfills

$$u_1(t) - u_2(t) = e^{t\Delta}(u_{01} - u_{02}) - \int_0^t \text{div } e^{(t-s)\Delta}\mathbf{P}((u_1 \otimes u_1) - (u_2 \otimes u_2))ds. \quad (10.2.2)$$

Since

$$\|\text{div } e^{t\Delta}\mathbf{P}f\|_\infty \le Ct^{-1/2}\|f\|_\infty \qquad (10.2.3)$$

by [GIM], estimating (10.2.2) yields

$$\|u_1 - u_2\|_\infty(t) \le \|u_{01} - u_{02}\|_\infty + 2C\int_0^t (t-s)^{-1/2}(\|u_1\|_\infty$$
$$+ \|u_2\|_\infty)(\|u_1 - u_2\|_\infty(s))ds$$

so that

$$\sup_{0<\tau<t} \|u_1 - u_2\|_\infty(\tau) \le \|u_{01} - u_{02}\|_\infty + 4C2Mt^{1/2} \sup_{0<\tau<t} \|u_1 - u_2\|_\infty(\tau).$$

If $8CMt^{1/2} \le 1/2$, then we have

$$\sup_{0<\tau<t} \|u_1 - u_2\|_\infty(\tau) \le 2\|u_{01} - u_{02}\|_\infty.$$

We divide $[0, T]$ into $[0, T_1], [T_1, T_2], \ldots, [T_N, T_0]$ so that the length of each time interval is less than $(16CM)^{-2}$ and repeat this argument on each interval to get (i).

If $\{u_{0k}\}$ is the Cauchy sequence in $L_\sigma^\infty(\mathbf{R}^n)$, $\{u_k\}$ is the Cauchy sequence in $L^\infty(\mathbf{R}^n \times (0, T))$ (= the space of all essentially bounded functions in $\mathbf{R}^n \times (0, T)$) by (i). Thus, sending k to infinity in (10.2.1) yields (ii) if we note (10.2.3). $\qquad \square$

We now recall the notion of almost periodicity in the sense of Bohr [AG], [Co]. Let f be in $BUC(\mathbf{R}^n)$. We say that f is almost periodic if the set

$$\Sigma_f = \{f(\cdot + \xi)|\xi \in \mathbf{R}^n\} \subset L^\infty(\mathbf{R}^n)$$

is relatively compact in $L^\infty(\mathbf{R}^n)$, i.e., any sequence in Σ_f has a convergent subsequence in $L^\infty(\mathbf{R}^n)$. (Actually, uniformly continuity assumption is redundant. In fact, if f is bounded continuous and Σ_f is relatively compact in $L^\infty(\mathbf{R}^n)$, then f must be uniformly continuous [AG], [Co]). If f is periodic in $(x_1, \ldots, x_n)$, i.e.,

$$f(x + \eta) = f(x) \text{ for all } x \in \mathbf{R}^n$$

for some $\eta = (\eta_1, \ldots, \eta_n), \eta_i > 0, i = 1, \ldots, n$, then Σ_f is identified with a torus $\prod_{i=1}^n (\mathbf{R}/\eta_i\mathbf{Z})$ so any periodic function is almost periodic. A finite sum of periodic continuous functions in L^∞ is almost periodic. If an infinite sum of periodic continuous functions converges in $L^\infty(\mathbf{R}^n)$, then it is almost periodic.

We are now in position to state our main result.

THEOREM 10.2.4 *Assume that $u_0 \in BUC_\sigma(\mathbf{R}^n)$ is almost periodic. Let u be the bounded (smooth) solution of the Navier-Stokes equation in $\mathbf{R}^n \times (0, T)$ with initial data u_0 in Proposition 10.2.1. Then $u(\cdot, t)$ is almost periodic as a function of $\mathbf{R}^n$ for all $t \in (0, T)$.*

Proof. Let $u(t) = u(\cdot, t)$ be the solution (in Proposition 10.2.1) of the Navier-Stokes equation with initial data u_0. We denote the mapping $u_0 \longmapsto u(t)$ by $S(t)$.

Since the Navier-Stokes equations are invariant under translation in the space variable, the solution u_η with initial data $u_{0\eta}(x) = u_0(x + \eta)$, $\eta \in \mathbf{R}^n$ fulfills $u_\eta(x, t) = u(x + \eta, t)$. Thus $S(t)$ maps Σ_{u_0} onto $\Sigma_{u(t)}$. Since

$$\sup_{0 < t < T} \|u_\eta\|_\infty(t) = \sup_{0 < t < T} \|u\|_\infty(t)$$

is independent of η, Lemma 10.1 implies that $S(t)$ is a well-defined continuous mapping from the closure $\bar{\Sigma}_{u_0}$ onto $\bar{\Sigma}_{u(\cdot,t)}$. Thus, if $\bar{\Sigma}_{u_0}$ is relatively compact, so is $\bar{\Sigma}_{u(\cdot,t)}$. We now conclude that $u(\cdot, t)$ is almost periodic. $\qquad \square$

10.3 THREE-DIMENSIONAL NAVIER-STOKES EQUATIONS IN A ROTATING FRAME

We shall show persistence of almost periodicity for the Cauchy problem (10.1.2)–(10.1.3). For this problem it seems impossible to establish wellposedness for L^∞-initial data. We recall a result of existence of a unique local solution for (10.1.2)–(10.1.3) [GIMM](see also [S]). Let $\mathcal{W}$ be

$$\mathcal{W} = \{u \in L^\infty_\sigma(\mathbf{R}^3) | \partial u/\partial x_3 = 0 \text{ in } \mathbf{R}^3 \text{ in the sense of distribution}\}.$$

An element of this space is often called a two-dimensional three-component divergence free-vector field. We need to recall a Besov space,

$$\dot{B}^0_{\infty,1}(\mathbf{R}^3) = \{f \in \mathcal{S}'(\mathbf{R}^3) | f = \sum_{j=-\infty}^{\infty} \varphi_j * f \text{ in } \mathcal{S}'(\mathbf{R}^3),$$

$$\|f\|_{\dot{B}^0_{\infty,1}} := \sum_{j=-\infty}^{\infty} \|\varphi_j * f\|_\infty < \infty\},$$

where $\{\varphi_j\}_{j=-\infty}^\infty$ is the Littlewood-Paley decomposition satisfying

$$\hat{\varphi}_j(\xi) = \hat{\varphi}_0(2^{-j}\xi) \in C_0^\infty(\mathbf{R}^3), \quad \text{supp } \hat{\varphi}_0 \subset \{1/2 \le |\xi| \le 2\},$$

$$\sum_{j=-\infty}^{\infty} \hat{\varphi}_j(\xi) = 1(\xi \ne 0);$$

here $\hat{\varphi}$ denotes the Fourier transform of φ and $C_0^\infty(\mathbf{R}^n)$ is the space of all smooth functions with compact support in $\mathbf{R}^n$.

We need to prepare several function spaces. We say that $u \in L^\infty_\sigma(\mathbf{R}^3)$ admits vertical averaging if

$$\lim_{L \to \infty} \frac{1}{2L} \int_{-L}^{L} u(x_1, x_2, x_3) dx_3 = \bar{u}(x_1, x_2)$$

exists almost everywhere (a.e.). The vector field $\bar{u}$ is called the vertical average of $u(x_1, x_2, x_3)$. Let $L_{\sigma,a}^\infty$ be the topological direct sum of the form

$$L_{\sigma,a}^\infty = \mathcal{W} \oplus \mathcal{B}^0$$

with

$$\mathcal{B}^0 = \{u \in \dot{B}_{\infty,1}^0 \cap L_\sigma^\infty | \bar{u}(x_1, x_2) = 0 \text{ a.e.} (x_1, x_2) \in \mathbf{R}^2\}.$$

We often consider a smaller space,

$$X = BUC_{\sigma,a} := \{u \in L_{\sigma,a}^\infty(\mathbf{R}^3) | u \in BUC(\mathbf{R}^3)\}$$
$$= \{u = u_1 + u_2 | u_1 \in \mathcal{W} \cap BUC, u_2 \in \mathcal{B}^0\}.$$

The second identity follows from the fact that $\mathcal{B}^0 \subset BUC$. The space X is equipped with the norm

$$\|u\|_X = \|u_1\|_{L^\infty} + \|u_2\|_{\dot{B}_{\infty,1}^0}$$

and it is a Banach space. We shall fix $\Omega \in \mathbf{R}$ below.

PROPOSITION 10.3.1 ([GIMM]). *Assume that $u_0 \in X$. There exists $T_0 > 0$ and a unique classical solution $(u, \nabla p)$ of (10.1.2)–(10.1.3) such that*

(i) *$u \in C([0, T_0), X)$ for any $\delta > 0$ and u is smooth in $\mathbf{R}^3 \times (0, T_0)$.*
(ii)

$$\frac{\partial p}{\partial x_\ell} = \frac{\partial}{\partial x_\ell} \sum_{i,j=1}^3 R_i R_j u^i u^j - \Omega R_\ell (R_2 u^1 - R_1 u^2), \ell = 1, 2, 3.$$

Moreover, $t^{1/2} \nabla u \in C([0, T_0]; L_\sigma^\infty(\mathbf{R}^3))$ and $T_0(\|u_0\|_X^2 + 1) \geq c_0$ with a positive constant c_0.

PROPOSITION 10.3.2 (Continuity with respect to initial data). *Let u_k be the bounded (smooth) solution in $\mathbf{R}^3 \times (0, T)$ of the Navier-Stokes equations (10.1.3) in a rotating frame with initial data $u_{0k} \in X$ for $k = 1, 2, \ldots$ Assume that $\sup_{0 \leq t < T} \|u_k\|_\infty(t) \leq M$. Then the following two properties hold:*

(i) *$\|u_1(\cdot, t) - u_2(\cdot, t)\|_\infty \leq C \|u_{01} - u_{02}\|_X, t \in (0, T)$ with C depending only on T and M.*
(ii) *If $\{u_{0k}\}_{k=1}^\infty$ converges to u_0 in X, then the solution u with initial data u_0 exists for $\mathbf{R}^3 \times (0, T)$ and it is the uniform limit in $\mathbf{R}^3 \times (0, T)$.*

Proof of Proposition 10.3.2. This is easy to prove by applying arguments in [GIMM]. The solution in [GIMM] is the mild solution of an integral equation, so u_k fulfills

$$u_k(t) = e^{-A(\Omega)t} u_{0k} - \int_0^t \text{dive}^{-A(\Omega)(t-s)\Delta} \mathbf{P}(u_k \otimes u_k) ds$$

with $A(\Omega) = -\Delta + \Omega \mathbf{P} J \mathbf{P}$, where $Ja = e_3 \times a$ for $a \in \mathbf{R}^3$. We estimate the difference $u_1 - u_2$ similarly to [GIMM, (4.8)] and obtain

$$\|u_1 - u_2\|_\infty(t) \leq \|u_{01} - u_{02}\|_X + 2CM \int_0^t (t-s)^{-1/2} \|u_1 - u_2\|_\infty(s) ds,$$

with a constant $C > 0$. As before, we apply the Gronwall inequality [GMS] to get

$$\|u_1 - u_2\|_\infty(t) \leq C_1 \|u_{01} - u_{02}\|_X e^{C_2 t}, \quad t \in (0, T),$$

with some positive constants C_1, C_2 depending only on C and M. This yields (i). From (i) it follows (ii) as before. $\square$

We are now in position to state our main result.

THEOREM 10.3.3 *Assume that $u_0 \in X$ is almost periodic. Let u be the bounded (smooth) solution of the Navier-Stokes equations (10.1.3) in a rotating frame in $\mathbf{R}^3 \times (0, T)$ with initial data u_0 in Proposition 10.3.1. Then $u(\cdot, t)$ is almost periodic as a function of $\mathbf{R}^n$.*

The proof parallels that of Theorem 10.2.4 by setting the solution operator $S(t)$: $u_0 \mapsto u(\cdot, t)$ if we use the next lemma.

LEMMA 10.3.4 *A function $f \in \dot{B}^0_{\infty,1}(\mathbf{R}^n)$ is almost periodic if and only if Σ_f is relatively compact in $f \in \dot{B}^0_{\infty,1}(\mathbf{R}^n)$.*

Proof. It suffices to prove "only if" part since $\dot{B}^0_{\infty,1} \subset L^\infty$ is continuous. Suppose that $\Sigma_f (\subset \dot{B}^0_{\infty,1})$ is relatively compact in $L^\infty(\mathbf{R}^n)$. Then any sequence $\{f_{n_k}\} \subset \Sigma_f$ has a convergent subsequence $\{f_\ell = f(\cdot + \eta_{k(\ell)})\} \to f_0$ in $L^\infty(\mathbf{R}^3)$ as $\ell \to \infty$. We note that

$$\|\varphi_i * f_\ell\|_\infty = \|\varphi_i * f\|_\infty \quad \text{for all} \quad i \in \mathbf{Z}.$$

to get

$$\|f_\ell - f_0\|_{\dot{B}^0_{\infty,1}} \leq \sum_{|i| \geq N} \|\varphi_i * (f_\ell - f_0)\|_\infty$$

$$+ \sum_{|i| \leq N-1} \|(f_\ell - f_0) * \varphi_i\|_\infty \leq 2 \sum_{|i| \geq N} \|\varphi_i * f\|_\infty + C_N \|f_\ell - f\|_\infty$$

with $C_N = 2N \|\varphi_0\|_{L^1}$. Sending $\ell \to \infty$, we observe that

$$\lim_{\ell \to \infty} \sup \|f_\ell - f_0\|_{\dot{B}^0_{\infty,1}} \leq 2 \sum_{|i| \geq N} \|\varphi_i * f\|_\infty.$$

Since $f \in \dot{B}^0_{\infty,1}$, the right-hand side tends to zero as $N \to \infty$ so $f_\ell \to f_0$ and $f_0 \in \dot{B}^0_{\infty,1}$. $\square$

Remark 10.3.5 We note that $\Omega e_3 \times u$ restricted to divergence-free vector fields is a skew-symmetric operator in (10.1.3). The fast singular oscillating limit (large Ω) of the 3D Navier-Stokes equations in a rotating frame (10.1.3) with almost periodic initial data (10.1.2) will be considered elsewhere. Global regularity for large Ω of solutions of the three-dimensional Navier-Stokes equations in a rotating frame with initial data on arbitrary periodic lattices and in bounded cylindrical domains in $\mathbb{R}^3$ was proven in [BMN1], [BMN2], [BMN3], and [MN] without any conditional assumptions on the properties of solutions at later times. The method of proving global regularity for large fixed Ω is based on the analysis of fast singular oscillating limits (singular limit $\Omega \to +\infty$), nonlinear averaging, and cancellation of oscillations in the nonlinear interactions for the vorticity field. It uses harmonic analysis tools of Littlewood-Paley dyadic decomposition and lemmas on restricted convolutions to prove global regularity of the limit-resonant three-dimensional Navier-Stokes equations which holds without any restriction on the size of initial data and strong convergence theorems for large Ω.

10.4 AVERAGING PROPERTY

As proved in [AG], [Co] an almost periodic function has the mean value at least for functions of one variable. We shall study a class of functions having its mean value.

DEFINITION 10.4.1 *Let D be a bounded C^1 domain in $\mathbf{R}^n$ containing the origin. Let χ_D be its characteristic function, i.e., $\chi_D(x) = 1$ if $x \in D$ and $\chi_D(x) = 0$ if $x \notin D$. Let $\chi_D^R(x) = \chi_D(x/R)R^{-n}|D|^{-1}$ for $R > 0$, where $|D|$ denotes the Lebesgue measure of D. (By definition, $\int_{\mathbf{R}^n} \chi_D^R dx = 1$.) A function $f \in L^\infty(D)$ is said to have its D-mean value if $\chi_D^R * f$ converges to a constant c uniformly in $\mathbf{R}^n$ as $R \to \infty$. The constant c is called D-mean value of f.*

Example 10.4.2 For any $\xi \in \mathbf{R}^n$ the function $e^{i\xi x}$ has its mean value for any D. This is trivial if $\xi = 0$, so we may assume $\xi \neq 0$. By rotation of coordinates we may assume $\xi_1 \neq 0$ for $\xi = (\xi_1, \ldots, \xi_n)$. Integrating by parts, we observe that

$$(\chi_D^R * f)(x) = \frac{1}{R^n|D|} \int_{RD} f(x-y)dy = -\frac{1}{R^n|D|} \int_{\partial(RD)} n_1 \frac{e^{i\xi(x-y)}}{i\xi_1} d\mathcal{H}^{n-1}(y),$$

where $\mathbf{n} = (n_1, \ldots, n_n)$ is the outer unit normal of $\partial(RD) = R(\partial D) = \{Rx | x \in \partial D\}$ and $d\mathcal{H}^{n-1}$ is the area element. Thus, $\|\chi_D^R * f\|_\infty \to 0$ as $R \to \infty$. Thus, unless $\xi = 0$, the mean value of $e^{i\xi x}$ equals zero. Since one can prove that an almost periodic function is a uniform limit of Bochner-Fejer trigonometric polynomials as proved in [AG], [Co] (for $n = 1$), an almost periodic function has its D-mean value for all D. (Note that a uniform limit of $\{f_m\}$ always has its D-mean value if f_m has its D-mean value.)

Evidently, even if a function has its D-mean value for any D, this does not imply the function is almost periodic. For example $f(x) = e^{iax}e^{-\varepsilon x^2}$, $x \in \mathbf{R}$, $a \neq 0$, $\varepsilon > 0$ has mean value zero, but it is not at all almost periodic. However, $f \in \dot{B}_{\infty,1}^0$ if ε is taken sufficiently small so that the support of $\hat{f}$ is away from the origin. This implies that an element of $\dot{B}_{\infty,1}^0$ is not necessarily almost periodic though it has mean value zero for any D.

LEMMA 10.4.3 *A function $f \in \dot{B}_{\infty,1}^0$ has D-mean value zero for any D.*

Proof. It suffices to prove that $\varphi_j * f$ has D-mean value zero for all $j \in \mathbf{Z}$ since $f_m \to f$ in $\dot{B}_{\infty,1}^0$ implies $f_m \to f$ in L^∞ for $f_m = \Sigma_{|j| \leq m} \varphi_j * f$. Let $\{\hat{\psi}_\ell\}_{\ell=1}^N \subset C_0^\infty(\mathbf{R}^n)$ be a partition of unity of the support of φ_j and supp $\hat{\psi}_\ell$ does not intersect the plane $\{\xi_k = 0\}$ for some $k = 1, \ldots, n$ (k may depend on ℓ). Then there is $\rho_\ell \in \mathcal{S}$ such that $\hat{\psi}_\ell = i\xi_k \,\hat{\rho}_\ell$. Thus we observe   .

$$\psi_\ell * \varphi_j * f = \partial_k(\rho_\ell * \varphi_j * f)$$

to get

$$F_R(x) := (\chi_D^R * \psi_\ell * \varphi_j * f)(x) = -\frac{1}{R^n |D|} \int_{\partial(RD)} n_k(\rho_\ell * \varphi_j * f)(x - y) d\mathcal{H}^{n-1}(y).$$

We estimate F_R to get

$$|F_R(x)| \le \frac{1}{R^n |D|} \|\rho_\ell * \varphi_j * f\|_\infty \mathcal{H}^{n-1}(\partial(RD)) \to 0$$

since $\rho_\ell * \varphi_j \in L^1$ is independent of R. Thus,

$$\varphi_j * f = \sum_{\ell=1}^N \psi_\ell * \varphi_j * f$$

has D-mean value zero. The proof is now complete. $\square$

An element of BUC having its mean value for some D may not have mean value for another D. Here is an example. We consider

$$f(x) = \frac{x}{\sqrt{x^2 + 1}} (\cos \log \sqrt{x^2 + 1} - \sin \log \sqrt{x^2 + 1}),$$

which is the derivative of

$$g(x) = \sqrt{x^2 + 1} \cos \log \sqrt{x^2 + 1}.$$

If $D = (-1, 1)$, the mean value exists and equals zero by uniform continuity of f. If $D = (-1/2, 3/2)$, the mean value does not exist. Indeed, $(\chi_D^R * f)(0) = \frac{1}{R}(g(3R/2) - g(R/2))$ does not converge as $R \to \infty$.

Our goal in this section is to prove that existence of mean value is preserved for the Navier-Stokes flow.

THEOREM 10.4.4 *Assume that u_0 has D-mean value $c \in \mathbf{R}^n$. Then the solution u of the Navier-Stokes equation with initial data u_0 (in Proposition 10.2.1) has D-mean value c for all $t \in (0, T_0)$.*

Proof. Since u solves (10.2.1), i.e.,

$$u(t) = e^{t\Delta} u_0 - \int_0^t \operatorname{div}(e^{(t-s)\Delta} \mathbf{P}(u \otimes u)) ds, \qquad (10.4.1)$$

it suffices to prove that $e^{t\Delta} u_0$ has D-mean value c and that the second term $F(t)$ of (10.4.1) belongs to $\dot{B}_{\infty,1}^0$ if we note that any element of $\dot{B}_{\infty,1}^0$ has D-mean value zero (Lemma 10.3). We shall prove these facts in the next lemmas. $\square$

LEMMA 10.4.5 *(i) If $a \in BUC(\mathbf{R}^n)$ has D-mean value $c \in \mathbf{R}$, then $f * a$ has D-mean value $c \int_{\mathbf{R}^n} f\,dx$ provided that $f \in L^1(\mathbf{R}^n)$. In particular, $e^{t\Delta} a$ has D-mean value c. (ii) If $u \in L^\infty(\mathbf{R}^n \times (0, T))$, then $F(t) \in \dot{B}_{\infty,1}^0(\mathbf{R}^n)$.*

Proof. (i) This is clear since

$$\|\chi_D^R * f * a - f * c\|_\infty = \|(\chi_D^R * a - c) * f\|_\infty \le \|f\|_{L^1}\|\chi_D^R * a - c\|_\infty \to 0$$

as $R \to \infty$.

(ii) We shall recall an estimate

$$\|\nabla e^{t\Delta} f\|_{\dot{B}_{\infty,1}^0} \le \frac{C}{t^{1/2}}\|f\|_{\dot{B}_{\infty,\infty}^0} \tag{10.4.2}$$

found, for example, in [I]. (Here the space $\dot{B}_{\infty,\infty}^0 (\subset S')$ is defined as the dual space of

$$\dot{B}_{1,1}^0 = \{f \in S' | f = \sum_{j=-\infty}^{\infty} f * \varphi_j \text{ in } S',$$

$$\|f\|_{\dot{B}_{1,1}^0} = \sum_{j=-\infty}^{\infty} \|\varphi_j * f\|_{L^1} < \infty\}.)$$

Using this estimate for $F(t)$, we observe that

$$\|F(t)\|_{\dot{B}_{\infty,1}^0} \le \int_0^t \frac{C}{(t-s)^{1/2}} |\mathbf{P}| \|u\|_\infty^2(s)ds,$$

where $|\mathbf{P}|$ is the operator norm in $\dot{B}_{\infty,\infty}^0$, which is finite (see, e.g., [A]); here we invoked the property that $\|f\|_{\dot{B}_{\infty,\infty}^0} \le C'\|f\|_\infty$. This estimate yields that $F(t) \in \dot{B}_{\infty,1}^0$. $\qquad\square$

Remark 10.4.6 (i) The estimate (10.4.2) also implies (10.2.3) if we note that $\|f\|_\infty \le \|f\|_{\dot{B}_{\infty,1}^0}$. (ii) A similar result holds for the Navier-Stokes equation in a rotating frame. In this case we have to assume that $\mathcal{W}$ component of the initial data u_0 has D-mean value.

Acknowledgment

This work is partly supported by the Grant-in-Aid for formation of COE "Mathematics of Nonlinear Structures via Singularities" (Hokkaido University). The work of the first author is partly supported by the Grant-in-Aid for Scientific Research, No.14204011, the Japan Society of the Promotion of Science. Much of the work of the first author was done when he was a faculty member of the Department of Mathematics at Hokkaido University. The work of the second and the third authors is partially supported by the AFOSR Contract FG9620-02-1-0026 and the CRDF Contract RU-M1-2596-ST-04.

REFERENCES

[AG] N. I. Akhiezer and I. M. Glazman (1961), *Theory of Linear Operators in Hilbert Space*, F. Ungar Pub. Co., New York.

[A]   H. Amann (1997), Operator-valued fourier multipliers, vector-valued Besov spaces, and applications, *Math. Nachr.*, 186, 5–56.

[BMN1]   A. Babin, A. Mahalov, and B. Nicolaenko (2001), 3D Navier-Stokes and Euler equations with initial data characterized by uniformly large vorticity, *Indiana University Mathematics Journal*, 50, 1–35.

[BMN2]   A. Babin, A. Mahalov, and B. Nicolaenko (1999), Global regularity of the 3D rotating Navier-Stokes equations for resonant domains, *Indiana University Mathematics Journal*, 48, No. 3, 1133–1176.

[BMN3]   A. Babin, A. Mahalov, and B. Nicolaenko (1997), Regularity and integrability of the 3D Euler and Navier-Stokes equations for uniformly rotating fluids, *Asympt. Anal.*, 15, No. 2, 103–150.

[CK]   J. R. Cannon and G. H. Knightly (1970), A note on the Cauchy problem for the Navier-Stokes equations, *SIAM J. Appl. Math.*, 18, 641–644.

[Ca]   M. Cannone (1995), *Ondelettes, Praproduits et Navier-Stokes*, ed. Diderot Arts et Sciences Paris–New York–Amsterdam.

[CaM]   M. Cannone and Y. Meyer (1995), Littlewood-Paley decomposition and Navier-Stokes equations, *Methods and Applications of Analysis*, 2, 307–319.

[Co]   C. Corduneanu (1968), *Almost Periodic Functions*, Interscience Publishers, New York.

[FMRT]   C. Foias, O. Manley, R. Rosa, and R. Temam (2001), *Navier-Stokes Equations and Turbulence, Encyclopedia of Mathematics and Its Applications*, vol. 83, Cambridge University Press, Cambridge, U.K.

[GIMM]   Y. Giga, K. Inui, A. Mahalov, and S. Matsui (2004), Navier-Stokes equations in a rotating frame in $\mathbb{R}^3$ with initial data nondecaying at infinity, *Hokkaido Math. J.*, 35, 321–364.

[GIM]   Y. Giga, K. Inui, and S. Matsui (1999), On the Cauchy problem for the Navier-Stokes equations with nondecaying initial data, *Quaderni di Matematika*, 4, 28–68.

[GMS]   Y. Giga, S. Matsui, and O. Sawada (2001), Global existence of two-dimensional Navier-Stokes flow with nondecaying initial velocity, *J. Math. Fluid Mech.*, 3, 302–315.

[I]   K. Inui, *Rotating Navier-Stokes Equations with Initial Data Nondecreasing at Infinity*, Sūrikaisekikenkyūsho Kōkyūroku, to appear.

[KT]   H. Koch and D. Tataru (2001), Well-posedness for the Navier-Stokes equations, *Advances in Mathematics*, 157, 22–35.

[MN]   A. Mahalov and B. Nicolaenko (2003), Global regularity of the 3D Navier-Stokes equations with weakly aligned large initial vorticity, *Russian Math. Surveys*, 58, No. 2 (350), 287–318.

[S]   O. Sawada (2004), The Navier-Stokes flow with linearly growing initial velocity in the whole space, *Bol. Soc. Paran. Mat.*, 22, 75–95.

[VF]   M. I. Vishik and A. V. Fursikov (1988), *Mathematical Problems of Statistical Hydromechanics*, Kluwer, Academic Publisher.

Chapter Eleven

Longtime Decay Estimates for the Schrödinger Equation on Manifolds

I. Rodnianski and T. Tao

11.1 INTRODUCTION

Let $(M, g) = (\mathbf{R}^3, g)$ be a compact perturbation of Euclidean space[1] $\mathbf{R}^3$, thus M is $\mathbf{R}^3$ endowed with a smooth metric g which equals the Euclidean metric outside of a Euclidean ball $B(0, R_0) := \{x \in \mathbf{R}^3 : |x| \leq R_0\}$ for some fixed R_0. We consider smooth solutions to the Schrödinger equation

$$u_t = -iHu, \qquad (11.1)$$

where for each time t, $u(t) : M \to \mathbf{C}$ is a Schwartz function on M, and H is the Hamiltonian operator

$$H := -\frac{1}{2}\Delta_M,$$

where $\Delta_M := \nabla^j \nabla_j$ is the Laplace-Beltrami operator (with ∇^j denoting covariant differentiation with respect to the Levi-Civita connection, in contrast with the Euclidean partial derivatives ∂_j). Note that H is positive definite and self-adjoint with respect to the natural inner product

$$\langle u, v \rangle_{L^2(M)} := \int_M u(x)\overline{v(x)}\, dg(x), \qquad (11.2)$$

where $dg := \sqrt{\det g_{ij}(x)}dx$ is the standard volume element induced by the metric g. In fact, the spectrum of H consists entirely of absolutely continuous spectrum on the positive real axis $[0, +\infty)$, in particular H has no eigenvalues or resonances at any energy. In particular, H enjoys a standard functional calculus on $L^2(M)$, the Hilbert space associated to the inner product, and one can define the homogeneous Sobolev norms $\|u\|_{\dot{H}^s(M)} := \|H^{s/2}u\|_{L^2(M)}$ on M using fractional powers of H for all $-1 \leq s \leq 1$. (One can of course define these norms for other s also, but there

---

[1] The analysis we give here also extends to higher dimensions $n > 3$, and are in fact slightly easier in those cases, but for simplicity of exposition we restrict our attention to the physically important three-dimensional case and to compact perturbations. The hypothesis that M is topologically $\mathbf{R}^3$ is technical but in any event is forced upon us by the nontrapping hypothesis, which we introduce below; see [14].

are some technicalities when s is too negative that we do not wish to address here.) Thus, for instance, $\|u\|_{\dot{H}^0(M)} = \|u\|_{L^2(M)}$ and

$$\|u\|^2_{\dot{H}^1(M)} = \langle u, Hu \rangle_{L^2(M)} = \frac{1}{2} \int_M |\nabla u|_g^2 \, dg. \tag{11.3}$$

It is well known (see [2]) that for any time t_0 and any Schwartz initial data u_0 there exists a unique global-in-time Schwartz solution $u : \mathbf{R} \times M \to \mathbf{C}$ to (11.1) with initial data $u(0) = u_0$; indeed, we have $u(t) = e^{-itH}u_0$. In this paper we develop a quantitative variant of Enss's method to obtain a new *global-in-time* local smoothing estimate for such solutions; to avoid needless technicalities we shall always restrict ourselves to Schwartz solutions. The methods here extend to more general classes of Hamiltonians than those considered here; for instance, they can handle asymptotically flat manifolds of dimension $n \geq 3$ as well as short-range potentials, provided that there are no resonances or eigenfunctions at zero; however, to simplify the exposition, we have chosen to restrict attention to the simple case of zero potential and compact perturbations of three-dimensional Euclidean space. We shall pursue more general Hamiltonians in [9] using a somewhat different method (based on limiting absorption principles).

Let us begin by recalling some earlier results. In the case where M is Euclidean space $\mathbf{R}^n$, so that $H = H_0 := -\frac{1}{2}\Delta_{\mathbf{R}^3}$ is just the free Hamiltonian, then we have the well-known global-in-time local smoothing estimate (see [3], [11], [15]), which we shall phrase as

$$\int_{-\infty}^{\infty} \|\langle x \rangle^{-1/2-\sigma} \nabla e^{-itH_0}u_0\|^2_{L^2(\mathbf{R}^3)}$$
$$+ \|\langle x \rangle^{-3/2-\sigma} e^{-itH_0}u_0\|^2_{L^2(\mathbf{R}^3)} \, dt \leq C_\sigma \|u_0\|^2_{\dot{H}^{1/2}(\mathbf{R}^3)} \tag{11.4}$$

for any $\sigma > 0$ and any Schwartz initial data f, where $\langle x \rangle := (1 + |x|^2)^{1/2}$. It is well known that the condition on σ is sharp (there is a logarithmic divergence in the left-hand side when $\sigma = 0$).

Now we suppose that $(M, g) = (\mathbf{R}^3, g)$ is a compact perturbation of Euclidean space obeying the nontrapping condition. It is known (see [2], [5]) that one has the *local-in-time* local smoothing estimate

$$\int_I \|\langle x \rangle^{-1/2-\sigma} \nabla e^{-itH}u_0\|^2_{L^2(M)}$$
$$+ \|\langle x \rangle^{-3/2-\sigma} e^{-itH}u_0\|^2_{L^2(M)} \, dt \leq C_{\sigma,I,M} \|u_0\|^2_{\dot{H}^{1/2}(M)} \tag{11.5}$$

for all *compact* time intervals $I \subset \mathbf{R}$, where the constant on the right-hand side is allowed to depend on I, if and only if the manifold M is *nontrapping*, i.e., every geodesic $s \mapsto x(s)$ in M eventually goes to spatial infinity, as has a theory for the Schrödinger equation as $s \to \pm\infty$. To understand why the nontrapping condition is necessary, observe that the localization in time means that the low and medium energies are easily controlled (for instance, by using the Sobolev embedding $\dot{H}^{1/2}(M) \subseteq L^3(M)$ and Hölder's inequality, noting that at low and medium energies $\nabla e^{-itH}u_0$ is also in $\dot{H}^{1/2}(M)$), and so one only needs to understand the evolution of the high energies, which evolve semiclassically. The semiclassical limit of the estimate (11.5) is the estimate $\int_{\mathbf{R}} \langle x(s) \rangle^{-1-2\sigma} |\dot{x}(s)|^2 \, ds \leq C_{\sigma,M} |\dot{x}(0)|$ for any geodesic $s \mapsto x(s)$, which is easily seen to hold if and only if the manifold is nontrapping (this also explains the requirement that $\sigma > 0$). Of course, one needs semiclassical tools such as pseudo-differential operators and the positive commutator method in order to make this argument rigorous; see [2], [5] (or Section 11.6) for more details.

The main result of this paper is to unify the global-in-time Euclidean estimate (11.4) with the local-in-time manifold estimate (11.5) as follows.

THEOREM 11.1.1 *Let M be a smooth compact perturbation of $\mathbf{R}^3$ which is non-trapping and which is smoothly diffeomorphic to $\mathbf{R}^3$. Then, for any Schwartz solution $u(t, x)$ to (11.1) and any $\sigma > 0$, we have*

$$\int_{\mathbf{R}} \| \langle x \rangle^{-1/2-\sigma} \nabla e^{-itH} u_0 \|^2_{L^2(M)} \tag{11.6}$$
$$+ \| \langle x \rangle^{-3/2-\sigma} e^{-itH} u_0 \|^2_{L^2(M)} \, dt \leq C_{\sigma,M} \| u_0 \|^2_{\dot{H}^{1/2}(M)}.$$

In other words, the constant $C_{\sigma,I,M}$ in (11.5) can be taken to be independent of the interval I.

As mentioned earlier, this is not the strongest result that one could obtain with this method; our purpose here is merely to illustrate a model example in which the method applies. By using different methods (in particular the limiting absorption principle) we were able to obtain results for more general Hamiltonians; see [9]. However, we believe the method we present here still has some merit; in particular, it is conceptually straightforward and seems to have some hope of generalizing to more "time-dependent" or "nonlinear" situations in which spectral theory tools are less useful. For instance, the methods here were already used in [13] to obtain new results on the asymptotic behavior of focusing nonlinear Schrödinger equations, at least in the spherically symmetric case.

We now informally discuss the proof of the theorem. First note that this estimate is already proven when M is Euclidean space $\mathbf{R}^3$, and we are only considering manifolds M that are compact perturbations of Euclidean space. Thus it is reasonable to expect that the only difficulties in proving (11.6) will arise from the compact region $B(0, R_0)$, and indeed by modifying the proof of (11.4) we will be able to show that a global-in-time local smoothing estimate in a slight enlargement $B(0, R)$ of $B(0, R_0)$ will automatically imply the full estimate (11.6). Thus we may (heuristically, at least) restrict our attention to a ball such as $B(0, R)$.

Next, we use the spectral theorem to decompose the evolution (11.1) into low-energy, medium-energy, and high-energy components. The high-energy components turn out to be treatable by the same positive commutator arguments used to prove the local-in-time estimate (11.5). From heuristic viewpoint, this is because high-energy components propagate very quickly and thus only linger in the compact region $B(0, R)$ for a very short period of time, after which they escape into the Euclidean region of M and never return to $B(0, R)$ again. Of course, the nontrapping hypothesis is essential here. At a more technical level, the reason why we can adapt the arguments used to prove (11.5) is that the error terms generated by the positive commutator method are lower order than the main term, and can thus be absorbed by the main term in the high-energy regime even when the time interval I is unbounded.

. The low-energy components are easy to treat, but for a different reason, namely that there is an "uncertainty principle" that shows that solutions that have extremely low energy cannot be concentrated entirely in the compact region K, and thus the low energies cannot be the dominant component to the local smoothing estimate inside this region K. This is ultimately a reflection of the well-known fact that a Hamiltonian $H = -\frac{1}{2}\Delta_M$ with no potential does not have any resonances or bound states at zero or negative energies.

The most interesting component to treat is the medium energies regime, which requires new methods. These are energies that are not high enough to behave semi-classically and escape the compact region K by means of the nontrapping hypothesis, but which are not low enough to escape the compact region K by means of the uncertainty principle. Fortunately, there is a third mechanism by which we can force solutions of the Schrödinger equation to escape the compact region K, namely the Ruelle-Amrein-Georgescu-Enss (RAGE) theorem. The point is that (as is well known) H contains no embedded eigenfunctions in the medium-energy portion of the spectrum (or indeed anywhere in the spectrum), and so the RAGE theorem then ensures that any given solution to (11.1) must eventually vacate the region K after some time T. We recall the abstract version of the RAGE Theorem (see, e.g., [8]).

THEOREM 11.1.2 (RAGE) *Let H be a self-adjoint operator on a Hilbert space $\mathcal{H}$. If $C : \mathcal{H} \to \mathcal{H}$ is a compact operator and u_0 lies in the continuous subspace of H, then*

$$\lim_{T\to\infty} \frac{1}{T} \int_0^T \|Ce^{-itH}u_0\|_H^2 = 0.$$

The RAGE theorem played a crucial role in Enss's approach to scattering for Schrödinger operators $H = H_0 + V$, [6]. In fact, our treatment of medium energies can be viewed as a quantitative version of Enss's method in a sense that we use both the RAGE type inequality and the decomposition into incoming and outgoing waves, also a major part of Enss's work, but derive an a priori estimate, as opposed to a qualitative result about completeness of the wave operators. The other major difference is that in scattering theory, density arguments allow one to consider *only* medium energies and compactly supported data, whereas we must necessarily treat all energy ranges and allow our data to have arbitrary support.

One might object that the RAGE theorem is "qualitative" in nature, in that the time T required for a solution to leave K depends on the choice of solution and thus need not be uniform. However, because we have localized the solution in both frequency (to medium energies) and position (to the region K), and because we can use linearity to normalize the $H^{1/2}(M)$ norm of u, the solution is in fact effectively contained in a compact region of (quantum) phase space. Because of this, one can make the time T required for a solution to leave K to be uniform for all medium-energy solutions u.

There is, however, still a remaining difficulty for medium-energy solutions, which is that once a solution leaves the compact region K one needs to ensure that it does not return back to K, since a solution that periodically left and then returned to K would eventually contribute an infinite left-hand side to (11.6). To resolve this we introduce a quantitative version of Enss's method. The starting point is the observation that, in the exterior of the domain K, any function can be decomposed into "outgoing" and "incoming" components (very roughly speaking, this corresponds to the spectral projections $\chi_{[0,+\infty)}(i\partial_r)$ and $\chi_{(-\infty,0]}(i\partial r)$ where ∂_r is the radial derivative, although we shall not perform these projections directly due to the singular behavior of the operator $i\partial_r$). Outgoing components will evolve toward spatial infinity as $t \to +\infty$, whereas incoming components will evolve toward spatial infinity as $t \to -\infty$. In particular, in both cases the solution will not encounter the compact region K, and the evolution is essentially Euclidean in nature.

Now suppose that a medium-energy solution to (11.1) is localized to K at some time t_0. By the quantitative RAGE theorem, at some later time $t_0 + T$ the solution has mostly vacated the region K. By Enss's decomposition it can have either outgoing or incoming components. But one can show that there is almost no incoming component, because if we evolved backward in time from $t_0 + T$ back to t_0 we see that the incoming component would have evolved back to a region far away from K, and thus be orthogonal to the initial data. Thus, at time $t_0 + T$, the solution consists almost primarily of outgoing components. But then, by the preceding discussion this means that the solution will continue to radiate to spatial infinity for times after t_0 and thus never return to K. To summarize, we have shown that any component of a medium-energy solution that is located in K at time t_0 will eventually radiate to spatial infinity as $t \to +\infty$; a similar argument also handles the $t \to -\infty$ evolution. Combining this with the finite energy of u (note that in the medium energy regime all Sobolev norms are equivalent), one can obtain the estimate (11.6).

The global-in-time local smoothing estimate in Theorem 11.1.1 has a number of consequences; for instance, by combining it with the arguments of Staffilani and Tataru [12] one can obtain global-in-time Strichartz estimates for compact nontrapping perturbations of Euclidean space, which then can be used to transfer some local and global existence results for nonlinear Schrödinger equations on Euclidean space, to the setting of compact nontrapping perturbations. We will not discuss these (fairly standard) generalizations here, but see [9] for further discussion.

11.2 NOTATION

Throughout this paper, the manifold M, the radius R_0, and the exponent σ will be fixed. All constants C will be allowed to depend on σ and M (and hence on R_0). If C needs to depend on other parameters, we will indicate this by subscripts.

Suppose $A_0, A_1, \ldots, A_k$ are real parameters. We use $O_{A_1,\ldots,A_k}(1)$ to denote any quantity depending on $A_1, \ldots, A_k$ (and possibly some other quantities), which is bounded in magnitude by some constant $C_{A_1,\ldots,A_k}$. For some fixed c (usually $c = 0$ or $c = \infty$), we also use $o_{A_0 \to c; A_1,\ldots,A_k}(1)$ to denote a quantity depending on $A_0, A_1, \ldots, A_k$ (and possibly some other parameters), which is bounded in magnitude by some quantity $F_{A_1,\ldots,A_k}(A_0)$, such that

$$\lim_{A_0 \to c} F_{A_1,\ldots,A_k} = 0 \text{ for all choices of } A_1, \ldots, A_k.$$

Often k will be zero, in which case the above notations simply read $O(1)$ and $o_{A_0 \to c}(1)$, respectively. We also abbreviate $O_{A_1,\ldots,A_k}(1)X$ and $o_{A_0 \to c; A_1,\ldots,A_k}(1)X$ as $O_{A_1,\ldots,A_k}(X)$ and $o_{A_0 \to c; A_1,\ldots,A_k}(X)$, respectively.

In the absence of parentheses, we read operators from right to left. Thus, for instance, $\nabla f g$ denotes the function $\nabla(fg)$ rather than $(\nabla f)g$.

We use the usual summation conventions on indices, and use g^{jk} to denote the dual metric to g_{jk} on the cotangent bundle, and use g to raise and lower indices in the usual manner. We use ∇_j, ∇^j to denote the usual covariant derivatives with respect to the Levi-Civita connection on M; these can be applied to any tensor field; we use $|\nabla f|_g = \sqrt{\nabla^j f \overline{\nabla_j f}}$ to denote the magnitude of the gradient with respect to the metric g, and $|\nabla f|$ to denote the Euclidean magnitude of the gradient. Since $\nabla g = 0$, the indices of covariant derivatives can be raised and lowered freely; thus, for instance, $\Delta_M = \nabla^j \nabla_j = \nabla_j \nabla^j$. Also, these covariant derivatives are antiselfadjoint with respect to the inner product (11.2), and thus we can integrate by parts using these derivatives freely.

As the operator H is self-adjoint and has spectrum on $[0, \infty)$, we can construct spectral multipliers $f(H)$ for any measurable function $f : [0, \infty) \to \mathbf{C}$ of at most polynomial growth; in particular, we can define fractional powers $H^{s/2}$ and $(1 + H)^{s/2}$, as well as Schrödinger propagators e^{-itH} and Littlewood-Paley type operators on H. These spectral multipliers commute with each other, and are bounded on L^2 if their symbol f is bounded.

11.3 OVERVIEW OF PROOF

We now begin the proof of Theorem 11.1.1. The first step will be to show that one can freely pass back and forth between the slowly decaying weight $\langle x \rangle^{-1/2-\sigma}$ in (11.6) and a suitably chosen compactly supported weight φ. By shrinking σ as necessary we may assume that $0 < \sigma \ll 1$.

Fix any compact time interval $[0, T]$, and let $K(T)$ be the best constant for which the inequality

$$\int_0^T \| \langle x \rangle^{-1/2-\sigma} |\nabla e^{-itH} u_0|_g \|_{L^2(M)}^2$$
$$+ \| \langle x \rangle^{-3/2-\sigma} e^{-itH} u_0 \|_{L^2(M)}^2 \, dt \leq K(T)^2 \|u_0\|_{\dot{H}^{1/2}(M)}^2 \tag{11.7}$$

holds for any Schwartz function u_0. From the (local-in-time) local smoothing theory in [2], [5] we already know that $K(T)$ is finite for each T. Our task is to show that $K(T)$ is bounded independently of T; the negative times can then be handled by time reversal symmetry.

Recall that M is equal to Euclidean space in the exterior region $|x| > R_0$. To take advantage of this, let us fix $\varphi : M \to \mathbf{R}$ to be a smooth function that equals 1 when

$|x| \leq 4R_0$ and equals 0 when $|x| \geq 8R_0$. We then define the localized quantity $K_\varphi(T)$ to be the best constant, such that

$$\int_0^T \||\varphi|\nabla e^{-itH} u_0|_g\|_{L^2(M)}^2 \, dt \leq K_\varphi(T)^2 \|u_0\|_{\dot{H}^{1/2}(M)}^2 \tag{11.8}$$

holds for any Schwartz solution u to (11.1). It is clear that $K_\varphi(T) \leq C_\varphi K(T)$. In Section 11.4 we shall establish the converse inequality

$$K(T) \leq C_\varphi + C_\varphi K_\varphi(T) \tag{11.9}$$

for all times $T > 0$, where the constants C_φ depend on φ but not on T.

In light of (11.9), we see that to bound $K(T)$ it suffices to bound the localized quantity $K_\varphi(T)$. We make the technical remark that $K_\varphi(T)$ is required only to control first derivatives of $e^{-itH} u_0$, and not $e^{-itH} u_0$ directly. This will be important in the low-freqency analysis later on.

The next step is energy decomposition into the very low energy, medium energy, and very high energy portions of the evolution. Let $0 < \varepsilon_0 \ll 1$ be a small parameter (depending on φ) to be chosen later, and decompose $1 = P_{\text{lo}} + P_{\text{med}} + P_{\text{hi}}$, where P_{lo}, P_{med}, P_{hi} are the spectral multipliers

$$P_{\text{lo}} := \chi(H/\varepsilon_0); \quad P_{\text{med}} = \chi(\varepsilon_0 H) - \chi(H/\varepsilon_0); \quad P_{\text{hi}} := 1 - \chi(\varepsilon_0 H),$$

and $\chi : \mathbf{R} \to \mathbf{R}$ is a bump function supported on $[-1, 1]$ which equals 1 on $[-1/2, 1/2]$. We shall prove the following three propositions, in Sections 11.5, 11.7, 11.6, respectively:

PROPOSITION 11.3.1 (Low-energy estimate). *For any $T > 0$, we have*

$$\int_0^T \|\varphi\nabla P_{\text{lo}} e^{-itH} u_0\|_{L^2(M)}^2 \, dt \leq o_{\varepsilon_0 \to 0; \varphi}(K(T)^2) \|u_0\|_{\dot{H}^{1/2}(M)}^2.$$

PROPOSITION 11.3.2 (Medium-energy estimate). *For any $T > 0$, any time-step $\tau \geq 1$, and radius $R \geq 10R_0$, we have*

$$\int_0^T \|\varphi\nabla P_{\text{med}} e^{-itH} u_0\|_{L^2(M)}^2 \, dt \leq (C_{\varphi,\varepsilon_0,R,\tau} + o_{R \to \infty; \varphi, \varepsilon_0}(K(T)^2)$$
$$+ o_{\tau \to \infty; \varphi, \varepsilon_0, R}(K(T)^2)) \|u_0\|_{\dot{H}^{1/2}(M)}^2.$$

PROPOSITION 11.3.3 (High-energy estimate). *For any $T > 0$, we have*

$$\int_0^T \|\varphi\nabla P_{\text{hi}} e^{-itH} u_0\|_{L^2(M)}^2 \, dt \leq o_{\varepsilon_0 \to 0; \varphi}(K(T)^2) \|u_0\|_{\dot{H}^{1/2}(M)}^2.$$

Combining these three propositions using the triangle inequality and using the definition (11.7) of $K_\varphi(T)$, we see that

$$K_\varphi(T)^2 \leq o_{\varepsilon_0 \to 0, \varphi}(K(T)^2) + (C_{\varphi,\varepsilon_0,R,\tau}$$
$$+ o_{R \to \infty; \varphi, \varepsilon_0}(K(T)^2) + o_{\tau \to \infty; \varphi, \varepsilon_0, R}(K(T)^2)) + o_{\varepsilon_0 \to 0; \varphi}(K(T)^2)$$

for any $\varepsilon_0 > 0$, $\tau \geq 1$, and $R \geq 1$. If we choose τ sufficiently large depending on φ, ε_0 and R, and R sufficiently large depending on φ, ε_0, we conclude that

$$K_\varphi(T)^2 \leq C_{\varphi,\varepsilon_0} + o_{\varepsilon_0 \to 0; \varphi}(K(T)^2);$$

combining this with (11.9) we obtain

$$K(T)^2 \leq C_{\varphi,\varepsilon_0} + o_{\varepsilon_0 \to 0; \varphi}(K(T)^2);$$

letting ε_0 be sufficiently small depending on φ and recalling that $K(T)$ is finite, we conclude that $K(T) \leq C_\varphi$ for all time T, which gives Theorem 11.1.1.

It remains to prove (11.9) and Propositions 11.3.1, 11.3.2, 11.3.3. This will be done in the following sections.

11.4 PHYSICAL SPACE LOCALIZATION

We first prove (11.9). Let $u = e^{-itH} u_0$ be a Schwartz solution to (11.1). Let $T > 0$. We will allow all constants C to depend on φ and will no longer mention this dependence explicitly. We normalize $\|u_0\|_{\dot{H}^{1/2}(M)} = 1$, which by unitarity of e^{-itH} and spectral calculus implies that

$$\sup_{t \in \mathbf{R}} \|u(t)\|_{\dot{H}^{1/2}(M)} = 1. \tag{11.10}$$

Our task is to show that

$$\int_0^T \| \langle x \rangle^{-1/2-\sigma} \nabla u \|^2_{L^2(M)} + \| \langle x \rangle^{-3-\sigma} u \|^2_{L^2(M)} \, dt$$
$$\leq (C + C K_\varphi(T)^2) \|u_0\|^2_{\dot{H}^{1/2}(M)}.$$

Note that it will not be relevant whether we measure the magnitude of ∇u using the metric g or the Euclidean metric as they only differ by at most a constant.

From (11.7) we already have

$$\int_0^T \int_{|x| \leq 3R_0} |\nabla u|^2 \, dx dt \leq C K_\varphi(T)^2. \tag{11.11}$$

From an easy Poincaré inequality argument we also can show that

$$\int_{|x| \leq 3R_0} |u|^2 \, dx \leq C \left(\int_{|x| \leq 4R_0} |\nabla u|^2 \, dx + \int_{3R_0 \leq |x| \leq 4R_0} |u|^2 \, dx \right).$$

Thus it will suffice to work in the Euclidean region $|x| > 3R_0$ and prove that

$$\int_0^T \int_{|x| > 3R_0} |x|^{-3-2\sigma} |\nabla u|^2 + |x|^{-1-2\sigma} |u|^2 \, dx dt \leq C + C K_\varphi(T)^2. \tag{11.12}$$

We now invoke the positive commutator method. Let A be an arbitrary linear operator on Schwartz functions. From the self-adjoint nature of H, we observe the *Heisenberg identity*

$$\frac{d}{dt} \langle Au(t), u(t) \rangle_{L^2(M)} = \langle i[H, A]u(t), u(t) \rangle_{L^2(M)}, \tag{11.13}$$

where $i[H, A] = i(HA - AH)$ is the Lie bracket of H and A; integrating this in t and using (11.10) and the duality of $\dot{H}^{1/2}(M)$ and $\dot{H}^{-1/2}(M)$, we obtain

$$\left| \int_0^T \langle i[H, A]u(t), u(t) \rangle_{L^2(M)} \, dx \right| \leq C \|A\|_{\dot{H}^{1/2}(M) \to \dot{H}^{-1/2}(M)}. \tag{11.14}$$

The *positive commutator method* is based on choosing A so that $i[H, A]$ is mostly positive definite, in order to extract useful information out of (11.14).

Let us now set A equal to the self-adjoint first-order operator

$$A = -ia_{,k}\partial_k - i\partial_k a_{,k},$$

where $a : M \to \mathbf{R}$ is the function $a := \chi(|x| - \varepsilon|x|^{1-\varepsilon})$, where χ is a smooth cutoff supported on the region $|x| > 2R_0$, which equals 1 when $|x| \geq 3R_0$, and $0 < \varepsilon \ll 1$ is a sufficiently small constant depending on R_0, $a_{,k}$ denotes the Euclidean derivative of a in the e_k direction, and we are summing indices in the usual manner. Since $\nabla a = O(1)$ and $\nabla^2 a = O(1/|x|)$, we observe from (11.3) and the classical Hardy inequality $\|u/|x|\|_{L^2(\mathbf{R}^3)} \leq C\|u\|_{\dot{H}^1(\mathbf{R}^3)}$ that A maps $\dot{H}^1(M)$ to $L^2(M)$, and by self-adjointness also maps $L^2(M)$ to $\dot{H}^{-1}(M)$. By interpolation we conclude that A maps $\dot{H}^{1/2}$ to $\dot{H}^{-1/2}$. Also, since $H = -\frac{1}{2}\partial_j\partial_j$ on the support of χ, we can compute

$$\begin{aligned} i[H, A] &= -\frac{1}{2}[\partial_j\partial_j, a_{,k}\partial_k + \partial_k a_{,k}] \\ &= -\frac{1}{2}(a_{,kjj}\partial_k + 2a_{,kj}\partial_{jk} + \partial_k a_{,kjj} + 2\partial_k a_{,kj}\partial_j) \\ &= -2\partial_j a_{,kj}\partial_k - \frac{1}{2}a_{,jjkk}, \end{aligned}$$

and thus from (11.14) and an integration by parts we conclude that

$$\left| \int_0^T \int_{|x|>R_0} 2a_{,jk}\partial_k u \overline{\partial_j u} - \frac{1}{2}\Delta_{\mathbf{R}^3}^2 a|u|^2 \, dxdt \right| \leq C. \tag{11.15}$$

Let us first consider the portion of (11.15) on the region $|x| > 3R_0$, for which $a = |x| - \varepsilon|x|^{1-\varepsilon}$. Then a computation shows that a has some convexity if ε is sufficiently small; indeed, in this region we have

$$a_{,jk}\partial_k u \overline{\partial_j u} \geq c_\varepsilon \frac{|\nabla u|^2}{|x|^{1+\varepsilon}}; \quad \Delta^2 a \leq -c_\varepsilon/|x|^{3+\varepsilon}$$

for some $c_\varepsilon > 0$. Invoking (11.15) and using (11.11) to estimate the region where $2R_0 \leq |x| \leq 3R_0$, we conclude that

$$2\int_0^T \int_{|x|>3R_0} \frac{|\nabla u|^2}{|x|^{1+\varepsilon}} + c_\varepsilon \frac{|u|^2}{|x|^{2+\varepsilon}} \, dxdt \leq C_\varepsilon + C_\varepsilon K_\varphi(T)^2$$

$$+ \frac{1}{2}\int_0^T \int_{|x|\leq 3R_0} \Delta_{\mathbf{R}^3}^2 a|u|^2 \, dxdt.$$

To conclude (11.12), it thus suffices by (11.11) to establish the following fixed-time estimate:

LEMMA 11.4.1 *If $\varepsilon > 0$ is sufficiently small, then there exists a constant $C > 0$ such that*

$$\int_{|x| \le 3R_0} \Delta_{\mathbf{R}^3}^2 a |f|^2 \le C \int_{|x| \le 3R_0} |\nabla f|^2$$

for all smooth functions f. (Note that the left-hand side can be negative.)

Proof. It is possible to establish this from Poincaré inequality and the Green's function computation below (the main point being that $\Delta^2 a$ has negative mean), but we shall use a compactness argument instead. Let $\delta > 0$ be a small number to be chosen later. It clearly suffices to show that there exists $C > 0$ such that

$$\int_{|x| \le 3R_0} (\delta + \Delta_{\mathbf{R}^3}^2 a)|f|^2 \le \delta \int_{|x| \le 3R_0} |f|^2 + C \int_{|x| \le 3R_0} |\nabla f|^2$$

for all smooth functions f. We have chosen to use the Euclidean measure dx here, but one could equally well run the following argument using the measure dg.

Suppose for contradiction that the above estimate failed. Then we can find a sequence f_n of smooth functions with the normalization

$$\delta \int_{|x| \le 3R_0} |f_n|^2 + n \int_{|x| \le 3R_0} |\nabla f_n|^2 = 1,$$

such that

$$\limsup_{n \to \infty} \int_{|x| \le 3R_0} (\delta + \Delta^2 a)|f_n|^2 > 1.$$

By Rellich compactness we can find a subsequence f_{n_j} of f_n that converges in L^2 to a limiting object $f \in L^2(B(0, 3R_0))$, and then by Fatou's lemma

$$\int_{|x| \le 3R_0} |\nabla f|^2 = 0.$$

In other words, f is equal to a constant on the region $\langle x \rangle \le 3R_0$. But by Green's formula we have

$$\int_{|x| \le 3R_0} \delta + \Delta^2 a = O(\delta) + \int_{|x| = 3R_0} \frac{d}{dn} \Delta a \, dS$$

$$= O(\delta) + 4\pi(3R_0)^2 \left(-\frac{1}{(3R_0)^2} + O(\varepsilon) \right) = -4\pi + O(\varepsilon) + O(\delta)$$

which is negative if ε and δ are chosen sufficiently small. This is a contradiction, and the claim follows.

11.5 LOW-ENERGY ESTIMATE

Now we prove the low-energy estimate in Proposition 11.3.1, which is the easiest of the three propositions to prove, especially in the model case when there is no potential V, and the manifold M is a compact perturbation of $\mathbf{R}^3$. The idea here is to exploit the uncertainty principle to extract some gain from the spatial projection φ and the frequency projection P_{lo}, but in order to do this we need to somehow exploit the fact that H contains no resonances or bound states at the zero energy. For technical reasons (having to do with our use of the homogeneous norm $\dot{H}^{1/2}(M)$ instead of the inhomogeneous norm $H^{1/2}(M)$), we shall need the following nonstandard formulation of this nonresonance property.

PROPOSITION 11.5.1 (Laplace-Beltrami operators have no resonance). *Suppose that* $f : M \to \mathbf{C}$ *is a measurable, weakly differentiable function such that*

$$\int_M \langle x \rangle^{-3-2\sigma} |f(x)|^2 + \langle x \rangle^{-1-2\sigma} |\nabla f|^2 \, dg(x) < \infty$$

for some $K < \infty$, and such that $Hf \equiv 0$ in the sense of distributions. Then (if $\sigma > 0$ is sufficiently small) f is a constant.

Proof. We may take f to be real. The condition $Hf = 0$, combined with the local square-integrability of f, implies that f is in fact smooth thanks to elliptic regularity (and the smoothness of g). Since $Hf = 0$ and $H = H_0 = -\frac{1}{2}\Delta_{\mathbf{R}^3}$ outside of $B(0, R_0)$, the function $\Delta_{\mathbf{R}^3} f$ is a smooth, compactly supported function. Thus, if we set $F := \frac{1}{4\pi|x|} * \Delta_{\mathbf{R}^3} f$, i.e., the convolution of $\Delta_{\mathbf{R}^3} f$ with the fundamental solution of the Euclidean Laplacian, then $f - F$ is harmonic, F decays like $O(1/\langle x \rangle)$, and $|\nabla F|$ decays like $O(1/\langle x \rangle^2)$. From hypothesis and the triangle inequality, we then have

$$\int_M \langle x \rangle^{-3-2\sigma} |f(x) - F(x)|^2 \, dg(x) < \infty.$$

Since $f - F$ is harmonic, we conclude (e.g., from the mean-value theorem applied to $f - F$ and its first derivative) that $f - F$ is constant, thus $f - F = c$. By subtracting this constant from f we may in fact take $F = f$. This now shows that $f(x) = O(1/\langle x \rangle)$ and $\nabla f(x) = O(1/\langle x \rangle^2)$. But then we can justify the computation

$$\int_M |\nabla f|_g^2 \, dx = -\int_M f(x) \nabla^j \nabla_j f(x) \, dg(x)$$

$$= \int_M 2f(x) Hf(x) \, dg(x) = 0$$

by inserting a suitable smooth cutoff to a large ball $B(0, R)$ and then letting $R \to \infty$; we omit the standard details. But then we have $\nabla f = 0$ and we are done. $\quad\blacksquare$

Using this fact and a compactness argument, we now conclude

PROPOSITION 11.5.2 (Poincaré-type inequality). *Let $f : M \to \mathbf{R}$ be a Schwartz function. Let φ be as in previous sections. Then for any $\varepsilon > 0$ we have*

$$\|\varphi |\nabla f|\|_{L^2(M)} \leq o_{\varepsilon \to 0}(1)(\|\langle x \rangle^{-1/2-\sigma} \nabla f\|_{L^2(M)} + \|\langle x \rangle^{-3/2-\sigma} f\|_{L^2(M)}$$
$$+ \varepsilon^{-1} \|\langle x \rangle^{-1/2-\sigma} \nabla Hf\|_{L^2(M)} \tag{11.16}$$
$$+ \varepsilon^{-1} \|\langle x \rangle^{-3/2-\sigma} Hf\|_{L^2(M)}).$$

Proof. Suppose for contradiction that Proposition 11.5.2 was false. Then there exists a $\delta > 0$, a sequence $\varepsilon_n > 0$ converging to zero, and Schwartz functions f_n such that

$$\||\varphi|\nabla f_n|\|_{L^2(M)} > \delta(\|\langle x\rangle^{-1/2-\sigma}\nabla f_n\|_{L^2(M)} + \|\langle x\rangle^{-3/2-\sigma} f_n\|_{L^2(M)}$$
$$+ \varepsilon_n^{-1}\|\langle x\rangle^{-1/2-\sigma}\nabla H f_n\|_{L^2(M)}$$
$$+ \varepsilon_n^{-1}\|\langle x\rangle^{-3/2-\sigma} H f_n\|_{L^2(M)}).$$

Without loss of generality we may assume that f_n is not identically zero, and then we can normalize the expression in parentheses to equal 1. Thus,

$$\|\langle x\rangle^{-1/2-\sigma}\nabla f_n\|_{L^2(M)}, \|\langle x\rangle^{-3/2-\sigma} f_n\|_{L^2(M)} \leq 1;$$
$$\|\langle x\rangle^{-1/2-\sigma}\nabla H f_n\|_{L^2(M)} + \|\langle x\rangle^{-3/2-\sigma} H f_n\|_{L^2(M)} \leq \varepsilon_n; \qquad (11.17)$$
$$\||\varphi|\nabla f|\|_{L^2(M)} \geq \delta.$$

Next, we establish weighted $\dot{H}^2$ bounds on f_n via a Bochner identity. From (11.17) and Cauchy-Schwartz we have

$$\int_M \langle x\rangle^{-1-2\sigma}\mathrm{Re}((\nabla^j f_n)\overline{\nabla_j H f_n}) \leq \varepsilon_n = O(1).$$

We substitute $H = -\frac{1}{2}\nabla^k\nabla_k$. Since M is flat outside of the compact set $B(0, R_0)$, we have

$$\nabla_j H f_n = -\frac{1}{2}\nabla_j\nabla^k\nabla_k f_n = -\frac{1}{2}\nabla^k\nabla_j\nabla_k f_n + O(|\nabla f_n|)\chi_{B(0,R_0)}.$$

Using this and (11.17), we obtain

$$-\int_M \langle x\rangle^{1-2\sigma}\mathrm{Re}((\nabla^j f_n)\nabla^k\nabla_j\nabla_k\overline{f_n}) \leq C.$$

Integrating by parts we obtain

$$\int_M \langle x\rangle^{1-2\sigma}\mathrm{Re}((\nabla^k\nabla^j f_n)\nabla_j\nabla_k\overline{f_n})$$
$$+ (\nabla^k\langle x\rangle^{1-2\sigma})\mathrm{Re}((\nabla^j f_n)\nabla^k\nabla_j\nabla_k\overline{f_n}) \leq C.$$

Since $\mathrm{Re}((\nabla^j f_n)\nabla^k\nabla_j\nabla_k\overline{f_n}) = \frac{1}{2}\nabla_k|\nabla f_n|_g^2$, we can integrate by parts once more to obtain

$$\int_M \langle x\rangle^{1-2\sigma}|\mathrm{Hess}(f_n)|_g^2 \leq C + \frac{1}{2}\int_M (\Delta_M\langle x\rangle^{1-2\sigma})|\nabla f_n|_g^2.$$

Since $\Delta_M\langle x\rangle^{1-2\sigma} = O(\langle x\rangle^{-1-2\sigma})$, the integral on the right-hand side is $O(1)$ by (11.17). Thus,

$$\int_M \langle x\rangle^{1-2\sigma}|\mathrm{Hess}(f_n)|^2 \leq C. \qquad (11.18)$$

From this, (11.17), and Rellich compactness we see that the sequence f_n, when localized smoothly to any ball $B(0, R)$, is contained in a compact subset of $H^1(B(0, R))$. From this and the usual Arzela-Ascoli diagonalization argument, we may extract a subsequence of f_n that converges locally in H^1 to some limit f. From (11.17) and Fatou's lemma we then see that

$$\| \langle x \rangle^{-1/2-\sigma} \nabla f \|_{L^2(M)}, \ \| \langle x \rangle^{-3/2-\sigma} f \|_{L^2(M)} \leq 1; \quad \| \varphi | \nabla f | \|_{L^2(M)} \geq \delta \quad (11.19)$$

and that $Hf = 0$ in the sense of distributions. But then by Proposition 11.5.1, f is constant. But this contradicts the last estimate in (11.19), and we are done.

We can now quickly prove Proposition 11.3.1. Applying (11.7) with u_0 replaced by $P_{lo}u_0$ and $HP_{lo}u_0$, and using some spectral theory to estimate the right-hand side, we obtain

$$\int_0^T \| \langle x \rangle^{-1/2-\sigma} \nabla P_{lo} e^{-itH} u_0 \|_{L^2(M)}^2 + \| \langle x \rangle^{-3/2-\sigma} P_{lo} e^{-itH} u_0 \|_{L^2(M)}^2 \, dt$$

$$\leq CK(T)^2 \| u_0 \|_{\dot{H}^{1/2}(M)}^2$$

and

$$\int_0^T \| \langle x \rangle^{-1/2-\sigma} \nabla H P_{lo} e^{-itH} u_0 \|_{L^2(M)}^2 + \| \langle x \rangle^{-3/2-\sigma} P_{lo} e^{-itH} u_0 \|_{L^2(M)}^2 \, dt$$

$$\leq C \varepsilon_0^2 K(T)^2 \| u_0 \|_{\dot{H}^{1/2}(M)}^2;$$

note that P_{lo} and H are spectral multipliers and hence commute with each other and with e^{-itH}. Applying Proposition 11.5.2, Proposition 11.3.1 follows.

11.6 HIGH-ENERGY ESTIMATE

We now prove Proposition 11.3.3, which is the next easiest of the three propositions. This case resembles the local-in-time theory of Craig-Kappeler-Strauss [2] and Doi [5], and indeed our main tool here will be the positive commutator method applied to a certain pseudo-differential operator, exploiting the nontrapping hypothesis to ensure that the symbol of the pseudo-differential operator increases along geodesic flow. As we shall now be working in the high-energy setting, we will not need to take as much care with lower-order terms as in previous sections. For similar reasons, we will not need to use the homogeneous Sobolev spaces $\dot{H}^s(M)$, relying instead on the more standard (and more stable) inhomogeneous Sobolev spaces $H^s(M)$. The argument here is in fact quite general and would work on any asymptotically conic manifold with a short-range metric perturbation and a short-range potential.

It will be convenient to use the *scattering pseudo-differential calculus*, which is an extension of the standard pseudo-differential calculus that keeps track of the decay of the symbol at infinity. We briefly summarize the relevant features of this calculus here, referring the reader to [2] for more complete details. For any $m, l \in \mathbf{R}$, we define a *symbol* $a : T^*M \to \mathbf{C}$ of order (m, l) to be any smooth function obeying the bounds

$$|\nabla_x^\alpha \nabla_\xi^\beta a(x, \xi)| \leq C_{\alpha, \beta} \langle \xi \rangle^{m - |\beta|} \langle x \rangle^{-l - |\alpha|};$$

the function $a(x, \xi) = \langle x \rangle^{-l} \langle \xi \rangle^m$ is a typical example of such a symbol. Note that we assume that each derivative in x gains a power of $\langle x \rangle$, in contrast to the standard symbol calculus in which no such gain is assumed. We let $S^{m,l}(\overline{M})$ denote the space of such symbols. Given any such symbol $a \in S^{m,l}(\overline{M})$, we can define an associated pseudo-differential operator $A = \mathrm{Op}(a)$ by the usual Kohn-Nirenberg quantization formula

$$\mathrm{Op}(a)u(x) := (2\pi)^{-n} \int e^{i\langle x-y,\xi\rangle} a(x, \xi) u(y)\, dy\, d\xi.$$

We sometimes denote a by $\sigma(A)$ and refer to it as the *symbol* of A. Heuristically speaking, we have $A = \sigma(A)(x, \frac{1}{i}\nabla_x)$. We refer to the class of pseudo-differential operators of order (m, l) as $\Psi_{\mathrm{sc}}^{m,l}$. Also, if $h : \mathbf{R} \to \mathbf{C}$ is any spectral symbol of order $m/2$, the corresponding spectral multiplier $h(H)$ is a pseudo-differential operator of order $(m, 0)$. In particular, $(1 + H)^{m/2}$ has order $(m, 0)$, and the Littlewood-Paley type operators P_{lo}, P_{med}, P_{hi} have order $(0, 0)$. We caution, however, that the Schrödinger propagators e^{-itH} are not pseudo-differential operators.

The composition of an operator $A = \mathrm{Op}(a)$ of order (m, l) with an operator of $B = \mathrm{Op}(b)$ order (m', l') is an operator AB of order $(m + m', l + l')$, whose symbol $\sigma(AB)$ is equal to $\sigma(A)\sigma(B)$ plus an error of order $(m + m' - 1, l + l' + 1)$; note the additional gain of 1 in the decay index l, which is not present in the classical calculus. Similarly, the commutator $i[A, B]$ will be an operator of order $(m + m' - 1, l + l' - 1)$ with symbol $\sigma(i[A, B])$ equal to the Poisson bracket

$$\{\sigma(A), \sigma(B)\} := \nabla_x \sigma(A) \cdot \nabla_\xi \sigma(B) - \nabla_\xi \sigma(A) \cdot \nabla_x \sigma(A),$$

plus an error of order $(m + m' - 2, l + l' + 2)$. We shall write the above facts schematically as

$$\sigma(AB) = \sigma(A)\sigma(B) + O(S^{m+m'-1,l+l'+1});$$
$$\sigma(i[A, B]) = \{\sigma(A), \sigma(B)\} + O(S^{m+m'-2,l+l'+2}),$$

or equivalently as

$$\mathrm{Op}(a)\,\mathrm{Op}(b) = \mathrm{Op}(ab) + O(\Psi_{\mathrm{sc}}^{m+m'-1,l+l'+1}),$$
$$i[\mathrm{Op}(a), \mathrm{Op}(b)] = \mathrm{Op}(\{a, b\}) + O(\Psi_{\mathrm{sc}}^{m+m'-2,l+l'+2}).$$

In particular, since H has order $(2, 0)$ and has principal symbol $\frac{1}{2}|\xi|^2_{g(x)}$ plus lower-order terms of order $(1, 1)$ and $(0, 2)$, we see that if $a \in S^{m,l}$, then we have

$$i[H, \mathrm{Op}(a)] = \mathrm{Op}(Xa) + O(\Psi_{\mathrm{sc}}^{m,l+2}),$$

where Xa denotes the derivative of a along geodesic flow in the cotagent bundle T^*M.

Associated with the scattering calculus are the weighted Sobolev spaces $H^{m,l}(M)$ defined (for instance) by

$$\|u\|_{H^{m,l}(M)} := \|\langle x\rangle^l (1 + H)^{m/2} u\|_{L^2(M)}$$

(many other equivalent expressions for this norm exist, of course); when $l = 0$ this corresponds to the usual Sobolev space $H^m(M)$. It is easy to verify that a scattering pseudo-differential operator of order (m, l) maps $H^{m',l'}(M)$ to $H^{m'-m,l'+l}(M)$ for any m', l'.

In [2] (see also [5]) it was shown that the nontrapping hypothesis on M allows one to construct a real-valued symbol $a \in S^{1,0}$ (depending on φ) which was nondecreasing along geodesic flow, $Xa \geq 0$, and in fact obeyed the more quantitative estimate

$$Xa(x, \xi) = \varphi(x)|\xi|_g^2 + |b|^2$$

for some symbol b of order $(1, 1/2 - \sigma)$. In Euclidean space, an example of such a symbol is $C_\varphi \frac{x}{\langle x \rangle} \cdot \xi$ for some sufficiently large constant C_φ. Quantizing this, we obtain

$$i[H, A] = \nabla^j \varphi(x) \nabla_j + B^* B + O(\Psi_{sc}^{1,2-2\sigma}),$$

where $A := \mathrm{Op}(a)$ is a symbol of order $(1, 0)$, and $B := \mathrm{Op}(b)$ is a symbol of order $(1, 1/2 - \sigma)$. We then apply the self-adjoint projection $P_{hi} = P_{hi}^*$ to both sides, and observe that this commutes with H, to obtain

$$i[H, P_{hi}^* A P_{hi}] = P_{hi}^* \nabla^j \varphi(x) \nabla_j P_{hi} + P_{hi}^* B^* B P_{hi} + P_{hi} O(\Psi_{sc}^{1,2-2\sigma}) P_{hi}.$$

Applying (11.14), and integrating by parts (discarding the positive term $B^* B$, and using that $\Psi_{sc}^{1,2-2\sigma}$ maps $H^{1/2,-1+\sigma}$ to $H^{-1/2,1-\sigma}$), we obtain

$$\int_0^T \int_M \varphi |\nabla e^{-itH} u_0|_g^2 \, dg dt \leq C \|P_{hi}^* A P_{hi}\|_{\dot{H}^{1/2}(M) \to \dot{H}^{-1/2}(M)} \|u_0\|_{\dot{H}^{1/2}(M)}^2$$

$$+ C \int_0^T \|P_{hi} e^{-itH} u_0\|_{H^{1/2,-1+\sigma}}^2 \, dt.$$

Since A is of order $(1, 0)$, and P_{hi} maps the homogeneous Sobolev spaces to their inhomogeneous counterparts, we obtain a bound of the form

$$\|P_{hi}^* A P_{hi}\|_{\dot{H}^{1/2}(M) \to \dot{H}^{-1/2}(M)} \leq C_\varphi.$$

To finish the proof of Proposition 11.3.3, it thus suffices to show that

$$\int_0^T \|P_{hi} e^{-itH} u_0\|_{H^{1/2,-1+\sigma}}^2 \, dt \leq o_{\varepsilon_0}(1) \|u(0)\|_{\dot{H}^{1/2}(M)}^2.$$

On the other hand, applying (11.7) to $H^{-j} P_{hi} u(0)$ for $j = 0, 1$ we have

$$\int_0^T \|\nabla H^{-j} e^{-itH} P_{hi} u_0\|_{H^{0,-1/2-\sigma}}^2$$

$$+ \|H^{-j} e^{-itH} P_{hi} u_0\|_{H^{0,-3/2-\sigma}}^2 \, dt \leq C K(T)^2 \|u(0)\|_{\dot{H}^{1/2}(M)}^2$$

for $j = 0, 1$. Thus fixing t and setting $f := H^{-1} e^{-itH} P_{hi} u_0$ it will suffice to prove the fixed-time estimate

$$\|Hf\|_{H^{1/2,-1+\sigma}(M)} \leq o_{\varepsilon_0 \to 0}(1)$$

$$\times \sum_{j=0}^1 \left(\|\nabla H^j f\|_{H^{0,-1/2-\sigma}(M)} + \|H^j f\|_{H^{0,-3/2-\sigma}(M)} \right).$$

It suffices to verify this for real-valued f. By Rellich compactness, the space $H^{1,-1/2-\sigma}(M)$ embeds compactly into $H^{1/2,-1+\sigma}(M)$. Since $P_{hi} f \to 0$ as $\varepsilon_0 \to 0$ for each individual f, we thus see by compactness that it suffices to show that

$$\|Hf\|_{H^{1,-1/2-\sigma}(M)} \le C(\|\nabla Hf\|_{H^{0,-1/2-\sigma}(M)} + \|Hf\|_{H^{0,-3/2-\sigma}(M)}$$
$$+ \|\nabla f\|_{H^{0,-1/2-\sigma}(M)} + \|f\|_{H^{0,-3/2-\sigma}(M)}). \tag{11.20}$$

The top order term of $\|Hf\|_{H^{1,-1/2-\sigma}(M)}$ is already controlled by the right-hand side of (11.20), so it suffices to control the lower-order term $\|Hf\|_{H^{0,-1/2-\sigma}(M)}$. But an integration by parts allows us to write

$$\int_M \langle x \rangle^{-1-2\sigma} Hf Hf \, dg$$
$$= \frac{1}{2} \int_M \langle x \rangle^{-1-2\sigma} \left(\nabla^j f \nabla_j Hf \, dg - (1+2\sigma)\frac{\nabla^j \langle x \rangle}{\langle x \rangle} \nabla^j f Hf \right).$$

The claim then follows from the Cauchy-Schwartz inequality. This concludes the proof of Proposition 11.3.3.

11.7 MEDIUM ENERGY ESTIMATE

We now turn to the medium-energy estimate, Proposition 11.3.2, which is the hardest of the three propositions. Neither the uncertainty principle nor the nontrapping condition will be of much use here. Instead our tools[2] will be a RAGE-type theorem, exploiting the fact that H has no embedded eigenvalues, to propagate the solution away from the origin after a long time τ (though the energy localization shows the solution will not move *too* far away from the origin in bounded time), combined with a decomposition of phase space into incoming and outgoing waves (cf. Enss's method, [6], [10]), and several applications of Duhamel's formula. Because we have eliminated the high frequencies, we will enjoy an approximate finite speed of propagation law for the solution (but the upper bound of the speed is quite large, being roughly $O(\varepsilon_0^{-1/2})$); and because we have eliminated the low energies, we will not encounter a frequency singularity when we decompose into incoming and outgoing waves, although again we will pick up some negative powers of ε_0. We shall compensate for these ε_0 losses by using the RAGE theorem to gain a factor of $o_{\tau \to \infty; \varepsilon_0, R}(1)$ within a distance R from the origin, and to also gain a factor of $o_{R \to \infty; \varepsilon_0}(1)$ when one is farther than R from the origin. The reader should view the comparative magnitudes of ε_0, R, τ according to the relationship $1/\varepsilon_0 \ll R \ll \tau$, which is of course the most interesting case of Proposition 11.3.2.

It is instructive at this point to recall the basic features of Enss's method. Define the wave operator $W_+ = s-\lim_{t \to +\infty} e^{itH} e^{-itH_0}$. Enss's method is concerned with establishing *completeness* of W_+, i.e., showing that the range of W_+ coincides with the continuous subspace of $H = H_0 + V$. We assume otherwise so that there exists ϕ_0 not in the range of W_+. Density arguments allow us to consider ϕ_0 with compact support and medium energies only. We then evolve ϕ_0 by e^{-itH} and claim that we

---

[2] One can also proceed via Kato's theory of H-smooth operators, using the limiting absorption principles obtained in [1]. We will pursue this approach in detail in [9].

can find a sequence of times t_n and a decomposition

$$e^{-it_n H}\phi_0 = \phi_n = \phi_{n,out} + \phi_{n,in} + \tilde{\phi}_n. \tag{11.21}$$

Compare this with the decomposition into F_{loc}, F_{glob} in (11.24) and a further decomposition of F_{glob} into the outgoing/incoming waves in (11.34)) with the properties that

$$\|(W_+ - 1)\phi_{n,out}\|_{L^2}, \|\phi_{n,in}\|_{L^2}, \|\tilde{\phi}_n\|_{L^2} = o_{n\to\infty}(1).$$

If we had such a decomposition, we could conclude from the L^2 boundedness of W_+ that

$$\|W_+ e^{it_n H_0}\phi_0 - \phi_0\|_{L^2} = o_{n\to\infty}(1),$$

and thus ϕ_0 lies in the range of W_+. The desired decomposition (11.21) can be found, for instance, in [10], and we sketch it as follows. The local component $\tilde{\phi}_n$ can be set, for instance, to $\tilde{\phi}_n := \chi_{|x|\leq n}\phi_n$; its convergence to zero is a consequence of the RAGE theorem (if the times t_n are chosen appropriately). The global component $(1 - \chi_{|x|\leq n})\phi_n$ is partitioned into functions $\phi_{n,\alpha}$ supported in unit balls with centers at the lattice points $\alpha \in \mathbf{R}^n$. One can then define the

$$\phi_{n,\alpha,out} = \int_{\xi,y} e^{i(x-y)\cdot\xi} m(\xi)\phi_{n,\alpha}(y)\, d\xi\, dy,$$

$$\phi_{n,\alpha,in} = \int_{\xi,y} e^{i(x-y)\cdot\xi}(1 - m(\xi))\phi_{n,\alpha}(y)\, d\xi\, dy,$$

where $m(\xi)$ is a smooth multiplier localizing to the region $\angle(\xi,\alpha) \leq 3\pi/2$. Since it is expected that a compact support function ϕ_0 propagated by $e^{-it_n H}$ should become mostly outgoing in the region $|x| \geq n$, we have that $\phi_{n,in} \to 0$, while the fact that $(W_+ - 1)\phi_{n,out} \to 0$ in L^2 follows since on the outgoing waves the evolution e^{-itH} is well approximated by the free flow e^{-itH_0} and in fact converges to it as $n \to \infty$. Our proof of Proposition 11.3.2 will follow in spirit the above construction, although we have to take additional care since we do not work with functions of compact support and need weighted estimates instead of L^2 bounds. On the other hand, we are still localized in energy, and so we will not be too concerned about losing or gaining too many derivatives (as we are able to lose factors of ε_0 or ε_0^{-1} in our estimates here). In particular, the metric perturbation H of H_0 is now of similar "strength" to a potential perturbation, and we will now be able to use H_0 as a reasonable approximant to H, at least when the solution is far away from the origin.

It is more convenient to work in the dual formulation. First observe that the claim is easy when $T \leq \tau$, since $P_{med}e^{-itH}$ maps $\dot{H}^{1/2}(M)$ to $H^1(M)$ thanks to the frequency localization of P_{med}. Similarly, the $\int_0^\tau$ portion of the integral is easy to deal with. Thus, we may assume that $T > \tau$ and reduce to proving

$$\int_\tau^T \|\varphi\nabla P_{med}e^{-itH}u_0\|_{L^2(M)}^2\, dt \leq (C_{\varepsilon_0,R,\tau} + o_{R\to\infty;\varepsilon_0}(K(T))$$

$$+ o_{\tau\to\infty;\varepsilon_0,R}(K(T)))\|u_0\|_{\dot{H}^{1/2}(M)}^2,$$

which after dualization (and shifting t by τ) becomes

$$\left\| \int_0^{T-\tau} e^{itH} e^{i\tau H} P_{\text{med}} \nabla_j \varphi F^j (t + \tau)\, dt \right\|_{\dot{H}^{-1/2}(M)}$$

$$\leq (C_{\varepsilon_0, R, \tau} + o_{R \to \infty; \varepsilon_0}(K(T)) + o_{\tau \to \infty; \varepsilon_0, R}(K(T))) \quad (11.22)$$

$$\times \left(\int_0^{T-\tau} \| F(t + \tau) \|_{L^2(M)}^2\, dt \right)^{1/2}$$

for any vector field $F(t)$ defined on the time interval $[\tau, T]$ which is Schwartz for each time t.

Fix F. We split

$$e^{itH} e^{i\tau H} P_{\text{med}} \nabla_j \varphi F^j (t + \tau) = e^{itH} F_{\text{loc}}(t) + e^{itH} F_{\text{glob}}(t), \quad (11.23)$$

where

$$\begin{aligned} F_{\text{loc}}(t) &:= \varphi_R e^{i\tau H} P_{\text{med}} \nabla_j \varphi F^j (t + \tau); \\ F_{\text{glob}}(t) &:= (1 - \varphi_R) e^{i\tau H} P_{\text{med}} \nabla_j \varphi F^j (t + \tau). \end{aligned} \quad (11.24)$$

11.7.1 Term 1. Contribution of the Local Part

Consider first the contribution of the local term $F_{\text{loc}}(t)$. To control this term we use the following local decay result:

PROPOSITION 11.7.1 (RAGE theorem). *Let φ_R be a bump function supported on $B(0, 2R)$, which equals 1 on $B(0, R)$. For all Schwartz vector fields f^j, we have*

$$\| \varphi_R e^{i\tau H} P_{\text{med}} \nabla_j \varphi f^j \|_{L^2(M)} \leq o_{\tau \to \infty; R, \varepsilon_0}(\| f \|_{L^2(M)}).$$

Proof. Observe that we have the crude bound

$$\| \varphi_R e^{i\tau H} P_{\text{med}} \nabla_j \varphi f^j \|_{L^2(M)} \leq C_{\varepsilon_0} \| f \|_{H^{-1,-1}(M)}$$

since $P_{\text{med}} \nabla \varphi$ maps $H^{-1,-1}$ to L^2, and $\varphi_R e^{i\tau H}$ is bounded on L^2. Since the unit ball of $L^2(M)$ is precompact in $H^{-1,-1}$, it thus suffices to prove the estimate

$$\| \varphi_R e^{i\tau H} P_{\text{med}} \nabla_j \varphi f^j \|_{L^2(M)} \leq o_{\tau \to \infty; R, f, \varepsilon_0}(1)$$

for all Schwartz vector fields f, the point being that the compactness allows us to ignore the dependence of the constants on f. Fix $\tilde{f}$, φ, R. Since f is Schwartz, the curve $\{\tilde{\varphi} e^{i\tau H} P_{\text{med}} \nabla_j \varphi f^j : \tau \in \mathbf{R}\}$ is bounded in $H^{1,1}(M)$ (for instance) and hence precompact in L^2. Thus, to prove the above strong convergence, it will suffice to prove the weak convergence result

$$|\langle \tilde{\varphi} e^{i\tau H} P_{\text{med}} \nabla_j \varphi f^j, \psi \rangle| \leq o_{\tau \to \infty; R, f, \varepsilon_0, \psi}(1)$$

for all Schwartz functions ψ.

Fix ψ. Since the spectrum of H is purely absolutely continuous,[3] we see from the Riemann-Lebesgue lemma and the spectral theorem that

$$\langle \varphi_R e^{i\tau H} P_{\text{med}} \nabla \varphi f, \psi \rangle = \langle e^{i\tau H} P_{\text{med}} \nabla_j \varphi f^j, \varphi_R \psi \rangle$$

$$= 0_{\tau \to \infty; P_{\text{med}} \nabla_j \varphi f^j, \varphi_R \psi} (1)$$

and the claim follows.

From this proposition we see in particular that

$$\| F_{loc}(t) \|_{L^2(M)} \leq 0_{\tau \to \infty; R, \varepsilon_0}(1) \| F(t + \tau) \|_{L^2(M)}.$$

From dualizing the second part of (11.7) we have

$$\left\| \int_0^T e^{itH} \langle x \rangle^{-3/2-\sigma} G(t) \, dt \right\|_{\dot{H}^{-1/2}(M)} \tag{11.25}$$

$$\leq K(T) \left(\int_0^T \| G(t) \|_{L^2(M)}^2 \, dt \right)^{1/2}$$

for any G. If we truncate the time interval to $T - \tau$, substitute $G(t) := \langle x \rangle^{3/2+\sigma} F_{loc}(t)$, and take advantage of the spatial localization of F_{loc} to $B(0, 2R)$, we thus have

$$\left\| \int_0^{T-\tau} e^{itH} F_{loc}(t) \, dt \right\|_{\dot{H}^{-1/2}(M)}$$

$$\leq 0_{\tau \to \infty; \varepsilon_0, R}(K(T)) \left(\int_0^{T-\tau} \| F(t + \tau) \|_{L^2(M)}^2 \, dt \right)^{1/2}.$$

Thus the first term in (11.23) is acceptable.

11.7.2 Term 2. Contribution of the Global Part

To conclude the proof of Proposition 11.3.2, it suffices to establish the estimate

$$\left\| \int_0^{T-\tau} e^{itH} F_{\text{glob}}(t) \, dt \right\|_{\dot{H}^{-1/2}(M)} \tag{11.26}$$

$$\leq (C_{\varepsilon_0, R, \tau} + 0_{R \to \infty; \varepsilon_0}(K(T))) \left(\int_0^{T-\tau} \| F(t + \tau) \|_{L^2(M)}^2 \, dt \right)^{1/2}$$

for the global term $F_{\text{glob}}(t)$.

Let us first show an associated estimate for the *free* propagator e^{itH_0}, namely

$$\left\| \int_0^{T-\tau} e^{itH_0} F_{\text{glob}}(t) \, dt \right\|_{\dot{H}^{-1/2}(M)} \tag{11.27}$$

$$\leq C_{\varepsilon_0, R, \tau} \left(\int_0^{T-\tau} \| F(t + \tau) \|_{L^2(M)}^2 \, dt \right)^{1/2}.$$

---

[3] Actually, the argument would still work well if H had some singular continuous spectrum, except that τ must now be averaged over an interval such as $[\tau_0, 2\tau_0]$, but the reader may verify that the arguments below will continue to work with this averaging. The spectral fact that is really being used here is that H contains no embedded eigenfunctions at medium frequencies, since such eigenfunctions would certainly contradict (11.6).

Note that our constants are allowed to depend on τ, as we will no longer need to place a factor of $K(T)$ on the right-hand side.

To prove (11.27), first observe that by dualizing the second part of (11.4) we have

$$\left\| \int_0^{T-\tau} e^{itH_0} F_{\text{glob}}(t)\, dt \right\|_{\dot{H}^{-1/2}(M)} \leq C \left(\int_0^{T-\tau} \| F_{\text{glob}}(t) \|_{H^{0,2}(M)}^2\, dt \right)^{1/2}$$

(for instance), so it will suffice to show that

$$\| F_{\text{glob}}(t) \|_{H^{0,2}(M)} \leq C_{\varepsilon_0, R, \tau} \| F(t+\tau) \|_{L^2(M)}$$

for all $0 \leq t \leq T - \tau$. On the other hand, we know from inspection of the symbol of $P_{\text{med}} \nabla_j \varphi$ that

$$\| P_{\text{med}} \nabla_j \varphi F^j(t+\tau) \|_{H^{2,2}(M)} \leq C_{\varepsilon_0} \| F(t+\tau) \|_{L^2(M)}$$

(for instance), so it will suffice by (11.24) to show that

$$\| e^{i\tau H} \|_{H^{2,2}(M) \to H^{0,2}(M)} \leq C_\tau.$$

But this can be easily established by standard energy methods.[4] This proves (11.27). Thus to prove (11.26) it suffices to show that

$$\left\| \int_0^{T-\tau} (e^{itH} - e^{itH_0}) F_{\text{glob}}(t)\, dt \right\|_{\dot{H}^{-1/2}(M)}$$

$$\leq o_{R \to \infty; \varepsilon_0} (1 + K(T)) \left(\int_0^{T-\tau} \| F(t+\tau) \|_{L^2(M)}^2\, dt \right)^{1/2}. \qquad (11.28)$$

At this stage it is necessary to decompose $F_{\text{glob}}(t)$ further into "incoming" and "outgoing" components, which roughly correspond to the regions of phase space where $x \cdot \xi < 0$ and $x \cdot \xi > 0$, respectively. Semiclassically, we expect $F_{\text{glob}}(t)$ to be supported almost entirely in the "incoming" region of phase space, since it is currently far away from the origin, but came by propagating a localized function backwards in time from $t + \tau$. However, if this function is supported in the incoming region of phase space, then by moving farther backward in time by t it should move even farther away from the origin, and in particular it should evolve much like the Euclidean flow (i.e., it should become small when $e^{itH} - e^{itH_0}$ is applied).

We now make this intuition precise. The first step is to formalize the decomposition into incoming and outgoing waves. We first take advantage of the fact that the spectral support of P_{med} vanishes near zero. From (11.24) we have

$$F_{\text{glob}}(t) = (1 - \varphi_R) H^2 \tilde{F}_{\text{glob}}(t), \qquad (11.29)$$

---

[4] For instance, if $f \in H^{2,2}(M)$, one can first establish uniform bounds on $e^{i\tau H} f$ in $H^{2,0}(M)$ by spectral methods, then use energy methods to control $e^{i\tau H} f$ in $H^{1,1}(M)$, and then finally in $H^{0,2}(M)$, losing polynomial factors of τ in each case. One could also argue using positive commutator methods based on (11.13) for such operators as $A = \langle x \rangle^4$. We omit the details.

where

$$\tilde{F}_{\text{glob}}(t) := e^{i\tau H} \tilde{P}_{\text{med}} \nabla_j \varphi F^j(t + \tau) \tag{11.30}$$

and $\tilde{P}_{\text{med}} := H^{-2} P_{\text{med}}$. This factor of H^2 we have extracted from P_{med} shall be helpful for managing the very low frequencies in the proof of Proposition 11.7.2 below, which would otherwise cause a significant problem for this portion of the evolution (at least in three dimensions; this step appears to be unnecessary in five and higher dimensions, for reasons similar to why resonances do not occur in those dimensions).

We now need the following phase space decomposition associated to the *Euclidean* flow e^{-itH_0}.

PROPOSITION 11.7.2 (Phase space decomposition). *There exist operators* P_{in}, P_{out} *such that*

$$(1 - \varphi_R) H^2 = (1 - \varphi_R) H^2 P_{\text{in}} + (1 - \varphi_R) H^2 P_{\text{out}} \tag{11.31}$$

and for which we have the estimates[5]

$$\|e^{isH_0}(1 - \varphi_R) H^2 P_{\text{in}} f\|_{H^{2,-8}(M)} \leq C(R^2 + |s|)^{-1-\sigma} \|f\|_{H^{20}(M)} \tag{11.32}$$

and

$$\|\langle x \rangle^{3/2+\sigma}(1 - \varphi_R) H^2 P_{\text{out}} e^{isH_0} f\|_{L^2(M)}$$
$$\leq C(R^2 + |s|)^{-1-\sigma} \|f\|_{H^{18,11}(M)} \tag{11.33}$$

for any time $s > 0$ *and all Schwartz* f.

The proof of this proposition is a straightforward application of the principle of stationary phase and can be justified heuristically by appealing to the intuition of microlocal analysis and the uncertainty principle. It is, however, a little technical and will be deferred to the next section. Assuming it for the moment, let us conclude the proof of Proposition 11.3.2. It suffices to prove (11.26). From (11.29) and Proposition 11.3.2 we can split F_{glob} as

$$F_{\text{glob}}(t) = (1 - \varphi_R) H^2 P_{\text{in}} \tilde{F}_{\text{glob}}(t) + (1 - \varphi_R) H^2 P_{\text{out}} \tilde{F}_{\text{glob}}(t), \tag{11.34}$$

and treat the components separately.

11.7.3 Term 2(a). Contribution of the Global Incoming Part

We now control the contribution of the incoming component of (11.34) to (11.26). We use Duhamel's formula to write

$$(e^{itH} - e^{itH_0})(1 - \varphi_R) H^2 P_{\text{in}} \tilde{F}_{\text{glob}}(t)$$
$$= i \int_0^t e^{i(t-s)H}(H - H_0) e^{isH_0}(1 - \varphi_R) H^2 P_{\text{in}} \tilde{F}_{\text{glob}}(t)\, ds$$

---

[5] The decay weights here should not be taken too seriously; indeed, since we are assuming H to be a compactly supported perturbation of H_0 we have enormous flexibility with these weights.

and so it suffices to show that

$$\left\| \int_0^{T-\tau} \int_0^t e^{i(t-s)H}(H - H_0)e^{isH_0}(1 - \varphi_R)H^2 P_{\text{in}} \tilde{F}_{\text{glob}}(t) \, ds \, dt \right\|_{\dot{H}^{-1/2}(M)}$$
$$= o_{R \to \infty; \varepsilon_0}(K(T)) \left(\int_0^{T-\tau} \|F(t+\tau)\|_{L^2(M)}^2 \, dt \right)^{1/2}.$$

Substituting $t' := t - s$ using Minkowski's inequality, we can estimate

$$\left\| \int_0^{T-\tau} \int_0^t e^{i(t-s)H}(H - H_0)e^{isH_0}(1 - \varphi_R)H^2 P_{\text{in}} \tilde{F}_{\text{glob}}(t) \, ds \, dt \right\|_{\dot{H}^{-1/2}(M)}$$
$$\leq \int_0^{T-\tau} \left\| \int_0^{T-\tau-s} e^{it'H}(H - H_0)e^{isH_0} P_{\text{in}} \tilde{F}_{\text{glob}}(t' + s) \, dt' \right\|_{\dot{H}^{-1/2}(M)}.$$

Applying (11.25), it thus suffices to show that

$$\int_0^{T-\tau} \int_0^{T-\tau} \|\langle x \rangle^{3/2+\sigma}(H - H_0)e^{isH_0}(1 - \varphi_R)H^2 P_{\text{in}} \tilde{F}_{\text{glob}}(t' + s)\|_{L^2(M)} \, dt'$$
$$= o_{R \to \infty; \varepsilon_0}\left(\left(\int_0^{T-\tau} \|F(t+\tau)\|_{L^2(M)}^2 \, dt \right)^{1/2} \right). \qquad (11.35)$$

On the other hand, since $H - H_0$ is a compactly supported second-order operator, we have

$$\|\langle x \rangle^{3/2+\sigma}(H - H_0)e^{isH_0}(1 - \varphi_R)H^2 P_{\text{in}} \tilde{F}_{\text{glob}}(t' + s)\|_{L^2(M)}$$
$$\leq C\|e^{isH_0}(1 - \varphi_R)H^2 P_{\text{in}} \tilde{F}_{\text{glob}}(t' + s)\|_{H^{2,-8}(M)},$$

and so from (11.32) we thus have

$$\|\langle x \rangle^{3/2+\sigma}(H - H_0)e^{isH_0}(1 - \varphi_R)H^2 P_{\text{in}} \tilde{F}_{\text{glob}}(t' + s)\|_{L^2(M)}$$
$$\leq C(R^2 + |s|)^{-1-\sigma}\|\tilde{F}_{\text{glob}}(t' + s)\|_{H^{20}(M)}.$$

From (11.30) we note that $\|\tilde{F}_{\text{glob}}(t' + s)\|_{L^2(M)} \leq C_{\varepsilon_0}\|F\|_{L^2(M)}$ (noting that $\tilde{P}_{\text{med}}$ maps $H^{-1}(M)$ to $H^{20}(M)$ with an operator norm of C_{ε_0}). Thus we have

$$\int_0^{T-\tau-s} \|\langle x \rangle^{3/2+\sigma}(H - H_0)e^{isH_0}(1 - \varphi_R)H^2 P_{\text{in}} \tilde{F}_{\text{glob}}(t' + s)\|_{L^2(M)}$$
$$\leq C_{\varepsilon_0}(R^2 + |s|)^{-1-\sigma} \left(\int_0^{T-\tau} \|F(t+\tau)\|_{L^2(M)}^2 \, dt \right)^{1/2}$$

and the claim (11.35) follows upon integrating in s.

11.7.4 Term 2(b). Contribution of the Global Outgoing Part

We now control the contribution of the outgoing component of (11.34) to (11.26). From (11.25) we have

$$\left\| \int_0^{T-\tau} e^{itH}(1-\varphi_R)H^2 P_{\text{out}} \tilde{F}_{\text{glob}}(t) \, dt \right\|_{\dot{H}^{-1/2}(M)}$$

$$\leq K(T) \left(\int_0^{T-\tau} \|S(t)\|_{L^2(M)}^2 \, dt \right)^{1/2}$$

where

$$S(t) := \langle x \rangle^{3/2+\sigma}(1-\varphi_R)H^2 P_{\text{out}} \tilde{F}_{\text{glob}}(t).$$

From (11.4), a similar estimate holds when e^{itH} is replaced by the free flow e^{itH_0}, with $K(T)$ replaced by a constant C. Thus, to control this contribution to (11.26) it will suffice to show that

$$\left(\int_0^{T-\tau} \|S(t)\|_{L^2(M)}^2 \, dt \right)^{1/2} = o_{R\to 0;\varepsilon_0} \left(\int_0^{T-\tau} \|F(t+\tau)\|_{L^2(M)}^2 \, dt \right)^{1/2}. \quad (11.36)$$

To prove this, we first use (11.30) to expand

$$\begin{aligned}
\|S(t)\|_{L^2(M)}^2 &= \langle \langle x \rangle^{3/2+\sigma}(1-\varphi_R)H^2 P_{\text{out}} e^{itH} \tilde{P}_{\text{med}} \nabla_j \varphi F^j(t+\tau), S(t) \rangle \\
&= -\langle F^j(t+\tau), \varphi \nabla_j P_{\text{med}} e^{-i\tau H} W(t) \rangle \\
&\leq C \|F(t+\tau)\|_{L^2(M)} \|P_{\text{med}} e^{-i\tau H} W(t)\|_{H^{1,-20}(M)} \\
&\leq C_{\varepsilon_0} \|F(t+\tau)\|_{L^2(M)} \|e^{-i\tau H} W(t)\|_{H^{-20,-20}(M)}
\end{aligned} \quad (11.37)$$

by the support of φ and the smoothing properties of P_{med}, where

$$W(t) := P_{\text{out}}^* H^2 (1-\varphi_R) \langle x \rangle^{3/2+\sigma} S(t). \quad (11.38)$$

We use Duhamel's formula to write

$$e^{-i\tau H} W(t) = e^{-i\tau H_0} W(t) - i \int_0^\tau e^{-i(\tau-s)H}(H - H_0)e^{-isH_0} W(t) \, ds,$$

and apply $H^{-20,-20}$ norms on both sides, to obtain

$$\begin{aligned}
\|e^{-i\tau H} W(t)\|_{H^{-20,-20}(M)} &\leq \|e^{-i\tau H_0} W(t)\|_{H^{-20,-20}(M)} \\
&\quad + \int_0^\tau \|e^{-i(\tau-s)H}(H - H_0)e^{-isH_0} W(t) \, ds\|_{H^{-20,-20}(M)}.
\end{aligned}$$

Observe that the propagator $e^{-i(\tau-s)H}$ maps $H^{-20}(M)$ to $H^{-20}(M)$ and hence to $H^{-20,-20}(M)$, uniformly in τ and s. Also, since $H - H_0$ is a compactly supported operator, we see that $H - H_0$ maps $H^{-18,-11}(M)$ to $H^{-20}(M)$. We thus have

$$\begin{aligned}
\|e^{-i\tau H} W(t)\|_{H^1(B(0,C))} &\leq C \|e^{-i\tau H_0} W(t)\|_{H^{-18,-11}(M)} \\
&\quad + C \int_0^\tau \|e^{-isH_0} W(t)\|_{H^{-18,-11}(M)} \, ds,
\end{aligned}$$

and thus by (11.38) and the adjoint of (11.33)

$$\|e^{-i\tau H}W(t)\|_{H^1(B(0,C))} \leq C(R^2+\tau)^{-1-\sigma}\|S(t)\|_{L^2(M)}$$
$$+ C\int_0^\tau (R^2+s)^{-1-\sigma}\|S(t)\|_{L^2(M)}$$
$$\leq C_{\varepsilon_0} R^{-\sigma}\|S(t)\|_{L^2(M)}.$$

The point here is that we have obtained a nontrivial decay in R. Inserting this estimate back into (11.37), we see that

$$\|S(t)\|_{L^2(M)} \leq C_{\varepsilon_0} R^{-1}\|F(t+\tau)\|_{L^2(M)},$$

and (11.36) follows.

The proof of Theorem 11.1.1 is now complete, once we complete the proof of Proposition 11.7.2, which we do in the next section.

11.8 PROOF OF THE PHASE SPACE DECOMPOSITION

We now prove Proposition 11.7.2. The ideas here have some similarity with a decomposition used in [12]; related ideas were also used recently in [13]. The idea of using a decomposition into incoming and outgoing waves to analyze the perturbation theory of the free Laplacian H_0 of course goes back to Enss (and, in a different context, even earlier to the work of Lax-Phillips.) The phase space decomposition, developed by Enss [6] and refined by Simon [10], is based on the following construction. Let $\{\zeta_{\mathbf{j}}\}_{\mathbf{j}\in\mathbf{Z}^d}$ and $\{m_{\mathbf{j}}\}_{\mathbf{j}\in\mathbf{Z}^d}$ be smooth partitions of unity in $\mathbf{R}_x^d$ and $\mathbf{R}_\xi^d$, respectively, with the property that each of the functions $\zeta_{\mathbf{j}}(x)$ and $m_{\mathbf{j}}(\xi)$ is supported in the ball $B(\mathbf{j}, 2)$. Define the symbols (of the ΨDO's) of the projections P_{in} and P_{out} on the incoming and outgoing states:

$$P_{\text{in}} = \sum_{\mathbf{j}\cdot\mathbf{k}\leq 0} m_{\mathbf{j}}(\xi)\,\zeta_{\mathbf{k}}(x), \qquad P_{\text{out}} = \sum_{\mathbf{j}\cdot\mathbf{k}<0} m_{\mathbf{j}}(\xi)\,\zeta_{\mathbf{k}}(x)$$

The following estimate, crucial to Enss's method, reflects the expectation that the outgoing waves never come back to the region where they originate. For any $t \geq 0, N \geq 0$ and $\mathbf{j}\cdot\mathbf{k} > 0$ such that $|\mathbf{j}| \geq 3$, we have

$$\|e^{-itH_0}m_{\mathbf{j}}\zeta_{\mathbf{k}}f\|_{L^2(|x|<1/2(|\mathbf{k}|+t/4))} \leq C_N(1+t+|\mathbf{k}|)^{-N}\|f\|_{L^2(\mathbf{R}^d)}$$

(compare with (11.32) noting that the transformation $t \to -t$ corresponds to the change $P_{\text{in}} \to P_{\text{out}}$). Observe that the condition $|\mathbf{j}| \geq 3$ ensures that we deal only with the outgoing waves of velocities bounded away from zero in absolute value.

A continuous decomposition, based on coherent (Gaussian) states, was used by Davies, [4], while the outgoing/incoming waves defined via projections on positive/negative spectral subspace of the dilation operator $x \cdot \nabla + \nabla \cdot x$ were introduced by Mourre, [7].

We now return to our decomposition into the incoming/outgoing wave needed to prove Proposition 11.7.2. We first observe from dyadic decomposition that it suffices to prove the claim with $(1-\varphi_R)$ replaced by $(\varphi_{2R}-\varphi_R)$, since the original claim then follows by replacing R by $2^m R$ and summing the telescoping series over $m \geq 0$. By further decomposition, and some rotation and scaling, we may replace $\varphi_{2R} - \varphi_R$ by a smooth cutoff function $\psi = \psi_R$ supported on the ball $B := \{x : |x - Re_1| \leq R/100\}$, where e_1 is a basis vector of $\mathbf{R}^n$.

Our construction will be based on Euclidean tools such as the Fourier transform, so we shall now use the Euclidean inner product and Euclidean Lebesgue measure instead of the counterparts corresponding to the metric g. In particular, the operator H will no longer be self-adjoint, but this will not concern us as we shall soon break it down into components anyway. To reflect this change of perspective we shall write our manifold M now as $\mathbf{R}^n$.

We begin with the Fourier inversion formula

$$\psi H^2 f(x) = \psi H^2 \tilde{\psi} f(x) = \psi H^2 \int_{\mathbf{R}^n} \int_{\mathbf{R}^n} e^{2\pi i(x-y)\cdot\xi} \tilde{\psi}(y) f(y) \, dy d\xi,$$

valid for all Schwartz f, where $\tilde{\psi}$ is a smooth cutoff to the ball $\tilde{B} := \{x : |x - Re_1| \leq R/50\}$ which equals 1 on B. We then split the ξ integration into subspaces $\pm \xi_1 > 0$, defining

$$P_{\text{in}} f(x) = \int_{\xi_1 < 0} \int_{\mathbf{R}^n} e^{2\pi i(x-y)\cdot\xi} \tilde{\psi}(y) f(y) \, dy d\xi$$

and

$$P_{\text{out}} f(x) = \int_{\xi_1 > 0} \int_{\mathbf{R}^n} e^{2\pi i(x-y)\cdot\xi} \tilde{\psi}(y) f(y) \, dy d\xi,$$

where $\xi_1 := \xi \cdot e_1$ is the e_1 component of ξ_1. We remark that these operators are essentially Hilbert transforms in the e_1 direction; the multiplier is of course discontinuous in the ξ_1 variable, but we will never integrate by parts in this variable so this will not be a difficulty.

Clearly we have the decomposition (11.31). It remains to prove (11.33), (11.32).

11.8.1 Proof of (11.7.33)

We now prove the estimate (11.33). Since $H = H_0$ on the support of ψ, we may write

$$\langle x \rangle^{3/2+\sigma} \psi H^2 = R^{-5/2+\sigma} a(x)(R\nabla_x)^4$$

for some bounded functions tensor $a(x)$ supported on B. We shall think of $R\nabla_x$ as a normalized gradient on B.

In light of the above decomposition, it thus suffices to show that

$$\left\| (R\nabla_x)^4 \int_{\xi_1 > 0} \int_{\mathbf{R}^3} e^{2\pi i(x-y)\cdot\xi} \tilde{\psi}(y) e^{isH_0} f(y) \, dy d\xi \right\|_{L^2_x(B)}$$
$$\leq C R^{\frac{5}{2}-\sigma} (R^2 + s)^{-1-\sigma} \| f \|_{H^{18,11}(\mathbf{R}^3)}$$

for all all times $s > 0$.

We first dispose of the derivatives $(R\nabla_x)^4$. Each x derivative in $R\nabla_x$ hits the phase $e^{2\pi i(x-y)\cdot\xi}$, where it can be converted to a y derivative, which after integration by parts either hits $\tilde{\psi}$ or $e^{isH_0}f(y)$. Since partial derivatives commute with e^{isH_0}, we thus reduce to showing that

$$\left\|\int_{\xi>0}\int_{\mathbf{R}^3} e^{2\pi i(x-y)\cdot\xi}\psi^{(j_1)}(y)e^{isH_0}f^{(j_2)}(y)\,dy d\xi\right\|_{L_x^2(B)} \tag{11.39}$$
$$\leq CR^{\frac{5}{2}-\sigma}(R^2+s)^{-1-\sigma}\|f\|_{H^{18,11}(\mathbf{R}^3)}$$

whenever $0\leq j_1, j_2$ with $j_1+j_2=4$, where $\psi^{(j_1)}:=(R\nabla_x)^{j_1}\tilde{\psi}$ is a minor variant of the cutoff $\tilde{\psi}$, and $f^{(j_2)}:=(R\nabla_x)^{j_2}f$.

Applying a smooth cutoff to f, we can divide into two cases: the local case where f is supported on the ball $\{|x|\leq R/100\}$, or the global case where f is supported on the exterior ball $\{x\geq R/200\}$. Let us consider the global case first, which is rather easy and for which one can be somewhat careless with powers of R. We first observe from Plancherel (or the L^2 boundedness of the Hilbert transform) that

$$\left\|\int_{\xi>0}\int_{\mathbf{R}^3} e^{2\pi i(x-y)\cdot\xi}g(y)\,dy d\xi\right\|_{L_x^2(B)} \leq C\|g\|_{L^2(\mathbf{R}^3)}.$$

Thus to prove (11.39) in the global case it suffices to show that

$$\|\psi^{(j_1)}e^{isH_0}f^{(j_2)}\|_{L^2(\mathbf{R}^3)} \leq CR^{\frac{5}{2}-\sigma}(R^2+s)^{-1-\sigma}\|f\|_{H^{18,11}(\mathbf{R}^3)}.$$

We divide further into two subcases, the short-time case $s\leq R^2$ and the long-time case $s\geq R^2$. In the short-time case the claim follows since

$$\|\psi^{(j_1)}e^{isH_0}f^{(j_2)}\|_{L^2(\mathbf{R}^3)} \leq C\|e^{isH_0}f^{(j_2)}\|_{L^2(\mathbf{R}^3)}$$
$$= C\|f^{(j_2)}\|_{L^2(\mathbf{R}^3)}$$
$$\leq CR^{j_2}R^{-11}\|f\|_{H^{18,11}(\mathbf{R}^3)}$$
$$\leq CR^{-7}\|f\|_{H^{18,11}(\mathbf{R}^3)},$$

which is certainly acceptable (if σ is small enough). In the long time case we interpolate between the two bounds

$$\|\psi^{(j_1)}e^{isH_0}g\|_{L^2(\mathbf{R}^3)} \leq C\|e^{isH_0}g\|_{L^2(\mathbf{R}^3)} \leq C\|g\|_{L^2(\mathbf{R}^3)}$$

and

$$\|\psi^{(j_1)}e^{isH_0}g\|_{L^2(\mathbf{R}^3)} \leq CR^{n/2}\|e^{isH_0}g\|_{L^\infty(\mathbf{R}^3)}$$
$$\leq CR^{3/2}s^{-3/2}\|g\|_{L^1(\mathbf{R}^3)} \leq CR^{n/2}s^{-n/2}\|g\|_{H^{0,n/2+\sigma}(\mathbf{R}^3)}$$

to obtain

$$\|\psi^{(j_1)}e^{isH_0}f^{(j_2)}\|_{L^2(\mathbf{R}^3)} \leq C(R/s)^{-1-\sigma}\|f^{(j_2)}\|_{H^{0,1+C\sigma}}$$
$$\leq C(R/s)^{-1-\sigma}R^{j_2}R^{-11+1+C\sigma}\|f\|_{H^{18,11}(\mathbf{R}^3)},$$

which is certainly acceptable.

It remains to prove (11.39) in the local case, when f is supported on the ball $\{|x| \leq R/100\}$. In this case, expand the fundamental solution of e^{isH_0} to write

$$\int_{\mathbf{R}^3} \int_{\mathbf{R}^3} \chi_+(\xi) e^{2\pi i(x-y)\cdot\xi} \psi^{(j_1)}(y) e^{isH_0} f^{(j_2)}(y) \, dy d\xi = \int K_s(x,z) f^{(j_2)}(z) \, dz,$$

where

$$K_s(x,z) := Cs^{-3/2} \int_{\xi_1>0} \int_{\mathbf{R}^3} e^{2\pi i(x-y)\cdot\xi} \psi^{(j_1)}(y) e^{-i|y-z|^2/2s} \, dy d\xi.$$

Observe from all the spatial cutoffs that x, y are localized to the ball $\tilde{B}$, while z is localized to the ball $|z| \leq R/100$. Also, ξ is localized to the half-plane $\xi_1 > 0$. Our task is to show that

$$\left\| \int_{|x|\leq R/100} K_s(x,z) f^{(j_2)}(z) \, dz \right\|_{L^2(B)} \tag{11.40}$$

$$\leq C R^{\frac{5}{2}-\sigma} (R^2+s)^{-1-\sigma} \|f\|_{H^{18,11}(\mathbf{R}^3)}.$$

We split K_s into K_s^{hi} and K_s^{lo}, corresponding to the regions $|\xi| \geq R^{-1+\sigma}$ and $|\xi| < R^{-1+\sigma}$ of frequency space, respectively. The contribution of K_s^{hi} will be very small. Indeed, for any $|\xi| \geq R^{-1+\sigma}$, we can evaluate the y integral using stationary phase as follows. Observe that the y gradient of the phase

$$2\pi(x-y)\cdot\xi - |y-z|^2/2s$$

is equal to

$$-2\pi\xi - (y-z)/s.$$

From the localizations on y, z, and ξ we observe that this quantity has magnitude at least $\geq c(|\xi| + R/s)$. One can then do repeated integration by parts in the y_1 variable, gaining an R every time one differentiates the $\psi^{(j_1)}$ cutoff, to obtain the bound

$$\left| \int_{\mathbf{R}^3} e^{2\pi i(x-y)\cdot\xi} \psi^{(j_1)}(y) e^{-i|y-z|^2/2s} \, dy \right| \leq C_N R^3 (R|\xi| + R^2/s)^{-N}$$

for any N. Integrating over all $|\xi| \geq R^{-1+\sigma}$, we thus see that $K_s^{hi}(x,z)$ is bounded by $O_N((R^2+s)^{-N})$ for any N, and so this contribution to (11.40) is easily shown to be acceptable (using crude estimates on $f^{(j_2)}$).

Now we deal with the low-frequency case $|\xi| \leq R^{-1+\sigma}$. In this case we expand $f^{(j_2)}$ and integrate by parts to write

$$\left| \int_{|z|\leq R/100} K_s^{lo}(x,z) f^{(j_2)}(z) \, dz \right| \leq C R^{j_2} \left| \int_{|z|\leq R/100} \nabla_x^{j_2} K_s^{lo}(x,z) f(z) \, dz \right|$$

$$\leq C s^{-3/2} \int_{|z|\leq R/100} \int_{|\xi|\leq R^{-1+\sigma};\xi_1>0} \left| \int_{\mathbf{R}^3} e^{2\pi i(x-y)\cdot\xi} \psi^{(j_1)}(y) e^{-i|y-z|^2/2s} \, dy \right| R^{j_2} |\xi|^{j_2} |f(z)| \, d\xi dz.$$

We crudely bound $R^{j_2}|\xi|^{j_2}$ by $O(R^{4\sigma})$. From stationary phase we also observe the estimate

$$\left| \int_{\mathbf{R}^3} e^{2\pi i(x-y)\cdot\xi} \psi^{(j_1)}(y) e^{-i|y-z|^2/2s}\, dy \right| \le Cs^{3/2}$$

while by taking absolute values everywhere we also have the estimate

$$\left| \int_{\mathbf{R}^3} e^{2\pi i(x-y)\cdot\xi} \psi^{(j_1)}(y) e^{-i|y-z|^2/2s}\, dy \right| \le CR^3$$

and hence

$$\left| \int_{|z|\le R/100} K_s^{\mathrm{lo}}(x,z) f^{(j_2)}(z)\, dz \right|$$

$$\le Cs^{-3/2} R^{4\sigma} \int_{|z|\le R/100} \int_{|\xi|\le R^{-1+\sigma}} \min(s^{n/2}, R^3)|f(z)|\, d\xi dz$$

$$\le CR^{C\sigma} s^{-3/2} \min(s^{3/2}, R^3) R^{-3} \int_{|z|\le R/100} |f(z)|\, dz$$

$$\le CR^{C\sigma} s^{-3/2} \min(s^{3/2}, R^3) R^{-3/2} \|f\|_{L^2(\mathbf{R}^3)}$$

$$\le CR^{C\sigma} R^2 (R^2+s)^{-1-\sigma} R^{-3/2} \|f\|_{H^{18,11}(\mathbf{R}^3)},$$

which is certainly acceptable. This completes the proof of (11.33).

11.8.2 Proof of (11.32)

To conclude the proof of Proposition 11.7.2 it suffices to prove (11.32) (with $1 - \varphi_R$ replaced by ψ, of course). This is in a spirit similar to the proof of (11.33), although the steps will be in a somewhat permuted order. We begin by estimating

$$\|e^{isH_0}\psi H^2 P_{\mathrm{in}} f\|_{H^{2,-8}(\mathbf{R}^3)} \le \sum_{k=0}^{2} \|\langle x\rangle^{-8} \nabla_x^k e^{isH_0} \psi H^2 P_{\mathrm{in}} f\|_{L^2(\mathbf{R}^3)}$$

$$\le \sum_{k=0}^{2} \|\langle x\rangle^{-8} e^{isH_0} \nabla_x^k \psi H^2 P_{\mathrm{in}} f\|_{L^2(\mathbf{R}^3)}.$$

Now observe that

$$\nabla_x^k \psi H^2 = \sum_{j=0}^{4} R^{j-4} a_{j,k}(x) \nabla_x^{j+k}$$

for some smooth cutoff functions $a_{j,k}$ adapted to the ball B. Our task is thus to show that

$$\|\langle x\rangle^{-8} e^{isH_0} a_{j,k} \nabla_x^{j+k} P_{\mathrm{in}} f\|_{L^2(\mathbf{R}^3)} \le CR^{4-j}(R^2+|s|)^{-1-\sigma} \|f\|_{H^{20}(\mathbf{R}^3)}$$

for all $0 \le j \le 4$ and $0 \le k \le 2$.

Fix j, k. Let us first control the contribution of the weight $\langle x\rangle^{-8}$ arising from the region $|x| \ge R/100$. This contribution is dealt with differently in the short-time case $s \leq R^2$ and the long-time case $s > R^2$. In the short-time case $s \leq R^2$, this contribution is dealt with

$$\leq CR^{-8} \|e^{isH_0} a_{j,k} \nabla^{j+k} P_{\text{in}} f\|_{L^2(\mathbf{R}^3)}.$$

Since e^{isH_0} and $a_{j,k}$ are both bounded on L^2, we can bound this by

$$\leq CR^{-8} \|\nabla^{j+k} P_{\text{in}} f\|_{L^2(\mathbf{R}^3)}.$$

Now observe from Plancherel that P_{in} is bounded on H^{20} (since multiplication by $\tilde\psi$ is certainly bounded on H^{20}), and so we can bound this by $O(R^{-8}\|f\|_{H^{20}})$, which is acceptable. In the long-time case, we control the contribution instead by

$$\leq CR^{-8+\frac{3}{2}} \|e^{isH_0} a_{j,k} \nabla^{j+k} P_{\text{in}} f\|_{H^{0,-3/2}(\mathbf{R}^3)}. \tag{11.41}$$

Now we interpolate between the energy estimate

$$\|e^{isH_0} g\|_{L^2(\mathbf{R}^3)} = \|g\|_{L^2(\mathbf{R}^3)}$$

and the decay estimate

$$\begin{aligned}
\|e^{isH_0} g\|_{H^{0,-n/2-\sigma}(\mathbf{R}^3)} &\leq C \|e^{isH_0} g\|_{L^\infty(\mathbf{R}^3)} \\
&\leq Cs^{-3/2} \|g\|_{L^1(\mathbf{R}^3)} \\
&\leq Cs^{-3/2} \|g\|_{H^{0,n/3+\sigma}(\mathbf{R}^3)}
\end{aligned}$$

to obtain

$$\|e^{isH_0} g\|_{H^{0,-3/2}(\mathbf{R}^3)} \leq Cs^{-3/2+\sigma} \|g\|_{H^{0,3/2}(\mathbf{R}^3)}.$$

The operator $a_{j,k}$ maps L^2 to $H^{0,3/2}$ with a bound of $O(R^{3/2})$, so we can therefore bound (11.41) by

$$\leq CR^{-8+\frac{3}{2}} s^{-3/2+\sigma} R^{3/2} \|\nabla^{j+k} P_{\text{in}} f\|_{L^2(\mathbf{R}^3)}.$$

As in the long-term case we can bound $\|\nabla^{j+k} P_{\text{in}} f\|_{L^2(\mathbf{R}^3)}$ by $\|f\|_{H^{20}(\mathbf{R}^3)}$, and so this case is also acceptable.

It remains to control the contribution in the region $|x| \leq R/100$. We then expand out the fundamental solution of e^{isH_0}, and reduce to showing that

$$\begin{aligned}
s^{-3/2} \|\langle x\rangle^{-8} \int_{\mathbf{R}^3} \int_{\xi_1 < 0} \int_{\mathbf{R}^3} & e^{-i|x-y|^2/2s} a_{j,k}(y) \\
& \nabla_y^{j+k} e^{2\pi i(y-z)\cdot\xi} \psi(z) f(z)\, dz d\xi dy\|_{L^2_x(|x|\leq R/100)} \\
& \leq CR^{4-j} (R^2 + |s|)^{-1-\sigma} \|f\|_{H^{20}(\mathbf{R}^3)}.
\end{aligned} \tag{11.42}$$

Note that the x variable is localized to the ball $|x| \leq R/100$, while y and z are localized to the ball $\tilde B$, and ξ is localized to the half-space $\xi_1 < 0$. Once again, we split into the high frequencies $|\xi| \geq R^{-1+\sigma}$ and low frequencies $|\xi| < R^{-1+\sigma}$. In the case of the high frequencies, we move the y derivatives onto $\psi(z) f(z)$ by integration by parts, and reduce to showing that

$$s^{-3/2}\|\langle x\rangle^{-8}\int_{|\xi|\geq R^{-1+\sigma};\xi_1<0}\int_{\mathbf{R}^3}\left(\int_{\mathbf{R}^3}e^{-i|x-y|^2/2s}a_{j,k}(y)e^{2\pi i(y-z)\cdot\xi}\,dy\right)$$

$$g(z)\,dzd\xi\|_{L^2_x(|x|\leq R/100)}\leq CR^{4-j}(R^2+|s|)^{-1-\sigma}\|g\|_{L^2(\tilde{B})}$$

for some g. But the y derivative of the phase

$$-|x-y|^2/2s+2\pi(y-z)\cdot\xi$$

is

$$-(x-y)/s+2\pi\xi,$$

which has magnitude at least $c(|\xi|+R/s)$, by the localizations on x, y, ξ. Thus, by stationary phase we have

$$\int_{\mathbf{R}^3}e^{-i|x-y|^2/2s}a_{j,k}(y)e^{2\pi i(y-z)\cdot\xi}\,dy=O_N(R^3(|\xi|+R/s)^{-N})$$

for any $N>0$, and the claim is now easy to establish by crude estimates. Thus, it remains to prove (11.42) in the low-frequency case $|\xi|<R^{-1+\sigma}$. In this case we convert the ∇_y^{j+k} derivative to $O(|\xi|^{j+k})=O(R^{-(j+k)}R^{C\sigma})$, and thus estimate the left-hand side of this contribution to (11.42) by

$$\leq Cs^{-3/2}R^{-(j+k)}R^{C\sigma}R^{-3}\|\langle x\rangle^{-8}\int_{\mathbf{R}^3}\sup_{|\xi|\leq R^{-1+\sigma};\xi_1<0}$$

$$\left|\int_{\mathbf{R}^3}e^{-i|x-y|^2/2s}a_{j,k}(y)e^{2\pi i(y-z)\cdot\xi}\,dy\right|\psi(z)|f(z)|\,dz\|_{L^2_x(|x|\leq R/100)}.$$

But by stationary phase as before, we have the estimates

$$\left|\int_{\mathbf{R}^3}e^{-i|x-y|^2/2s}a_{j,k}(y)e^{2\pi i(y-z)\cdot\xi}\,dy\right|\leq C\min(s^{3/2},R^3),$$

and thus we can bound the previous expression by

$$\leq Cs^{-3/2}R^{-(j+k)}R^{C\sigma}R^{-3}\min(s^{3/2},R^3)\|\langle x\rangle^{-8}$$

$$\int_{\mathbf{R}^3}\psi(z)|f(z)|\,dz\|_{L^2(|x|\leq R/100)}.$$

We crudely bound the $L^2(|x|\leq R/100)$ norm of $\langle x\rangle^{-8}$ by $O(1)$, and use Cauchy-Schwartz we can bound the previous expression by

$$\leq Cs^{-3/2}R^{-(j+k)}R^{C\sigma}R^{-3}\min(s^{3/2},R^3)R^{3/2}\|f\|_{L^2(\mathbf{R}^3)},$$

which is acceptable (treating the $s\leq R^2$ and $s>R^2$ cases separately). The proof of Proposition 11.7.2 is now complete.

I.R. is a Clay Prize Fellow and supported in part by the NSF grant DMS-01007791. T.T. is supported in part by a grant from the Packard Foundation. We thank Nicolas Burq for helpful comments.

REFERENCES

[1] N. Burq, *Décroissance de l'énergie locale de l'équation des ondes pour le problème extérieur et absence de résonance au voisinage du réel*, Acta Math., 180, 1–29, 1998.

[2] W. Craig, T. Kappeler, and W. Strauss, *Microlocal dispersive smoothing for the Schrödinger equation*, Comm. Pure Appl. Math 48 (1995), 769–860.

[3] P. Constantin and J. C. Saut, *Effets régularisants locaux pour des équations disperives générales*, C. R. Acad. Sci. Paris, Sér. I. Math. 304 (1987), 407–410.

[4] E. B. Davies, *On Enss's approach to scattering theory*, Duke Math. J. 47 (1980), 171–185.

[5] S. Doi, *Smoothing effects of Schrödinger evolution groups on Riemannian manifolds*, Duke Math J. 82 (1996), 679–706.

[6] V. Enss, *Asymptotic completeness for quantum-mechanical potential scattering, I. Short-range potentials*, Comm. Math. Phys. 61 (1978), 285–291.

[7] E. Mourre, *Link between the geometrical and the spectral transformation approaches in scattering theory*, Comm. Math. Phys. 68 (1979), 91–94.

[8] M. Reed and B. Simon, *Methods of modern mathematical physics III: Scattering theory*, Academic Press, New York (1979).

[9] I. Rodnianski, T. Tao, *The limiting absorption principle on manifolds, and applications*, preprint.

[10] B. Simon, *Phase space analysis of simple scattering systems: Extensions of some work of Enss*, Duke Math. J. 46 (1979), 119–168.

[11] P. Sjölin, *Regularity of solutions to the Schrödinger equation*, Duke Math. J. 55 (1987), no. 3, 699–715.

[12] G. Staffilani and D. Tataru, *Strichartz estimates for a Schrödinger operator with nonsmooth coefficients*, to appear, Comm. Part. Diff. Eq.

[13] T. Tao, *On the asymptotic behavior of large radial data for a focusing nonlinear Schrödinger equation*, preprint.

[14] G. Thorbergsson, *Closed geodesics on non-compact Riemannian manifolds*, Math. Z. 159 (1978), no. 3, 249–258.

[15] L. Vega, *Schrödinger equations: Pointwise convergence to the initial data*, Proc. Amer. Math. Soc. 102 (1988), no. 4, 874–878.

Chapter Twelve

Dispersive Estimates for Schrödinger operators: A survey

W. Schlag

12.1 INTRODUCTION

The purpose of this note is to give a survey of some recent work on dispersive estimates for the Schrödinger flow

$$e^{itH} P_c, \qquad H = -\Delta + V \text{ on } \mathbb{R}^d, \ d \geq 1, \tag{12.1}$$

where P_c is the projection onto the continuous spectrum of H. V is a real-valued potential that is assumed to satisfy some decay condition at infinity. This decay is typically expressed in terms of the point-wise decay $|V(x)| \leq C\langle x \rangle^{-\beta}$, for all $x \in \mathbb{R}^d$ and some $\beta > 0$. Throughout this paper, $\langle x \rangle = (1 + |x|^2)^{\frac{1}{2}}$. Occasionally, we will use an integrability condition $V \in L^p(\mathbb{R}^d)$ (or a weighted variant thereof) instead of a point-wise condition. These decay conditions will also be such that H is asymptotically complete, i.e.,

$$L^2(\mathbb{R}^d) = L^2_{p.p.}(\mathbb{R}^d) \oplus L^2_{a.c.}(\mathbb{R}^d),$$

where the spaces on the right-hand side refer to the span of all eigenfunctions and the absolutely continuous subspace, respectively.

The dispersive estimate for (12.1) with which we will be most concerned is of the form

$$\sup_{t \neq 0} |t|^{\frac{d}{2}} \left\| e^{itH} P_c f \right\|_\infty \leq C \|f\|_1 \text{ for all } f \in L^1(\mathbb{R}^d) \cap L^2(\mathbb{R}^d). \tag{12.2}$$

Interpolating with the L^2 bound $\left\| e^{itH} P_c f \right\|_2 \leq C \|f\|_2$ leads to

$$\sup_{t \neq 0} |t|^{d(\frac{1}{2} - \frac{1}{p})} \left\| e^{itH} P_c f \right\|_{p'} \leq C \|f\|_p \text{ for all } f \in L^1(\mathbb{R}^d) \cap L^2(\mathbb{R}^d), \tag{12.3}$$

The author was partially supported by NSF grant DMS-0300081 and a Sloan fellowship. This article is based in part on a talk that the author gave at the PDE meeting at the Institute for advanced study in Princeton, N.J., in March 2004. The author is grateful to the organizers for the invitation to speak at that conference, as well as to the Clay Foundation and the IAS for their support. Also, he wishes to thank Fabrice Planchon for useful comments on a preliminary version of this article.

where $1 \le p \le 2$. It is well known that via a T^*T argument (12.3) gives rise to the class of Strichartz estimates

$$\left\| e^{itH} P_c f \right\|_{L_t^q(L_x^p)} \le C \| f \|_2, \quad \text{for all} \quad \frac{2}{q} + \frac{d}{p} = \frac{d}{2}, \ 2 < q \le \infty. \tag{12.4}$$

The endpoint $q = 2$ is not captured by this approach; see Keel and Tao [53].

In heuristic terms, for the free problem $V = 0$ the rate of decay $|t|^{-\frac{d}{2}}$ in (12.2) follows from L^2- conservation and the classical Newton law $\ddot{x} = 0$, which leads to the trajectories $x(t) = vt + x_0$. Mathematically, (12.2) follows from the explicit solution

$$(e^{-it\Delta} f)(x) = C_d \, t^{-\frac{d}{2}} \int_{\mathbb{R}^d} e^{-i \frac{|x-y|^2}{4t}} f(y) \, dy.$$

For general $V \ne 0$, no explicit solutions are available, and one needs to proceed differently.

If V is small and $d \ge 3$, then one can proceed perturbatively. We will give examples of such arguments in Section 12.2. A purely perturbative approach cannot work in the presence of bound states of H since those need to be removed. In other words, in the presence of bound states the nature of the spectral measure and/or resolvents of H becomes essential. Since it is well known that bound states can arise for arbitrarily small potentials in dimensions $d = 1, 2$; see Theorem XIII.11 in Reed and Simon [63], we conclude that a perturbative approach will necessarily fail in those dimensions. On the other hand, if $d = 3$ and V satisfies the Rollnik condition

$$\| V \|_{Roll}^2 := \int_{\mathbb{R}^6} \frac{|V(x)||V(y)|}{|x - y|^2} \, dxdy < \infty,$$

then Kato [52] showed that $-\Delta + V$ is unitarily equivalent with $-\Delta$ provided $4\pi \| V \|_{Roll} < 1$. Similar conditions are known for unitary equivalence if $d \ge 4$.

Dispersive estimates for large V and $d = 3$ were established by Rauch [62] and Jensen, Kato [48]. In contrast to (12.2), these authors measured the decay on weighted $L^2(\mathbb{R}^3)$, i.e., they proved that

$$\left\| w e^{itH} P_c w f \right\|_2 \le C |t|^{-\frac{3}{2}} \| f \|_2, \tag{12.5}$$

with $w(x) = e^{-\rho(x)}$ with some $\rho > 0$ and V exponentially decaying (Rauch) or $w(x) = \langle x \rangle^{-\sigma}$ for some $\sigma > 0$ and V decaying at a power rate (Jensen, Kato). In addition, they needed to assume that the resolvent of H has the property that

$$\limsup_{\lambda \to 0} \| w(H - (\lambda \pm i0))^{-1} w \|_{2 \to 2} < \infty. \tag{12.6}$$

This condition is usually referred to as *zero energy being neither an eigenvalue nor a resonance*. While it is clear what it means for zero to be an eigenvalue of H, the notion of a resonance depends on the norms relative to which the resolvent is required to remain bounded at zero energy; see (12.6). In the context of L^2 with power weights, which are most commonly used, one says that there is a resonance at zero iff there exists a distributional solution f of $Hf = 0$ with the property that $f \notin L^2(\mathbb{R}^3)$ but such that $\langle x \rangle^{-\sigma} f \in L^2(\mathbb{R}^3)$ for all $\sigma > \frac{1}{2}$. With this definition the following holds: (12.6) *is valid for* $w(x) = \langle x \rangle^{-\frac{1}{2}-\varepsilon}$ *for any* $\varepsilon > 0$ *iff zero is neither an eigenvalue nor a resonance.* The proof proceeds via the Fredholm alternative and the mapping properties of $(-\Delta + (\lambda + i0))^{-1}$ on weighted $L^2(\mathbb{R}^3)$ spaces; see Section 12.2. The notion of a resonance arises also in other dimensions, and we will discuss the cases $d = 1, 2$ in the corresponding sections below. If $|V(x)| \leq C\langle x \rangle^{-2-\varepsilon}$ with $\varepsilon > 0$ arbitrary, and $d \geq 5$, then H cannot have any resonances at zero energy. This is due to the fact that under these assumptions $(-\Delta)^{-1}V : L^2(\mathbb{R}^d) \to L^2(\mathbb{R}^d)$.

Rauch and Jensen, and Kato went beyond (12.5) by showing that if zero is an eigenvalue and/or a resonance, then (12.5) fails. In fact, they observed that if zero is a resonance but not an eigenvalue, then

$$C^{-1} < \sup_{\|f\|_2=1} \sup_{t \geq 1} |t|^{\frac{1}{2}} \left\| e^{itH} P_c f \right\|_2 < C < \infty.$$

Furthermore, this loss of decay can occur also if zero is an eigenvalue *even though* P_c *is understood to project away the corresponding eigenfunctions.* They obtained these results as corollaries of asymptotic expansions of e^{itH} as $t \to \infty$ on weighted L^2 spaces.

These asymptotic expansions are basically obtained as the Fourier transforms of asymptotic expansions of the resolvents (or rather, the imaginary part of the resolvents) around zero energy. In *odd* dimensions the latter are of the form, with $\Im z > 0$,

$$(-\Delta + V - z^2)^{-1} = z^{-2}A_{-2} + z^{-1}A_{-1} + A_0$$
$$+ zA_1 + O(z^2) \text{ as } z \to 0, \qquad (12.7)$$

where the O-term is understood in the operator norm on a suitable weighted L^2-space. These expansions can, of course, be continued to higher-order z^m, with the degree of the weights in L^2 needed to control the error $O(z^m)$ increasing with m. In addition, the decay of V needs to increase with m as well. The operator $-A_{-2}$ is the orthogonal projection onto the eigenspace of H, and A_{-1} is a finite rank operator related to both the eigenspace and the resonance functions. In odd dimensions, the free resolvent $(-\Delta + z^2)^{-1}$ is analytic for all $z \neq 0$ (and if $d \geq 3$ for all $z \in \mathbb{C}$), whereas in even dimensions the Riemann surface of the free resolvent is that of the logarithm. In practical terms, this means that (12.7) needs to include (inverse) powers of $\log z$ in even dimensions.

In [46] and [47], Jensen derived analogous expansions for the resolvent around zero energy (and thus for the evolution as $t \to \infty$) in dimensions $d \geq 4$. Resolvent expansion at thresholds for the cases $d = 1$ and $d = 2$ were treated by Bollé, Gesztesy, and Wilk [7], and Bollé, Gesztesy, and Danneels [5], [6]. However, their approach requires separate treatment of the cases $\int V \, dx = 0$ and $\int V \, dx \neq 0$. Moreover, for $d = 2$ only the latter case was worked out. A unified approach to resolvent expansions was recently found by Jensen and Nenciu in [49]. Their method can be applied to all dimensions, but in [49] the authors only present $d = 1, 2$

in detail, because for those cases novel results are obtained by their method. The method developed by Jensen and Nenciu was applied by Erdogan and the author for $d = 3$; see [28], [29], and by the author for $d = 2$; see [69]. A very general treatment of resolvent expansions as in (12.7) and of local L^2 decay estimates can be found in Murata's paper [57]. It is general in the sense that Murata states expansions in all dimensions, and covers the case of elliptic operators as well. However, his method is partially implicit in the sense that the coefficients of the singular powers in (12.7) depend on operators that are solutions of certain equations, but those equations are not solved explicitly.

The first authors to address (12.2) were Journeé, Soffer, and Sogge [51]. Under suitable decay and regularity conditions on V, and under the assumption that zero is neither an eigenvalue nor a resonance, they proved (12.2) for $d \geq 3$. In addition, they conjectured that (12.4) should hold for all V such that $|V(x)| \leq C\langle x\rangle^{-2-\varepsilon}$ with arbitrary $\varepsilon > 0$ and for which $-\Delta + V$ has neither an eigenvalue nor a resonance at zero energy.

The decay rate $\langle x\rangle^{-2-\varepsilon}$, which corresponds to $L^{\frac{d}{2}}(\mathbb{R}^d)$ integrability, plays a special role in dispersive estimates in particular, and the spectral theory of $-\Delta + V$ in general. On the one hand, potentials that decay more slowly than $|x|^{-2}$ at infinity can lead to operators with infinitely many negative bound states. On the other hand, in [13] and [14] Burq, Planchon, Stalker, and Tahvildar-Zadeh obtain Strichartz estimates for

$$i\partial_t u + \Delta u - \frac{a}{|x|^2}u = 0,$$

provided $a > -(d - 2)^2/4$ and $d \geq 2$, and they show that this condition is also necessary. Furthermore, for the case of the wave equation, it is known that pointwise decay estimates fail in the attractive case $a < 0$; see the work of Planchon, Stalker, and Tahvildar-Zadeh.

For $d = 3$ the assumptions on V in [51] are $|V(x)| \leq C\langle x\rangle^{-7-\varepsilon}$, $\hat{V} \in L^1(\mathbb{R}^3)$, and some small amount of differentiability of V. These requirements were subsequently relaxed by Yajima [84], [85], and [86], who proved much more, namely the L^p boundedness of the wave operators for $1 \leq p \leq \infty$. A different approach, which led to even weaker conditions on V, was found by Rodnianski and the author [64] (for small V), as well as by Goldberg and the author [35] (for large V). In addition, the aforementioned conjecture from [51] is proved in [64] (for large V).

Finally, Goldberg [34] proved that (12.2)—and not just (12.4)—holds for all V for which $|V(x)| \leq C\langle x\rangle^{-2-\varepsilon}$ with arbitrary $\varepsilon > 0$ and for which $-\Delta + V$ has neither an eigenvalue nor a resonance at zero energy. In fact, he only required a suitable L^p condition; see Section 12.2. In contrast, trying to adapt [35] to higher dimensions has led Goldberg and Visan [37] to show that for $d \geq 4$, (12.2) *fails* unless V has some amount of regularity, i.e., decay alone is insufficient for (12.2) to hold if $d \geq 4$. More precisely, they exhibit potentials $V \in C_{\mathrm{comp}}^{\frac{d-3}{2}-\varepsilon}(\mathbb{R}^d)$ for which the dispersive $L^1(\mathbb{R}^d) \to L^\infty(\mathbb{R}^d)$ decay with power $t^{-\frac{d}{2}}$ fails.

The first results for $d = 1$ are due to Weder [80], [78], [81]; see also Artbazar and Yajima [4]. These authors make use of the following explicit expression for the resolvent. If $\Im z > 0$, then

$$(-\partial_x^2 + V - z^2)^{-1}(x, y) = \frac{f_+(x, z) f_-(y, z)}{W(z)} \quad \text{if } x > y$$

and symmetrically if $x < y$. Here $f_\pm$ are the *Jost solutions* defined as solutions of

$$-f_\pm''(\cdot, z) + V f_\pm(\cdot, z) = z^2 f_\pm(\cdot, z)$$

with the asymptotics

$$f_+(x, z) \sim e^{ixz} \quad \text{as } x \to \infty$$
$$f_-(x, z) \sim e^{-ixz} \quad \text{as } x \to -\infty,$$

and $W(z) = W[f_+(\cdot, z), f_-(\cdot, z)]$ is their Wronskian. These Jost solutions are known to exist and have boundary values as $\Im z \to 0+$ as long as $V \in L^1(\mathbb{R})$ (in particular, this proves that the spectrum of H is purely a.c. on $(0, \infty)$ for such V). In order for these boundary values $f_\pm(\cdot, \lambda)$ to be continuous at $\lambda = 0$, one needs to require that $\langle x \rangle V(x) \in L^1(\mathbb{R})$. In that case, we say that *zero energy is a resonance* iff $W(0) = 0$. Note that the free case $V = 0$ has a resonance at zero energy, since then $f_\pm(\cdot, 0) = 1$. This condition is equivalent to the existence of a bounded solution f of $Hf = 0$ (in particular, zero cannot be an eigenvalue).

Using some standard properties of the Jost solutions; see [25], Goldberg and the author proved that

$$\|e^{itH} P_c f\|_{L^\infty(\mathbb{R})} \leq C|t|^{-\frac{1}{2}} \|f\|_{L^1(\mathbb{R})} \tag{12.8}$$

provided $\langle x \rangle V(x) \in L^1(\mathbb{R})$ and provided zero is not a resonance. Note that in terms of pointwise decay, this is in agreement with the $\langle x \rangle^{-2}$ threshold mentioned above. If zero is a resonance, then the same estimate holds for all V such that $\langle x \rangle^2 V(x) \in L^1(\mathbb{R})$. In Section 12.3 below, we present a variant of (12.8) with faster decay that seems to be new. It states that under sufficient decay on V and *provided zero is not a resonance,*

$$\|\langle x \rangle^{-1} e^{itH} P_c f\|_{L^\infty(\mathbb{R})} \leq C t^{-\frac{3}{2}} \|\langle x \rangle f\|_{L^1(\mathbb{R})} \tag{12.9}$$

for all $t > 0$. This estimate was motivated by the work of Murata [57] and Buslaev and Perelman [15], where such improved decay was obtained on $L^2(\mathbb{R})$ and with weights of the form $\langle x \rangle^{3.5+\varepsilon}$. It combines dispersive decay and the rate of propagation for H. However, to the best of the author's knowledge, (12.9) has not appeared before, and we therefore include a complete proof in Section 12.3. A version of (12.9) for the evolution of linearized nonlinear Schrödinger equations was crucial to the recent work [55] by Krieger and the author on stable manifolds for all supercritical NLS in one dimension.

Generally speaking, there is a very important difference between the one-dimensional dispersive bounds and those in other dimensions that have been proved so far, namely with regard to the constants. Indeed, in the one-dimensional case these constants exhibit an explicit dependence on the potential via the Jost solutions, which are solutions to a Volterra integral equation. On the other hand, in higher dimensions one resorts to a Fredholm alternative argument in order to invert the operator $H - (\lambda^2 \pm i0)$. This indirect argument is traditionally used to prove the so-called limiting absorption principle for the resolvent; see Agmon [1] and (12.25) below. Any constructive proof of such an estimate for the perturbed resolvent would be most interesting, as it would allow for quantitative constants in dispersive estimates. Such a result was achieved by Rodnianski and Tao; see [67] as well as their article in this volume. More generally, their work deals with dispersive estimates for the Schrödinger operator on $\mathbb{R}^n$ (or other manifolds) with variable metrics and is thus closely related to the subject matter of this article. Unfortunately, it is outside the scope of this review to discuss this exciting field of research. For example, see Bourgain [8], Doi [26], Burq, Gerard, and Tzvetkov [12] (as well as other papers by these authors), Hassell, Tao, and Wunsch [39], [40], Smith and Sogge [72], and Staffilani and Tataru [73].

For the wave equation with a potential, dispersive estimates have also been developed in recent years; see Cuccagna [20], Georgiev and Visciglia [31], Pierfelice [59], Planchon, Stalker, and Tahvildar-Zadeh [60], [61], d'Ancona and Pierfelice [22], as well as Stalker and Tahvildar-Zadeh [74]. The paper by Krieger and the author [56] establishes dispersive estimates for the wave equation with a potential in the presence of a zero energy resonance. This is necessary because of the particular nonlinear application they consider: the construction of a co-dimension one family of global solutions to the focusing H^1 critical wave equation in $\mathbb{R}^3$. There is some overlap with the results here, in particular with respect to certain bounds on the resolvent, but we will restrict ourselves to the Schrödinger equation. For Klein-Gordon, see Weder's work [79].

Much of the work in this paper has been motivated by nonlinear problems (see, e.g., Bourgain's book [11], in particular pages 17–27). In recent years there has been much interest in the asymptotic stability of standing waves of the focusing NLS

$$i\partial_t \psi + \Delta\psi + f(|\psi|^2)\psi = 0. \tag{12.10}$$

A "standing wave" here refers to a solution of the form $\psi(t, x) = e^{i\alpha^2 t}\phi(x)$, where $\alpha \neq 0$ and

$$\alpha^2\phi - \Delta\phi = f(\phi^2)\phi, \tag{12.11}$$

or any solution obtained from this one by applying the symmetries of the NLS, namely Galilei, scaling, and modulation (if the nonlinearity is critical, then there is one more symmetry by the name of pseudoconformal). Most work has been devoted to the standing wave generated by the ground state, i.e., a positive, decaying, solution of (12.11). In fact, such a solution must be radial and decay exponentially. Linearizing (12.10) around a standing wave yields a system of Schrödinger equations with non-selfadjoint matrix operator

$$\mathcal{H} = \begin{bmatrix} -\Delta + \alpha^2 - U & -W \\ W & \Delta - \alpha^2 + U \end{bmatrix} \tag{12.12}$$

and exponentially decaying, real-valued potentials U, W. In order to address the question of asymptotic stability of standing waves, one needs to study the spectrum of $\mathcal{H}$, as well as prove dispersive estimates for $e^{it\mathcal{H}}$ restricted to the stable subspace (which is defined as the range of a suitable Riesz projection). In the following sections we will mostly report on work on the *scalar* case rather than the system case. However, most of what is being said can be generalized to systems; see, e.g., [21], [65], [29], [70], [55]. Although it may seem that the exponential decay of the potential in (12.12) may simplify matters greatly, this turns out not to be the case. In fact, the method from the paper [35], which is concerned with weakening the decay assumptions on V in the scalar, three-dimensional case, has led to the resolution of some open questions about matrix operators as in (12.12); see [29], [70], [55].

12.2 DIMENSIONS THREE AND HIGHER

We start with a perturbative argument for small V that can be considered as a sketch of the method from [51]. As above, let $H = -\Delta + V$ and suppose $d \geq 3$. Define

$$M_0 = \sup_{0 \leq t} \sup_{\|f\|_{1 \cap 2}=1} \langle t \rangle^{\frac{d}{2}} \|e^{itH_0}f\|_{2+\infty},$$

$$M(T) = \sup_{0 \leq t \leq T} \sup_{\|f\|_{1 \cap 2}=1} \langle t \rangle^{\frac{d}{2}} \|e^{itH}f\|_{2+\infty}.$$

Here,

$$\|f\|_{1 \cap 2} = \|f\|_{L^1 \cap L^2}, \qquad \|f\|_{2+\infty} = \inf_{f_1+f_2=f} (\|f_1\|_2 + \|f_2\|_\infty).$$

Then the Duhamel formula

$$e^{itH} = e^{itH_0} + i \int_0^t e^{i(t-s)H_0} V e^{isH}\, ds$$

implies that

$$M(T) \leq M_0 + \langle T \rangle^{\frac{d}{2}} \int_0^T M_0 \langle t - s \rangle^{-\frac{d}{2}} \|V\|_{1 \cap \infty} M(T) \langle s \rangle^{-\frac{d}{2}}\, ds$$

$$\leq M_0 + C \|V\|_{1 \cap \infty} M_0 M(T).$$

Consequently, as long as

$$C \|V\|_{1 \cap \infty} M_0 \leq \frac{1}{2},$$

we obtain the bound

$$\sup_{T \geq 0} M(T) \leq 2M_0.$$

Note first that such an argument necessarily fails if $d = 1, 2$ due to the nonintegrability of $t^{-\frac{d}{2}}$ at infinity. Moreover, there are spectral reasons for this failure, which we outlined in the introduction. Second, we would like to point out that it equally applies to time-dependent potentials provided the evolution e^{itH} is replaced with the propagator of the associated Schrödinger equation. The inclusion of the space

L^2 allows us to deal with the singularity of $t^{-\frac{d}{2}}$ at $t = 0$ that arises in the $L^1 \to L^\infty$ estimate. In order to avoid it, Journeé, Soffer, and Sogge use the bound

$$\|e^{-itH_0} V e^{itH_0}\|_{p \to p} \le \|\hat{V}\|_1,$$

which holds uniformly in $1 \le p \le \infty$. This explains the origin of the condition $\hat{V} \in L^1$ in their paper.

The main difficulty in [51] is of course the fact that V is large. Let us first present an unpublished argument of Ginibre [32] in dimensions $d \ge 3$ that allows passing from the weighted (or local) decay (12.5) to global decay, albeit in the form of a $L^1 \cap L^2 \to L^2 + L^\infty$ estimate rather than the one in (12.2). Applying the Duhamel formula twice, we obtain

$$e^{itH} P_c = e^{itH_0} P_c + i \int_0^t e^{i(t-s)H_0} V P_c e^{isH_0} \, ds$$

$$- \int_0^t \int_0^s e^{i(t-s)H_0} V e^{i(s-\sigma)H} P_c V e^{i\sigma H_0} \, d\sigma \, ds . \tag{12.13}$$

As long as V decays sufficiently rapidly so as to absorb the weights w, i.e., such that

$$\|w^{-1} V\|_{L^1 \cap L^\infty} < \infty ,$$

we can combine the $L^1 \cap L^2 \to L^2 + L^\infty$ bound

$$\|e^{itH_0}\|_{2+\infty} \le C \langle t \rangle^{-\frac{d}{3}} \|f\|_{1 \cap 2} \tag{12.14}$$

with (12.5) as above, to conclude from (12.13) that (12.14) also holds for H. Here we also used that $P_c : L^1 \cap L^\infty \to L^1 \cap L^\infty$ which holds provided all eigenfunctions of H with negative eigenvalue belong to $L^1 \cap L^\infty$ (recall that zero is assumed not to be an eigenvalue). That property, however, follows from Agmon's exponential decay bound [2] and Sobolev imbedding, provided V also some small of regularity.

This argument, however, does not shed much light on the question of $L^1 \to L^\infty$ bounds (without assuming more regularity on V). The inclusion of L^2 is undesirable for a number of reasons, the main one being nonlinear applications. We therefore proceed differently, and first recall the small-potential argument from [64] in $d = 3$. For certain standard details we refer the reader to [64].

The starting point is the standard fact

$$e^{itH} P_{ac} = \int_0^\infty e^{it\lambda} E_{ac}(d\lambda), \tag{12.15}$$

where E_{ac} is the absolutely continuous part of the spectral resolution. Its density is given by

$$\frac{dE_{ac}(\lambda)}{d\lambda} = \frac{1}{2\pi i}[(H - (\lambda + i0))^{-1} - (H - (\lambda - i0))^{-1}]$$

on $\lambda > 0$. As already mentioned, Kato's theorem [52] insures that $E_{ac} = E$, provided

$$\|V\|_R^2 := \int_{\mathbb{R}^3 \times \mathbb{R}^3} \frac{|V(x)| |V(y)|}{|x - y|^2} \, dx \, dy < (4\pi)^2 . \tag{12.16}$$

Let $R_V(z) = (-\Delta + V - z)^{-1}$ and $R_0(z) = (-\Delta - z)^{-1}$. Then with V as in (12.16), for all $f, g \in L^2(\mathbb{R}^3)$ and $\varepsilon \geq 0$, one has the Born series expansion

$$\langle R_V(\lambda \pm i\varepsilon) f, g \rangle - \langle R_0(\lambda \pm i\varepsilon) f, g \rangle = \sum_{\ell=1}^{\infty} (-1)^{\ell} \langle R_0(\lambda \pm i\varepsilon)(V R_0(\lambda \pm i\varepsilon))^{\ell} f, g \rangle.$$

$$(12.17)$$

It is well known that the resolvent $R_0(z)$ for $\Im z \geq 0$ has the kernel

$$R_0(z)(x, y) = \frac{\exp(i\sqrt{z}|x - y|)}{4\pi |x - y|} \tag{12.18}$$

with $\Im(\sqrt{z}) \geq 0$. Then there is the following simple lemma that is basically an instance of stationary phase. For the proof we refer the reader to [64].

LEMMA 12.2.1 *Let ψ be a smooth, even bump function with $\psi(\lambda) = 1$ for $-1 \leq \lambda \leq 1$ and $\mathrm{supp}(\psi) \subset [-2, 2]$. Then for all $t \geq 1$ and any real a,*

$$\sup_{L \geq 1} \left| \int_0^{\infty} e^{it\lambda} \sin(a\sqrt{\lambda}) \, \psi(\frac{\sqrt{\lambda}}{L}) \, d\lambda \right| \leq C \, t^{-\frac{3}{2}} |a| \tag{12.19}$$

where C only depends on ψ.

In addition to (12.16), we will assume that

$$\|V\|_{\mathcal{K}} := \sup_{x \in \mathbb{R}^3} \int_{\mathbb{R}^3} \frac{|V(y)|}{|x - y|} \, dy < 4\pi. \tag{12.20}$$

In [64] this norm was introduced by the term *global Kato norm* (it is closely related to the well-known *Kato norm*; see Aizenman and Simon [3], [71]). The following lemma explains to some extent why condition (12.20) is needed. Iterated integrals as in (12.21) will appear in a series expansion of the spectral resolution of $H = -\Delta + V$. For the sake of completeness, and in order to show how these global Kato norms arise, we reproduce the simple proof from [64].

LEMMA 12.2.2 *For any positive integer k and V as above,*

$$\sup_{x_0, x_{k+1} \in \mathbb{R}^3} \int_{\mathbb{R}^{3k}} \frac{\prod_{j=1}^{k} |V(x_j)|}{\prod_{j=0}^{k} |x_j - x_{j+1}|} \sum_{\ell=0}^{k} |x_\ell - x_{\ell+1}| \, dx_1 \ldots dx_k \leq (k + 1) \|V\|_{\mathcal{K}}^k. \tag{12.21}$$

Proof. Define the operator $\mathcal{A}$ by the formula

$$\mathcal{A} f(x) = \int_{\mathbb{R}^3} \frac{|V(y)|}{|x - y|} f(y) \, dy.$$

Observe that the assumption (12.20) on the potential V implies that $\mathcal{A} : L^{\infty} \to L^{\infty}$ and $\|\mathcal{A}\|_{L^{\infty} \to L^{\infty}} \leq c_0$, where we have set $c_0 := \|V\|_{\mathcal{K}}$ for convenience. Denote by $\langle , \rangle$ the standard L^2 pairing. In this notation the estimate (12.21) is equivalent to proving that the operators $\mathcal{B}_k$ defined as

$$\mathcal{B}_k f = \sum_{m=0}^{k} < f, \mathcal{A}^{k-m} 1 > \mathcal{A}^m 1$$

are bounded as operators from $L^1 \to L^\infty$ with the bound

$$\|\mathcal{B}_k\|_{L^1 \to L^\infty} \le (k+1) c_0^k.$$

For arbitrary $f \in L^1$ one has

$$\|\mathcal{B}_k f\|_{L^\infty} \le \sum_{m=0}^{k} | < f, \mathcal{A}^{k-m} 1 > | \, \|\mathcal{A}^m 1\|_{L^\infty}$$

$$\le \sum_{m=0}^{k} \|\mathcal{A}^{k-m}\|_{L^\infty \to L^\infty} \|\mathcal{A}^m\|_{L^\infty \to L^\infty} \|f\|_{L^1}$$

$$\le \sum_{m=0}^{k} c_0^k \|f\|_{L^1} \le (k+1) c_0^k \|f\|_{L^1},$$

as claimed.   □

We are now in a position to prove the small V result from [64]. In [59], Perfelice obtained an analogous result for the wave equation.

THEOREM 12.2.1 *With $H = -\Delta + V$ and V satisfying the conditions (12.16) and (12.20), one has the bound*

$$\|e^{itH}\|_{L^1 \to L^\infty} \le C t^{-\frac{3}{2}}$$

in three dimensions.

Proof. Fix a real potential V as above, as well as any $L \ge 1$, and real $f, g \in C_0^\infty(\mathbb{R}^3)$. Then applying (12.17), (12.18), Lemma 12.2.1, and Lemma 12.2.2 in this order, we obtain

$$\sup_{L \ge 1} \left| \langle e^{itH} \psi(\sqrt{H}/L) f, g \rangle \right|$$

$$\le \sup_{L \ge 1} \left| \int_0^\infty e^{it\lambda} \, \psi(\sqrt{\lambda}/L) \langle E'(\lambda) f, g \rangle \, d\lambda \right|$$

$$= \sup_{L \ge 1} \left| \int_0^\infty e^{it\lambda} \, \psi(\sqrt{\lambda}/L) \Im \langle R_V(\lambda + i0) f, g \rangle \, d\lambda \right|$$

$$= \sup_{L \ge 1} \left| \int_0^\infty e^{it\lambda} \, \psi(\sqrt{\lambda}/L) \sum_{k=0}^{\infty} \Im \langle R_0(\lambda + i0)(V R_0(\lambda + i0))^k f, g \rangle \, d\lambda \right|.$$

To proceed, we now use the explicit form of the free resolvent. This yields

$$\leq \sum_{k=0}^{\infty} \int_{\mathbb{R}^6} |f(x_0)||g(x_{k+1})| \int_{\mathbb{R}^{3k}} \frac{\prod_{j=1}^{k} |V(x_j)|}{\prod_{j=0}^{k} 4\pi |x_j - x_{j+1}|} \cdot$$

$$\cdot \sup_{L\geq 1} \left| \int_0^{\infty} e^{it\lambda} \, \psi(\sqrt{\lambda}/L) \sin\left(\sqrt{\lambda} \sum_{\ell=0}^{k} |x_\ell - x_{\ell+1}|\right) d\lambda \right|$$

$$\times d(x_1, \ldots, x_k) \, dx_0 \, dx_{k+1}$$

$$\leq C t^{-\frac{3}{2}} \sum_{k=0}^{\infty} \int_{\mathbb{R}^6} |f(x_0)||g(x_{k+1})| \int_{\mathbb{R}^{3k}} \frac{\prod_{j=1}^{k} |V(x_j)|}{(4\pi)^{k+1} \prod_{j=0}^{k} |x_j - x_{j+1}|} \quad (12.22)$$

$$\times \sum_{\ell=0}^{k} |x_\ell - x_{\ell+1}| \, d(x_1, \ldots, x_k) \, dx_0 \, dx_{k+1}$$

$$\leq C t^{-\frac{3}{2}} \sum_{k=0}^{\infty} \int_{\mathbb{R}^6} |f(x_0)||g(x_{k+1})| \, (k+1)(\|V\|_{\mathcal{K}}/4\pi)^k \, dx_0 \, dx_{k+1}$$

$$\leq C t^{-\frac{3}{2}} \|f\|_1 \|g\|_1,$$

since $\|V\|_{\mathcal{K}} < 4\pi$. In order to pass to (12.22) one uses the explicit representation of the kernel of $R_0(\lambda + i0)$, see (12.18), which leads to a k-fold integral. Next, one interchanges the order of integration in this iterated integral. □

The next step is to remove the smallness assumption on V. This was done in [35] for potentials decaying like $|V(x)| \leq C\langle x \rangle^{-\beta}$ with $\beta > 3$. The proof required splitting the energies into the regions $[\lambda_0, \infty)$ (the "large energies") and $[0, \lambda_0]$ (the "small energies") where $\lambda_0 > 0$ is small. In the regime of large energies, one expands the resolvent R_V into a finite Born series,

$$R_V(\lambda^2 \pm i0) = \sum_{\ell=0}^{2m+1} R_0(\lambda^2 \pm i0)(-V R_0(\lambda^2 \pm i0))^{\ell}$$
$$+ R_0(\lambda^2 \pm i0)(V R_0(\lambda^2 \pm i0))^m V R_V(\lambda^2 \pm i0) \quad (12.23)$$
$$\times V(R_0(\lambda^2 \pm i0)V)^m R_0(\lambda^2 \pm i0),$$

where m is any positive integer. All but the last term (which involves R_V) are treated by the same argument in [64] that we sketched previously. To bound the contribution of the final term in (12.23), let $R_0^{\pm}(\lambda^2) := R_0(\lambda^2 \pm i0)$. Moreover, set

$$G_{\pm,x}(\lambda^2)(x_1) := e^{\mp i\lambda|x|} R_0(\lambda^2 \pm i0)(x_1, x) = \frac{e^{\pm i\lambda(|x_1-x|-|x|)}}{4\pi |x_1 - x|}.$$

Similar kernels appear already in Yajima's work [87] (see his high-energy section).

Hence, we are led to proving that

$$\left| \int_0^\infty e^{it\lambda^2} e^{\pm i\lambda(|x|+|y|)} \, (1 - \chi(\lambda/\lambda_0))\lambda \langle V R_V^\pm(\lambda^2) V (R_0^\pm(\lambda^2)V)^m G_{\pm,y}(\lambda^2), \right.$$
$$\left. (R_0^\mp(\lambda^2)V)^m G_{\pm,x}^*(\lambda^2) \rangle \, d\lambda \right| \lesssim |t|^{-\frac{3}{2}} \qquad (12.24)$$

uniformly in $x, y \in \mathbb{R}^3$. Here χ is a bump function that is equal to one on a neighborhood of the origin. The estimate (12.24) is proved by means of stationary phase and the *limiting absorption principle*. The latter refers to estimates of the form, with $\lambda > 0$ and $\sigma > \frac{1}{2}$,

$$\| R_0(\lambda^2 \pm i0) f \|_{L^{2,-\sigma}} \le C(\lambda) \| f \|_{L^{2,\sigma}}, \qquad (12.25)$$

where $L^{2,\sigma} = \langle x \rangle^{-\sigma} L^2$, see [1]. Similar estimates also hold for the derivatives of R_0 in λ. Moreover, $C(\lambda)$ decays powerlike with $\lambda \to \infty$. By means of the resolvent identity and arguments of Agmon and Kato, analogous estimates hold for $R_V(\lambda^2 \pm i0)$ (this essentially amounts to the absence of imbedded eigenvalues in the continuous spectrum). These properties insure that the integrand in (12.24), viz.,

$$a_{x,y}(\lambda) := (1 - \chi(\lambda/\lambda_0))\lambda \, \langle V R_V^\pm(\lambda^2) V (R_0^\pm(\lambda^2)V)^m G_{\pm,y}(\lambda^2),$$
$$(R_0^\mp(\lambda^2)V)^m G_{\pm,x}^*(\lambda^2) \rangle,$$

decays at least as fast as λ^{-2} (provided m is large) and is twice differentiable, say. Moreover, due to the presence of the functions $G_{\pm,y}$ and $G_{\pm,x}^*$ at the edges, one checks that if the critical point $\lambda_1 = \frac{|x|+|y|}{2t}$ of the phase falls into the support of this integrand, which requires $\lambda_1 \ge \lambda_0$, then the entire integral is bounded by

$$t^{-\frac{1}{2}} |a_{x,y}(\lambda_1)| \le C t^{-\frac{1}{2}} (\langle x \rangle \langle y \rangle)^{-1} \le C t^{-\frac{3}{2}},$$

as desired.

In the low-energy regime $\lambda \in [0, \lambda_0]$, one writes

$$\langle e^{itH} \chi(\sqrt{H}/\lambda_0) P_{a.c.} f, g \rangle$$
$$= \int_0^\infty e^{it\lambda^2} \lambda \, \chi(\lambda/\lambda_0) \langle [R_V(\lambda^2 + i0) - R_V(\lambda^2 - i0)]f, g \rangle \frac{d\lambda}{\pi i}$$

and proceeds via the resolvent identity

$$R_V^\pm(\lambda^2) = R_0^\pm(\lambda^2) - R_0^\pm(\lambda^2) V (I + R_0^\pm(\lambda^2)V)^{-1} R_0^\pm(\lambda^2).$$

Expanding $R_0^\pm(\lambda^2)$ around zero, the invertibility of $I + R_0^\pm(\lambda^2)V$ reduces to the invertibility of

$$S_0 := I + R_0^\pm(0)V. \qquad (12.26)$$

However, the latter is equivalent to zero energy being neither an eigenvalue nor a resonance. Writing $R_0^\pm(\lambda^2) = R_0(0) + B^\pm(\lambda)$, we conclude that

$$[I + R_0^\pm(\lambda^2)V]^{-1} = S_0^{-1}[I + B^\pm(\lambda) V S_0^{-1}]^{-1} =: S_0^{-1} \tilde{B}^\pm(\lambda).$$

Some elementary calculations based on the explicit form of the kernel of R_0 and the decay of V then reduce the $t^{-\frac{3}{2}}$ dispersive decay to the finiteness of

$$\int_{-\infty}^{\infty} \|[\chi_0(\tilde{B}^+)']^{\vee}(u)\|_{HS(-1^-,-2^-)} \, du \quad \text{and} \quad \int_{-\infty}^{\infty} \|[\chi_0\tilde{B}^+]^{\vee}(u)\|_{HS(-1^-,-2^-)} \, du,$$

where the norm is that of the Hilbert-Schmidt operators from $L^{-1-\varepsilon}(\mathbb{R}^3) \to L^{-2-\varepsilon}(\mathbb{R}^3)$. Expanding into a Neuman series

$$\tilde{B}^+(\lambda) = [I + B^+(\lambda)VS_0^{-1}]^{-1} = \sum_{n=0}^{\infty} (-B^+(\lambda)VS_0^{-1})^n$$

and making careful use of the explicit kernel

$$B^{\pm}(\lambda)(x, y) = \frac{e^{\pm i\lambda|x-y|} - 1}{4\pi|x-y|}$$

finishes the proof; see [35].

This argument was extended in various directions. First Yajima [88] and, independently, Erdogan and the author [28] have adapted it to the case of zero energy being an eigenvalue and/or a resonance. The difference is, of course, that in this case S_0 as in (12.26) is no longer invertible and $(I + R_0^{\pm}(\lambda^2)V)^{-1}$ involves singular powers of λ. Yajima uses the expansion from [48] for this purpose, whereas [28] uses the method from [49]. The latter is based on the symmetric resolvent identity and is therefore entirely situated in L^2 rather than weighted L^2. The following theorem is from [28]. Yajima proves the same, but assuming less decay on V.

THEOREM 12.2.2 *Assume that V satisfies $|V(x)| \le C\langle x \rangle^{-\beta}$ with $\beta > 10$ and assume that there is a resonance at energy zero but that zero is not an eigenvalue. Then there is a time-dependent rank one operator F_t such that*

$$\|e^{itH} P_{ac} - t^{-1/2}F_t\|_{1 \to \infty} \le Ct^{-3/2},$$

for all $t > 0$, and F_t satisfies

$$\sup_t \|F_t\|_{L^1 \to L^\infty} < \infty, \qquad \limsup_{t \to \infty} \|F_t\|_{L^1 \to L^\infty} > 0.$$

A similar result holds also in the presence of eigenvalues, but in general F_t is no longer of rank one.

The paper [29] extends these methods further, namely to the case of systems of the type that arise from linearizing NLS around a ground-state standing wave.

In another direction, Goldberg has improved on the method from [35] in several aspects. In [33], he proves that $V \in L^{\frac{3}{2}(1+\varepsilon)}(\mathbb{R}^3) \cap L^1(\mathbb{R}^3)$ suffices for the dispersive estimate (assuming, of course, that zero is neither an eigenvalue nor a resonance). This amount of integrability is analogous to the $\beta > 3$ point-wise decay from [35]. Goldberg's result requires a substitute for (12.25) on $L^p(\mathbb{R}^3)$ spaces, rather than weighted L^2 spaces. Such a substitute exists and is known to be related to the Stein-Tomas theorem in Fourier analysis; see [75]. It was first obtained for the free resolvent by Kenig, Ruiz, and Sogge [54] and extended to perturbed resolvents by Goldberg and the author [36], as well as Ionescu and the author [45]. For example, in $\mathbb{R}^3$ the bound from [54] takes the form

$$\|R_0(\lambda^2 + i\varepsilon)f\|_{L^4(\mathbb{R}^3)} \leq C\,\lambda^{-\frac{1}{2}}\|f\|_{L^{\frac{4}{3}}(\mathbb{R}^3)},$$

and in [36] it is proved that

$$\sup_{0<\varepsilon<1,\,\lambda\geq\lambda_0} \left\|(-\Delta + V - (\lambda^2 + i\varepsilon))^{-1}\right\|_{\frac{4}{3}\to 4} \leq C(\lambda_0, V)\,\lambda^{-\frac{1}{2}}. \tag{12.27}$$

for all real-valued $V \in L^p(\mathbb{R}^3) \cap L^{\frac{3}{2}}(\mathbb{R}^3)$, $p > \frac{3}{2}$ and every $\lambda_0 > 0$. Of course this requires absence of imbedded bound states in the continuous spectrum, which was proved for the same class of V by Ionescu and Jerison [44]. A very different approach from the one in [36] to estimates of the form (12.27) was found in [45], which is related to [68]. [45] applies to all dimensions $d \geq 2$ and quite general perturbations (including magnetic ones) of $-\Delta$, but it also does not rely on [44]. In fact, as in Agmon's classical paper [1] it is shown that the imbedded eigenvalues form a discrete set outside of which a bound as in (12.27) holds (albeit on somewhat different spaces). Moreover, this is obtained under the assumption that $V \in L^p(\mathbb{R}^d)$ for some $\frac{d}{2} \leq p \leq \frac{d+1}{2}$. The upper limit of $\frac{d+1}{2}$ here is natural in some ways, since Ionescu and Jerison have found a smooth, real-valued potential in $L^p(\mathbb{R}^d)$ for all $p > \frac{d+1}{2}$, which has an imbedded eigenvalue. The lower limit of $d/2$ is the usual one for self-adjointness purposes.

Returning to dispersive estimates, Goldberg [34] proved that even $V \in L^p(\mathbb{R}^3) \cap L^q(\mathbb{R}^3)$ with $p < \frac{3}{2} < q$ suffices for a dispersive estimate with the usual restriction on zero energy. Note that this is nearly critical with respect to the natural scaling of the Schrödinger equation in $\mathbb{R}^3$. One of his main observations for the low-energy argument was that for such V (and assuming that zero is neither an eigenvalue nor a resonance)

$$\sup_{\lambda\in\mathbb{R}} \left\|(I + VR_0^+(\lambda^2))^{-1}\right\|_{1\to 1} < \infty$$

(see [34] for further details). As far as high energies are concerned, Goldberg noticed that the Born series estimate from [64] can be improved so that the kth term is bounded by $(\lambda_1^{-\varepsilon}\|V\|)^k$ with $\|V\| = \max(\|V\|_p, \|V\|_q)$ as opposed to $(\|V\|_{\mathcal{K}}/4\pi)^k$. Choosing λ_1 large bound guarantees a convergent series.

12.3 THE ONE-DIMENSIONAL CASE

We will not repeat the discussion of the one-dimensional theorems from the introduction where the results from [80], [78], [4], or [35] were described. Rather, we would like to focus on a novel estimate that exploits the absence of a resonance by means of weights and obtains a better rate of decay. It was motivated by the work of Murata [57] and Buslaev and Perelman [15] on improved local L^2 decay in the absence of resonances in dimension one. Note that the weight $\langle x \rangle$ is optimal in the sense that it cannot be replaced with $\langle x \rangle^\tau$, $\tau < 1$.

THEOREM 12.3.1 *Suppose V is real-valued and $\|\langle x \rangle^4 V\|_1 < \infty$. Let $H = -\frac{d^2}{dx^2} + V$ have the property that zero energy is not a resonance. Then*

$$\|\langle x \rangle^{-1} e^{itH} P_{ac} f\|_\infty \leq C t^{-\frac{3}{2}} \|\langle x \rangle f\|_1$$

for all $t > 0$.

Proof. Let $\lambda_0 = \|\langle x \rangle V\|_1^2$ and suppose χ is a smooth cut-off such that $\chi(\lambda) = 0$ for $\lambda \leq \lambda_0$ and $\chi(\lambda) = 1$ for $\lambda \geq 2\lambda_0$. Recall that

$$R_0(\lambda \pm i0)(x) = \frac{\pm i}{2\sqrt{\lambda}} e^{\pm i|x|\sqrt{\lambda}}.$$

Hence,

$$\left| \langle R_0(\lambda + i0)(V R_0(\lambda + i0))^n f, g \rangle \right| \leq (2\sqrt{\lambda})^{-n-1} \|V\|_1^n \|f\|_1 \|g\|_1,$$

and the Born series

$$R_V(\lambda \pm i0) = \sum_{n=0}^{\infty} R_0(\lambda \pm i0)(-V R_0(\lambda \pm i0))^n \tag{12.28}$$

converges in the operator norm $L^1(\mathbb{R}) \to L^\infty(\mathbb{R})$ provided $\lambda > \lambda_0$. The absolutely continuous part of the spectral measure is given by

$$\langle E_{a.c.}(d\lambda) f, g \rangle = \left\langle \frac{1}{2\pi i} [R_V(\lambda + i0) - R_V(\lambda - i0)] f, g \right\rangle d\lambda.$$

Therefore, integrating by parts once yields

$$\langle e^{itH} \chi(H) f, g \rangle$$
$$= -(4\pi t)^{-1} \sum_{n=0}^{\infty} \int_{-\infty}^{\infty} e^{it\lambda^2} \frac{d}{d\lambda} \left[\chi(\lambda^2) \langle R_0(\lambda^2 + i0)(V R_0(\lambda^2 + i0))^n f, g \rangle \right] d\lambda,$$

$$\tag{12.29}$$

where we have first changed variables $\lambda \to \lambda^2$. Summation and integration may be exchanged because the Born series converges absolutely in the $L^1(d\lambda)$ norm, and the domain of integration is extended to $\mathbb{R}$ via the identity $R_0(\lambda^2 - i0) = R_0((-\lambda)^2 + i0)$. The kernel of $R_0(\lambda^2 + i0)(V R_0(\lambda^2 + i0))^n$ is given explicitly by the formula

$$R_0(\lambda^2 + i0)(V R_0(\lambda^2 + i0))^n(x, y)$$
$$= \frac{1}{(2\lambda)^{n+1}} \int_{\mathbb{R}^n} \prod_{j=1}^{n} V(x_j) e^{i\lambda(|x-x_1|+|y-x_n|+\sum_{k=2}^{n}|x_k-x_{k-1}|)} dx,$$

with $dx = dx_1 \dots dx_n$. Hence, in view of the derivative in (12.29),

$$\left|\langle e^{itH}\chi(H)f, g\rangle\right|$$

$$\leq C|t|^{-1}\sum_{n=0}^{\infty}(2\sqrt{\lambda_0})^{-n-1}\sup_{a\in\mathbb{R}}\left|\int_{-\infty}^{\infty}e^{i(t\lambda^2+a\lambda)}\chi(\lambda^2)\lambda^{-n-1}\lambda_0^{(n+1)/2}\,d\lambda\right|$$

$$\|\langle x\rangle V\|_1^n\,\|\langle x\rangle f\|_1\|\langle x\rangle g\|_1$$

$$+C|t|^{-1}\sum_{n=0}^{\infty}(2\sqrt{\lambda_0})^{-n-1}\sup_{a\in\mathbb{R}}\left|\int_{-\infty}^{\infty}e^{i(t\lambda^2+a\lambda)}\chi'(\lambda^2)\lambda^{-n}\lambda_0^{(n+1)/2}\,d\lambda\right| \tag{12.30}$$

$$\|V\|_1^n\,\|f\|_1\|g\|_1$$

$$\leq C(V)\,|t|^{-\frac{3}{2}}\|\langle x\rangle f\|_1\|\langle x\rangle g\|_1.$$

We used the dispersive bound for the one-dimensional Schrödinger equation to pass to (12.30), observing in particular that

$$\sup_{n\geq 0}\left\|[\chi(\lambda^2)\,\lambda^{-n-1}\lambda_0^{(n+1)/2}]^\vee\right\| < \infty,$$

where the norm refers to the total variation norm of measures.

It remains to consider small energies, i.e., those λ for which $\chi(\lambda^2) \neq 1$. In this case, we let $f_j(\cdot, \lambda)$ for $j = 1, 2$ be the Jost solutions. They satisfy

$$\left(-\frac{d^2}{dx^2} + V - \lambda^2\right)f_j(x, \lambda) = 0, \quad f_1(x, \lambda) \sim e^{ix\lambda}$$

$$\text{as } x \to \infty, \quad f_2(x, \lambda) \sim e^{-ix\lambda} \text{ as } x \to -\infty$$

for any $\lambda \in \mathbb{R}$. Furthermore, if $\lambda \neq 0$, then

$$f_1(\cdot, \lambda) = \frac{R_1(\lambda)}{T(\lambda)}f_2(\cdot, \lambda) + \frac{1}{T(\lambda)}f_2(\cdot, -\lambda),$$

$$f_2(\cdot, \lambda) = \frac{R_2(\lambda)}{T(\lambda)}f_1(\cdot, \lambda) + \frac{1}{T(\lambda)}f_1(\cdot, -\lambda), \tag{12.31}$$

where $T(\lambda) = \frac{-2i\lambda}{W(\lambda)}$ with $W(\lambda) = W[f_1(\cdot, \lambda), f_2(\cdot, \lambda)]$ and

$$R_1(\lambda) = -\frac{T(\lambda)}{2i\lambda}W[f_1(\cdot, \lambda), f_2(\cdot, -\lambda)], \quad R_2(\lambda) = \frac{T(\lambda)}{2i\lambda}W[f_1(\cdot, -\lambda), f_2(\cdot, \lambda)].$$

Then the jump condition of the resolvent R_V across the spectrum takes the form

$$\left(R_V(\lambda^2 + i0) - R_V(\lambda^2 - i0)\right)(x, y)$$

$$= \frac{|T(\lambda)|^2}{-2i\lambda}(f_1(x, \lambda)f_1(y, -\lambda) + f_2(x, \lambda)f_2(y, -\lambda)),$$

with $\lambda \geq 0$. Let us denote the distorted Fourier basis by

$$e(x, \lambda) = \frac{1}{\sqrt{2\pi}}\begin{cases} T(\lambda)f_1(\cdot, \lambda) & \text{if } \lambda \geq 0 \\ T(-\lambda)f_2(x, -\lambda) & \text{if } \lambda < 0 \end{cases}$$

(see Weder's papers [80] and [78] for more details on this basis). Then the evolution $e^{itH}(1 - \chi(H))P_{a.c.}$ can be written as

$$
\begin{aligned}
&\langle e^{itH}(1 - \chi(H))P_{a.c.}\phi, \psi \rangle \\
&= \frac{1}{2\pi i} \int_{\mathbb{R}^2} \int_0^\infty 2\lambda e^{it\lambda^2}(1 - \chi(\lambda^2))\big(R_V(\lambda^2 + i0) - R_V(\lambda^2 - i0)\big) \\
&\hspace{3cm} (x, y)\, d\lambda\, \bar{\phi}(x)\psi(y)\, dx dy \\
&= \int_{-\infty}^\infty e^{it\lambda^2}(1 - \chi(\lambda^2))\langle \psi, e(\cdot, \lambda)\rangle \langle e(\cdot, \lambda), \phi \rangle\, d\lambda.
\end{aligned}
\tag{12.32}
$$

Our assumption that zero energy is not a resonance implies that $T(\lambda) = \alpha\lambda + o(\lambda)$ where $\alpha \neq 0$. In particular, $T(0) = 0$ and $R_1(0) = R_2(0) = -1$. Integrating by parts in (12.32) therefore yields

$$
\begin{aligned}
\langle e^{itH}(1 - \chi(H))P_{a.c.}\phi, \psi \rangle &= -\frac{1}{4\pi it} \int_0^\infty e^{it\lambda^2} \partial_\lambda\Big[(1 - \chi(\lambda^2))|T(\lambda)|^2\lambda^{-1} \\
&\hspace{2cm} \langle \psi, f_1(\cdot, \lambda)\rangle \langle f_1(\cdot, \lambda), \phi \rangle\Big]\, d\lambda \\
&\quad -\frac{1}{4\pi it} \int_{-\infty}^0 e^{it\lambda^2} \partial_\lambda\Big[(1 - \chi(\lambda^2))|T(\lambda)|^2\lambda^{-1} \\
&\hspace{2cm} \langle \psi, f_2(\cdot, -\lambda)\rangle \langle f_2(\cdot, -\lambda), \phi \rangle\Big]\, d\lambda.
\end{aligned}
\tag{12.33}
$$

By symmetry, it will suffice to treat the integral involving $f_1(\cdot, \lambda)$. We distinguish three cases, depending on where the derivative ∂_λ falls. We start with the integral

$$
\int_0^\infty e^{it\lambda^2} \omega(\lambda) f_1(x, \lambda) f_1(y, -\lambda)\, d\lambda,
\tag{12.34}
$$

where we have set $\omega(\lambda) = \partial_\lambda[(1 - \chi(\lambda^2))|T(\lambda)|^2\lambda^{-1}]$. By the preceding, ω is a smooth function with compact support in $[0, \infty)$. As usual, we will estimate (12.34) by means of a Fourier transform in λ. Since we are working on a half-line, this will actually be a cosine transform. Let $\tilde{\omega}$ be another cut-off function satisfying $\omega\tilde{\omega} = \omega$. Then

$$
\left| \int_0^\infty e^{it\lambda^2} \omega(\lambda) f_1(x, \lambda) f_1(y, -\lambda)\, d\lambda \right|
$$
$$
\leq C|t|^{-\frac{1}{2}} \| [\omega f_1(x, \cdot)]^\vee \|_1 \| [\tilde{\omega} f_1(y, -\cdot)]^\vee \|_1.
\tag{12.35}
$$

It remains to estimate

$$
[\omega f_1(x, \cdot)]^\vee(u) := \int_0^\infty \cos(u\lambda)\omega(\lambda) f_1(x, \lambda)\, d\lambda
\tag{12.36}
$$

in L^1 relative to u. The second L^1-norm in (12.35) is treated the same way. We need to consider the cases $x \geq 0$ and $x \leq 0$ separately. In the former case,

$$[\omega f_1(x, \cdot)]^\vee (u) := \int_0^\infty \cos(u\lambda) e^{ix\lambda} \omega(\lambda) e^{-ix\lambda} f_1(x, \lambda) \, d\lambda$$

$$= \frac{1}{2} \int_0^\infty e^{i(x+u)\lambda} \omega(\lambda) e^{-ix\lambda} f_1(x, \lambda) \, d\lambda \qquad (12.37)$$

$$+ \frac{1}{2} \int_0^\infty e^{i(x-u)\lambda} \omega(\lambda) e^{-ix\lambda} f_1(x, \lambda) \, d\lambda.$$

If $||u| - |x|| \leq |x|$, then we simply estimate

$$|[\omega f_1(x, \cdot)]^\vee (u)| \leq C.$$

On the other hand, if $||u| - |x|| > |x|$, then we integrate by parts in (12.37):

$$[\omega f_1(x, \cdot)]^\vee (u) = -\frac{1}{2i(x+u)} \omega(0) f_1(x, 0) - \frac{1}{2i(x-u)} \omega(0) f_1(x, 0)$$

$$- \frac{1}{2i(x+u)} \int_0^\infty e^{i(x+u)\lambda} \partial_\lambda [\omega(\lambda) e^{-ix\lambda} f_1(x, \lambda)] \, d\lambda \qquad (12.38)$$

$$- \frac{1}{2i(x-u)} \int_0^\infty e^{i(x-u)\lambda} \partial_\lambda [\omega(\lambda) e^{-ix\lambda} f_1(x, \lambda)] \, d\lambda.$$

Since

$$\sup_{x \geq 0, \lambda} |\partial_\lambda^j [\omega(\lambda) e^{-ix\lambda} f_1(x, \lambda)]| \leq C(V),$$

for $j = 0, 1, 2$, it follows that

$$|[\omega f_1(x, \cdot)]^\vee (u)| \leq C \frac{|x|}{|x^2 - u^2|} + C(u + x)^{-2} + C(u - x)^{-2}.$$

The conclusion is that

$$\int_{\mathbb{R}} |[\omega f_1(x, \cdot)]^\vee (u)| \, du \leq C \langle x \rangle. \qquad (12.39)$$

To deal with $x \leq 0$, we use (12.31). Thus,

$$[\omega f_1(x, \cdot)]^\vee (u) = \int_0^\infty \cos(u\lambda) \omega(\lambda) \frac{R_1(\lambda) + 1}{T(\lambda)} f_2(x, \lambda) \, d\lambda \qquad (12.40)$$

$$+ \int_0^\infty \cos(u\lambda) \omega(\lambda) \frac{1}{T(\lambda)} (f_2(x, \lambda) - f_2(x, -\lambda)) \, d\lambda. \qquad (12.41)$$

Set $\omega_1 = \omega(\lambda) \frac{R_1(\lambda)+1}{T(\lambda)}$. Then (12.40) can be written as

$$\int_0^\infty \cos(u\lambda) \omega(\lambda) \frac{R_1(\lambda) + 1}{T(\lambda)} f_2(x, \lambda) \, d\lambda = \int_0^\infty \cos(u\lambda) e^{-ix\lambda} \omega_1(\lambda) e^{ix\lambda} f_2(x, \lambda) \, d\lambda.$$

Hence, it can be treated by the same arguments as (12.36) with $x \geq 0$. Indeed, simply use that

$$\sup_{x \leq 0, \lambda} |\partial_\lambda[\omega_1(\lambda)e^{ix\lambda} f_2(x, \lambda)]| \leq C(V).$$

On the other hand, (12.41) is the same (with ∂_2 being the partial derivative with respect to the second variable of f_2) as

$$\int_{-1}^{1} \int_0^\infty \cos(u\lambda)\omega(\lambda)\frac{\lambda}{T(\lambda)}\partial_2 f_2(x, \lambda\sigma)\, d\lambda d\sigma$$

$$= \int_{-1}^{1} \int_0^\infty \cos(u\lambda)e^{-i\lambda x\sigma} \omega_2(\lambda)\partial_2[e^{i\lambda x\sigma} f_2(x, \lambda\sigma)]\, d\lambda d\sigma \qquad (12.42)$$

$$- ix \int_{-1}^{1} \int_0^\infty \cos(u\lambda)e^{-i\lambda x\sigma} \omega_2(\lambda)e^{i\lambda x\sigma} f_2(x, \lambda\sigma)\, d\lambda d\sigma, \qquad (12.43)$$

where we have set $\omega_2(\lambda) = \omega(\lambda)\frac{\lambda}{T(\lambda)}$ (a smooth, compactly supported function in $[0, \infty)$). We will focus on the second integral (12.43), since the first one (12.42) is similar. We will integrate by parts in λ, but only on the set $|\sigma x \pm u| \geq 1$. Then

$$- ix \int_{-1}^{1} \int_0^\infty \cos(u\lambda)e^{-i\lambda x\sigma} \omega_2(\lambda)e^{i\lambda x\sigma} f_2(x, \lambda\sigma)\, d\lambda \chi_{[|\sigma x \pm u| \geq 1]}\, d\sigma$$

$$= \int_{-1}^{1} \frac{x}{2(-\sigma x + u)}\omega_2(0) f_2(x, 0)\, \chi_{[|\sigma x \pm u| \geq 1]}\, d\sigma$$

$$+ \int_{-1}^{1} \frac{x}{2(-\sigma x - u)}\omega_2(0) f_2(x, 0)\, \chi_{[|\sigma x \pm u| \geq 1]}\, d\sigma$$

$$+ \int_{-1}^{1} \frac{x}{2(-\sigma x + u)} \int_0^\infty e^{i(-\sigma x + u)\lambda}\partial_\lambda \qquad (12.44)$$

$$\times \left[\omega_2(\lambda)e^{ix\lambda\sigma} f_2(x, \lambda)\right] d\lambda\, \chi_{[|\sigma x \pm u| \geq 1]}\, d\sigma$$

$$+ \int_{-1}^{1} \frac{x}{2(-\sigma x - u)} \int_0^\infty e^{i(-\sigma x - u)\lambda}\partial_\lambda \qquad (12.45)$$

$$\times \left[\omega_2(\lambda)e^{ix\lambda\sigma} f_2(x, \lambda)\right] d\lambda\, \chi_{[|\sigma x \pm u| \geq 1]}\, d\sigma.$$

The first two integrals here (which are due to the boundary $\lambda = 0$) contribute

$$\int_{-1}^{1} \frac{x}{2(-\sigma x + u)}\omega_2(0) f_2(x, 0)\, \chi_{[|\sigma x \pm u| \geq 1]}\, d\sigma$$

$$+ \int_{-1}^{1} \frac{x}{2(\sigma x - u)}\omega_2(0) f_2(x, 0)\, \chi_{[|\sigma x \pm u| \geq 1]}\, d\sigma = 0,$$

where we performed a change of variables $\sigma \mapsto -\sigma$ in the second one. Integrating by parts one more time in (12.44) and (12.45) with respect to λ implies

$$\int_{-\infty}^{\infty} \left| x \int_{-1}^{1} \int_{0}^{\infty} \cos(u\lambda) e^{-i\lambda x\sigma} \omega_2(\lambda) e^{i\lambda x\sigma} f_2(x, \lambda\sigma) \, d\lambda \chi_{[|\sigma x \pm u| \geq 1]} \, d\sigma \right| du$$

$$\leq C \int_{-\infty}^{\infty} \int_{-1}^{1} \frac{|x|}{(-\sigma x + u)^2} \chi_{[|\sigma x \pm u| \geq 1]} \, d\sigma \, du$$

$$+ C \int_{-\infty}^{\infty} \int_{-1}^{1} \frac{|x|}{(-\sigma x - u)^2} \chi_{[|\sigma x \pm u| \geq 1]} \, d\sigma \, du \leq C \, |x|.$$

Finally, the cases $|\sigma x + u| \leq 1$ and $|\sigma x - u| \leq 1$ each contribute at most $C|x|$ to the u-integral. Hence,

$$\int_{-\infty}^{\infty} \left| x \int_{-1}^{1} \int_{0}^{\infty} \cos(u\lambda) e^{-i\lambda x\sigma} \omega_2(\lambda) e^{i\lambda x\sigma} f_2(x, \lambda\sigma) \, d\lambda \, d\sigma \right| du \leq C|x|.$$

Since (12.42) can be treated the same way (in fact, the bound is $O(1)$), we obtain

$$\int_{-\infty}^{\infty} \left| \int_{-1}^{1} \int_{0}^{\infty} \cos(u\lambda) \omega(\lambda) \frac{\lambda}{T(\lambda)} \partial_2 f_2(x, \lambda\sigma) \, d\lambda \, d\sigma \right| du \leq C\langle x \rangle.$$

In view of (12.39), (12.40), and (12.41),

$$\left\| [\omega f_1(x, \cdot)]^{\vee} \right\|_1 \leq C\langle x \rangle \quad \forall \, x \in \mathbb{R},$$

which in turn implies that

$$\left| \int_{0}^{\infty} e^{it\lambda^2} \omega(\lambda) f_1(x, \lambda) f_1(y, -\lambda) \, d\lambda \right| \leq C \, |t|^{-\frac{1}{2}} \langle x \rangle \langle y \rangle, \tag{12.46}$$

for all $x, y \in \mathbb{R}$; see (12.34). This is the desired estimate on (12.33), but only for the case when ∂_λ falls on the factors not involving f_1. We now consider the case when ∂_λ falls on $f_1(x, \lambda)$. The integral in which ∂_λ falls on $f_1(y, -\lambda)$ is analogous. Hence, we need to estimate

$$\int_{0}^{\infty} e^{it\lambda^2} (1 - \chi(\lambda^2)) |T(\lambda)|^2 \lambda^{-1} \partial_\lambda f_1(x, \lambda) \, f_1(y, -\lambda) \, d\lambda$$

$$= \int_{0}^{\infty} e^{it\lambda^2} (1 - \chi(\lambda^2)) T(-\lambda) \lambda^{-1} \partial_\lambda [T(\lambda) f_1(x, \lambda)] \, f_1(y, -\lambda) \, d\lambda \tag{12.47}$$

$$+ \int_{0}^{\infty} e^{it\lambda^2} (1 - \chi(\lambda^2)) T(-\lambda) T'(\lambda) \lambda^{-1} f_1(x, \lambda) \, f_1(y, -\lambda) \, d\lambda. \tag{12.48}$$

The integral in (12.48) is of the same form as that in (12.34). It therefore suffices to control (12.47). Let $\omega_3(\lambda) = (1 - \chi(\lambda^2)) T(-\lambda) \lambda^{-1}$. By the same reductions as before, we need to show that

$$\left\| [\omega_3 \partial_\lambda [T(\lambda) f_1(x, \cdot)]]^{\vee} \right\|_1 \leq C\langle x \rangle \quad \forall \, x \in \mathbb{R}.$$

Thus consider

$$\int_0^\infty \cos(u\lambda)\omega_3(\lambda)\partial_\lambda[T(\lambda)f_1(x,\lambda)]\,d\lambda$$

$$= ix \int_0^\infty \cos(u\lambda)e^{ix\lambda}\omega_3(\lambda)T(\lambda)e^{-ix\lambda}f_1(x,\lambda)\,d\lambda$$

$$+ \int_0^\infty \cos(u\lambda)e^{ix\lambda}\omega_3(\lambda)\partial_\lambda[T(\lambda)e^{-ix\lambda}f_1(x,\lambda)]\,d\lambda.$$

If $x \geq 0$, integrating by parts leads to

$$ix \int_0^\infty \cos(u\lambda)e^{ix\lambda}\omega_3(\lambda)T(\lambda)e^{-ix\lambda}f_1(x,\lambda)\,d\lambda \qquad (12.49)$$

$$= -\frac{ix}{2i(x+u)} \int_0^\infty e^{i(x+u)\lambda}\partial_\lambda\big[\omega_3(\lambda)T(\lambda)e^{-ix\lambda}f_1(x,\lambda)\big]\,d\lambda$$

$$- \frac{ix}{2i(x-u)} \int_0^\infty e^{i(x-u)\lambda}\partial_\lambda\big[\omega_3(\lambda)T(\lambda)e^{-ix\lambda}f_1(x,\lambda)\big]\,d\lambda$$

as well as

$$\int_0^\infty \cos(u\lambda)e^{ix\lambda}\omega_3(\lambda)\partial_\lambda[T(\lambda)e^{-ix\lambda}f_1(x,\lambda)]\,d\lambda \qquad (12.50)$$

$$= -\frac{1}{2i(x+u)}\omega_3(0)\partial_\lambda[T(\lambda)e^{-ix\lambda}f_1(x,\lambda)]\Big|_{\lambda=0}$$

$$- \frac{1}{2i(x-u)}\omega_3(0)\partial_\lambda[T(\lambda)e^{-ix\lambda}f_1(x,\lambda)]\Big|_{\lambda=0}$$

$$- \frac{1}{2i(x+u)} \int_0^\infty e^{i(x+u)\lambda}\partial_\lambda\big[\omega_3(\lambda)\partial_\lambda[T(\lambda)e^{-ix\lambda}f_1(x,\lambda)]\big]\,d\lambda$$

$$- \frac{1}{2i(x-u)} \int_0^\infty e^{i(x-u)\lambda}\partial_\lambda\big[\omega_3(\lambda)\partial_\lambda[T(\lambda)e^{-ix\lambda}f_1(x,\lambda)]\big]\,d\lambda.$$

Integrating by parts one more time in (12.49) implies

$$\left| ix \int_0^\infty \cos(u\lambda)e^{ix\lambda}\omega_3(\lambda)T(\lambda)e^{-ix\lambda}f_1(x,\lambda)\,d\lambda \right|$$

$$\leq C|x|(1+|x-u|)^{-2} + C|x|(1+|x+u|)^{-2}$$

uniformly in $x \geq 0$, whereas (12.50) is treated the same way as (12.38). One needs to use here that

$$\sup_{x\geq 0,\,\lambda} |\partial_\lambda^j[\omega_3(\lambda)e^{-ix\lambda}f_1(x,\lambda)]| \leq C(V),$$

for $j = 0,1,2,3$ which follows from $\|\langle x\rangle^4 V\|_1 < \infty$. Consequently, we have proved that

$$\int_{\mathbb{R}} \left| \int_0^\infty \cos(u\lambda)\omega_3(\lambda)\partial_\lambda[T(\lambda)f_1(x,\lambda)]\,d\lambda \right| du \leq C\langle x\rangle$$

uniformly in $x \geq 0$.

Next, we deal with the case $x \leq 0$. In view of (12.31),

$$T(\lambda) f_1(\cdot, \lambda) = R_1(\lambda) f_2(\cdot, \lambda) + f_2(\cdot, -\lambda).$$

This implies that

$$\int_0^\infty \cos(u\lambda)\omega_3(\lambda)\partial_\lambda[T(\lambda)f_1(x,\lambda)]\,d\lambda \tag{12.51}$$

$$= \int_0^\infty \cos(u\lambda)e^{-ix\lambda}\omega_3(\lambda)\partial_\lambda[R_1(\lambda)e^{ix\lambda}f_2(x,\lambda)]\,d\lambda$$

$$- ix \int_0^\infty \cos(u\lambda)e^{-ix\lambda}\omega_3(\lambda)R_1(\lambda)e^{ix\lambda}f_2(x,\lambda)\,d\lambda$$

$$+ \int_0^\infty \cos(u\lambda)e^{-ix\lambda}\omega_3(\lambda)\partial_\lambda[e^{ix\lambda}f_2(x,-\lambda)]\,d\lambda$$

$$+ ix \int_0^\infty \cos(u\lambda)e^{ix\lambda}\omega_3(\lambda)e^{-ix\lambda}f_2(x,-\lambda)\,d\lambda.$$

The two integrals that are not preceded by factors of ix are treated just as in (12.50). The only difference here is that the estimates are uniform in $x \leq 0$ rather than $x \geq 0$. On the other hand, the integrals preceded by ix need to be integrated by parts in λ. It is important to check that the boundary terms at $\lambda = 0$ do not contribute to this case. Indeed, these boundary terms are

$$\frac{x}{2(u-x)}\omega_3(0)R_1(0)f_2(x,0) - \frac{x}{2(u+x)}\omega_3(0)R_1(0)f_2(x,0)$$

$$- \frac{x}{2(u+x)}\omega_3(0)f_2(x,0) - \frac{x}{2(x-u)}\omega_3(0)f_2(x,0) = 0,$$

since $R_1(0) = -1$. Hence, integrating by parts leads to an expression similar to (12.49). The conclusion is that (12.51) satisfies

$$\int_{\mathbb{R}} \left| \int_0^\infty \cos(u\lambda)\omega_3(\lambda)\partial_\lambda[T(\lambda)f_1(x,\lambda)]\,d\lambda \right| du \leq C\langle x\rangle$$

uniformly in $x \leq 0$, and we are done. □

In [55] the same bound is proved for non-selfadjoint systems of the type that arise by linearizing NLS around a ground-state standing wave. It is crucial for proving the existence of stable manifolds for all supercritical NLS in one dimension.

In dimension one, there is some recent work of Cai [17] on dispersion for Hill's operator. More precisely, let $H = -\frac{d^2}{dx^2} + q$, where q is periodic and such that its spectrum has precisely one gap. It is well known that such q are characterized in terms of Weierstrass elliptic functions. As part of his Caltech Ph.D. thesis, Cai showed that for this H one always has

$$\|e^{itH}f\|_\infty \leq Ct^{-\frac{1}{4}}\|f\|_1, \qquad t \geq 1,$$

and that generically in the potential one can replace $\frac{1}{4}$ with $\frac{1}{3}$.

12.4 THE TWO-DIMENSIONAL CASE

The following two-dimensional dispersive estimate was obtained in [69].

THEOREM 12.4.1 *Let $V : \mathbb{R}^2 \to \mathbb{R}$ be a measurable function such that $|V(x)| \leq C(1+|x|)^{-\beta}$, $\beta > 3$. Assume in addition that zero is a regular point of the spectrum of $H = -\Delta + V$. Then*

$$\left\| e^{itH} P_{ac}(H) f \right\|_\infty \leq C|t|^{-1} \|f\|_1$$

for all $f \in L^1(\mathbb{R}^2)$.

The definition of zero being a regular point amounts to the following; see Jensen and Nenciu [49]: *Let $V \neq 0$ and set $U = \text{sign } V$, $v = |V|^{\frac{1}{2}}$. Let P_v be the orthogonal projection onto v and set $Q = I - P_v$. Finally, let*

$$(G_0 f)(x) := -\frac{1}{2\pi} \int_{\mathbb{R}^2} \log|x - y| \, f(y) \, dy.$$

Then zero is regular iff $Q(U + vG_0 v)Q$ is invertible on $QL^2(\mathbb{R}^2)$.

Jensen and Nenciu study $\ker[Q(U+vG_0 v)Q]$ on $QL^2(\mathbb{R}^2)$. It can be completely described in terms of solutions Ψ of $H\Psi = 0$. In particular, its dimension is at most three plus the dimension of the zero energy eigenspace; see Theorem 6.2 and Lemma 6.4 in [49]. The extra three dimensions here are called resonances. Hence, the requirement that zero is a regular point is the analog of the usual condition that zero is neither an eigenvalue nor a resonance of H. An equivalent characterization of a regular point was given in [6], albeit under the additional assumption that $\int_{\mathbb{R}^2} V(x) \, dx \neq 0$.

As far as the spectral properties of H are concerned, we note that under the hypotheses of Theorem 12.4.1 the spectrum of H on $[0, \infty)$ is purely, absolutely continuous, and that the spectrum is pure point on $(-\infty, 0)$ with at most finitely many eigenvalues of finite multiplicities. The latter follows for example from Stoiciu [76], who obtained Birman-Schwinger type bounds in the case of two dimensions.

Theorem 12.4.1 appears to be the first $L^1 \to L^\infty$ bound with $|t|^{-1}$ decay in $\mathbb{R}^2$. Yajima [87] and Jensen and Yajima [50] proved the $L^p(\mathbb{R}^2)$ boundedness of the wave operators under stronger decay assumptions on $V(x)$, but only for $1 < p < \infty$. Hence, their result does not imply Theorem 12.4.1. Local L^2 decay was studied by Murata [57], but he does not consider $L^1 \to L^\infty$ estimates.

The main challenge in two dimensions is, of course, the low-energy part. This is due to the fact that the free resolvent $R_0^\pm(\lambda^2) = (-\Delta - (\lambda^2 \pm i0))^{-1}$ has the kernel ($H_0^\pm$ being the Hankel functions)

$$R_0^\pm(\lambda^2)(x, y) = \pm\frac{i}{4} H_0^\pm(\lambda|x - y|),$$

which is singular at energy zero (which, just as in dimension one, expresses the fact that the free problem has a resonance at zero). It is a consequence of the asymptotic expansion of Hankel functions that for all $\lambda > 0$,

$$R_0^\pm(\lambda^2) = \left[\pm \frac{i}{4} - \frac{1}{2\pi}\gamma - \frac{1}{2\pi}\log(\lambda/2)\right]P_0 + G_0 + E_0^\pm(\lambda). \quad (12.52)$$

Here $P_0 f := \int_{\mathbb{R}^2} f(x)\,dx$, $G_0 f(x) = -\frac{1}{2\pi}\int_{\mathbb{R}^2}\log|x - y|\,f(y)\,dy$, and the error $E_0^\pm(\lambda)$ has the property that

$$\left\|\sup_{0<\lambda}\lambda^{-\frac{1}{2}}|E_0^\pm(\lambda)(\cdot,\cdot)|\right\| + \left\|\sup_{0<\lambda}\lambda^{\frac{1}{2}}|\partial_\lambda E_0^\pm(\lambda)(\cdot,\cdot)|\right\| \lesssim 1 \quad (12.53)$$

with respect to the Hilbert-Schmidt norm in $\mathcal{B}(L^{2,s}(\mathbb{R}^2), L^{2,-s}(\mathbb{R}^2))$ with $s > \frac{3}{2}$. These error estimates may seem artificial, but they allow for the least amount of decay on V. The following lemma from [69] contains the expansion of the perturbed resolvent around energy zero needed in the proof of Theorem 12.4.1. It displays an important idea from [49], namely to re-sum infinite series of powers of $\log\lambda$ into one function $h_\pm(\lambda)$. This feature is crucial for our purposes. Given $V \ne 0$, set $U = \operatorname{sign} V$, $v = |V|^{\frac{1}{2}}$. Let P_v be the orthogonal projection onto v and set $Q = I - P_v$. Finally, let $D_0 = [Q(U + vG_0v)Q]^{-1}$ on $QL^2(\mathbb{R}^2)$.

LEMMA 12.4.1 *Suppose that zero is a regular point of the spectrum of $H = -\Delta + V$. Then for some sufficiently small $\lambda_1 > 0$, the operators $M^\pm(\lambda) := U + vR_0^\pm(\lambda^2)v$ are invertible for all $0 < \lambda < \lambda_1$ as bounded operators on $L^2(\mathbb{R}^2)$, and one has the expansion*

$$M^\pm(\lambda)^{-1} = h_\pm(\lambda)^{-1}S + QD_0Q + E^\pm(\lambda), \quad (12.54)$$

where $h_+(\lambda) = a\log\lambda + z$, a is real, z complex, $a \ne 0$, $\Im z \ne 0$, and $h_-(\lambda) = \overline{h_+(\lambda)}$. Moreover, S is of finite rank and has a real-valued kernel, and $E^\pm(\lambda)$ is a Hilbert-Schmidt operator that satisfies the bound

$$\left\|\sup_{0<\lambda<\lambda_1}\lambda^{-\frac{1}{2}}|E^\pm(\lambda)(\cdot,\cdot)|\right\|_{HS} + \left\|\sup_{0<\lambda<\lambda_1}\lambda^{\frac{1}{2}}|\partial_\lambda E^\pm(\lambda)(\cdot,\cdot)|\right\|_{HS} \lesssim 1, \quad (12.55)$$

where the norm refers to the Hilbert-Schmidt norm on $L^2(\mathbb{R}^2)$. Finally, let $R_V^\pm(\lambda^2) = (-\Delta + V - (\lambda^2 \pm i0))^{-1}$. Then

$$R_V^\pm(\lambda^2) = R_0^\pm(\lambda^2) - R_0^\pm(\lambda^2)vM^\pm(\lambda)^{-1}vR_0^\pm(\lambda^2). \quad (12.56)$$

This is to be understood as an identity between operators $L^{2,\frac{1}{2}+\varepsilon}(\mathbb{R}^2) \to L^{2,-\frac{1}{2}-\varepsilon}(\mathbb{R}^2)$ for some sufficiently small $\varepsilon > 0$.

The low-energy part of the proof of Theorem 12.4.1 is based on a careful estimation of the contribution of each of the terms in (12.54) to R_V in (12.56) by means of the method of stationary phase; see [69].

Murata [57] discovered that under the assumptions of Theorem 12.4.1

$$\left\|we^{itH}P_{ac}(H)w\,f\right\|_2 \le C|t|^{-1}(\log t)^{-2}\|f\|_2,$$

provided $w(x) = \langle x \rangle^{-\sigma}$ with some sufficiently large $\sigma > 0$. In other words, he obtained improved local L^2 decay provided zero energy is regular. Needless to say, such improved decay is impossible for the $L^1 \to L^\infty$ bound, but a weighted $L^1 \to L^\infty$ estimate as in Theorem 12.3.1 with the improved $|t|^{-1}(\log t)^{-2}$ decay is quite possibly true but currently unknown. Due to the integrability of this decay at infinity, such a bound would be useful for the study of nonlinear asymptotic stability of (multi) solitons in dimension two.

12.5 TIME-DEPENDENT POTENTIALS

It seems unreasonable to expect a general theory of dispersion for the Schrödinger equation

$$i\partial_t \psi + \Delta \psi + V(t, \cdot)\psi = 0 \qquad (12.57)$$

for time-dependent potentials $V(t, \cdot)$. While the L^2 norm is preserved for real-valued V, it is well known that, in contrast to time-independent V, higher H^s norms can grow in this case; see, e.g., Bourgain [9], [10], and Erdogan, Killip, and Schlag [27].

The classical work of Davies [24], Howland [41], [42], [43], and Yajima [83] deals with scattering and wave operators in this context. Recall that if $U(t, s)$ denotes the evolution of (12.57) from time s to time t, then

$$W_\pm(s) = s - \lim_{t \to \pm\infty} e^{-i(t-s)\Delta} U(t, s)$$

are the wave operators (strictly speaking, the existence of these limits is usually referred to as *completeness*, but we are following Howland's terminology). In analogy to the treatment of time-dependent Hamiltonians in classical mechanics, Howland [42] develops a formalism for treating time-dependent potentials in which $K = -i\partial_t + H(t)$ is considered as a self-adjoint operator on the Hilbert space $L^2(-\infty, \infty; L^2(\mathbb{R}^d))$. He shows that the existence of $W_\pm$ is equivalent to the existence of the strong limits

$$\mathcal{W}_\pm := s - \lim_{\sigma \to \pm\infty} e^{i\sigma K_0} e^{-i\sigma K},$$

and that $\mathcal{W}_\pm$ is the same as multiplication by $W_\pm(t)$. Furthermore, following Kato [52], he formulates a condition which insures that the wave operators are unitary. He applies this to (12.57) with (real-valued) potentials

$$V \in L_t^{r+\varepsilon}(L_x^p) \cap L_t^{r-\varepsilon}(L_x^p), \quad r = \frac{2p}{2p-d}, \quad \frac{d}{2} < p \leq \infty, \ d > 1$$

to conclude that for such V the wave operators exist and are unitary. In [42], Howland obtained similar results for $d \geq 3$ potentials that are small at infinity (rather than vanishing).

Dispersive estimates were obtained by Rodnianksi and the author [64] for small but not necessarily decaying time-dependent potentials in $\mathbb{R}^3$, whereas the case of decaying V and dimensions ≥ 2 was studied by Naibo and Stepanov [58], and d'Ancona, Pierfelice, and Visciglia [23]. In particular, the result from [64] insures that in $\mathbb{R}^3$ and for small ε

$$i\partial_t \psi + \Delta\psi + \varepsilon F(t)V(x)\psi = 0$$

has the usual $t^{-\frac{3}{2}}$ dispersive $L^1 \to L^\infty$ decay for any real-valued trigonometric polynomial $F(t)$ (or more generally, any quasi-periodic analytic function $F(t)$) and V satisfying $\|V\|_{\mathcal{K}} < \infty$; see (12.20).

Another much studied case is that of *time-periodic* V; see [24], [43], and [83]. Suppose $T > 0$ is the smallest period of V. Then the theory of (12.57) reduces to that of the Floquet operator $\mathcal{U} = U(T, 0)$. The Floquet operator can exhibit bound states and the question arises as to the existence and ranges of the wave operators (the so-called completeness problem) as well as the structure of the discrete spectrum. These issues are addressed in the aforementioned references.

More recently, in [30], Galtbayar, Jensen, and Yajima show that on the orthogonal complement of the bound states of the Floquet operator, the solutions decay locally in $L^2(\mathbb{R}^3)$. In addition, O. Costin, R. Costin, Lebowitz, and Rohlenko [19], [18] have made a very detailed analysis of some special models with time-periodic potentials. More precisely, they have found and applied a criterion that ensures scattering of the wave function. On the level of the Floquet operator, this means that there is no discrete spectrum. It would be interesting to obtain dispersive estimates for these cases.

Another well-studied class of time-dependent potentials are the so-called charge transfer models. These are Hamiltonians of the form

$$H(t) = -\Delta + \sum_{j=1}^{m} V_j(\cdot - v_j t),$$

where $\{v_j\}_{j=1}^m$ are distinct velocities and V_j are well-localized potentials. They admit localized states that travel with each of these potentials and asymptotically behave like the sum of bound states of each of the "channel Hamiltonians"

$$H(t) = -\Delta + V_j(\cdot - v_j t).$$

Those are, of course, Galilei transformed bound states of the corresponding stationary Hamiltonians. Yajima [82] and Graf [38] proved that these Hamiltonians are asymptotically complete, i.e., that as $t \to \infty$, each state decomposes into a sum of wave functions associated with each of the channels, including the free channel.

Rodnianski, Soffer, and the author obtained dispersive estimates for these models in the spaces $L^1 \cap L^2 \to L^2 \cap L^\infty$. Later, Cai [16] as part of his Caltech thesis, removed L^2 from these bounds. Such estimates were needed in order to prove asymptotic stability of N-soliton solutions; see [66].

REFERENCES

[1] Agmon, S. *Spectral properties of Schrödinger operators and scattering theory.* Ann. Scuola Norm. Sup. Pisa Cl. Sci. (4) 2 (1975), no. 2, 151–218.

[2] Agmon, S. *Lectures on exponential decay of solutions of second-order elliptic equations: bounds on eigenfunctions of N-body Schrödinger operators.*

Mathematical Notes, 29. Princeton University Press, Princeton, NJ; University of Tokyo Press, Tokyo, 1982.

[3] Aizenman, M., Simon, B. *Brownian motion and Harnack inequality for Schrödinger operators.* Comm. Pure Appl. Math. 35 (1982), no. 2, 209–273.

[4] Artbazar, G., Yajima, K. *The L^p-continuity of wave operators for one dimensional Schrödinger operators.* J. Math. Sci. Univ. Tokyo 7 (2000), no. 2, 221–240.

[5] Bollé, D., Danneels, C., Gesztesy, F. *Scattering theory for one-dimensional systems with $\int dx\, V(x) = 0$.* J. Math. Anal. Appl. 122 (1987), no. 2, 496–518.

[6] Bollé, D., Danneels, C., Gesztesy, F. *Threshold scattering in two dimensions.* Ann. Inst. H. Poincaré Phys. Théor. 48 (1988), no. 2, 175–204.

[7] Bollé, D., Gesztesy, F., Wilk, S. F. *A complete treatment of low-energy scattering in one dimension.* J. Operator Theory 13 (1985), no. 1, 3–31.

[8] Bourgain, J. *Fourier transform restriction phenomena for certain lattice subsets and applications to nonlinear evolution equations. I. Schrödinger equations.* Geom. Funct. Anal. 3 (1993), no. 2, 107–156.

[9] Bourgain, J. *Growth of Sobolev norms in linear Schrödinger equations with quasi-periodic potential.* Comm. Math. Phys. 204 (1999), no. 1, 207–247.

[10] Bourgain, J. *On growth of Sobolev norms in linear Schrödinger equations with smooth time dependent potential.* J. Anal. Math. 77 (1999), 315–348.

[11] Bourgain, J. *Global solutions of nonlinear Schrödinger equations.* American Mathematical Society Colloquium Publications, 46. American Mathematical Society, Providence, RI, 1999.

[12] Burq, N., Gerard, P., Tzvetkov, N. *Strichartz inequalities and the nonlinear Schrödinger equation on compact manifolds.* Amer. J. Math. 126 (2004), no. 3, 569–605.

[13] Burq, N., Planchon, F., Stalker, J., Tahvildar-Zadeh, A. S. *Strichartz estimates for the wave and Schrödinger equations with the inverse-square potential.* J. Funct. Anal. 203 (2003), no. 2, 519–549.

[14] Burq, N., Planchon, F., Stalker, J., Tahvildar-Zadeh, A. S. *Strichartz estimates for the Wave and Schrödinger Equations with Potentials of Critical Decay,* Indiana Univ. Math. J., 53 (2004), no. 6, 1665–1680.

[15] Buslaev, V. S., Perelman, G. S. *Scattering for the nonlinear Schrödinger equation: States that are close to a soliton.* (Russian) Algebra i Analiz 4 (1992), no. 6, 63–102; translation in St. Petersburg Math. J. 4 (1993), no. 6, 1111–1142.

[16] Cai, K. *Fine properties of charge transfer models,* preprint 2003, to appear in Comm. PDE.

[17] Cai, K. *Dispersion for Schrödinger operators with one-gap periodic potentials on $\mathbb{R}^1$,* to appear in Dyn. Partial Differ. Equ.

[18] Costin, O., Costin, R. D., Lebowitz, J. L. *Time asymptotics of the Schrödinger wave function in time-periodic potentials.* J. Statist. Phys. 116 (2004), no. 1-4, 283–310.

[19] Costin, O., Lebowitz, J. L., Rokhlenko, A. *Decay versus survival of a localized state subjected to harmonic forcing: Exact results.* J. Phys. A 35 (2002), no. 42, 8943–8951.

[20] Cuccagna, S. *On the wave equation with a potential.* Comm. Partial Differential Equations 25 (2000), no. 7-8, 1549–1565.

[21] Cuccagna, S. *Stabilization of solutions to nonlinear Schrödinger equations.* Comm. Pure Appl. Math. 54 (2001), no. 9, 1110–1145.

[22] d'Ancona, P., Pierfelice, V. *On the wave equation with a large rough potential,* J. Funct. Anal. 227 (2005), no. 1, 30–77.

[23] d'Ancona, P., Pierfelice, V., Visciglia, N. *Some remarks on the Schrödinger equation with a potential in $L_t^r L_x^s$,* preprint 2003.

[24] Davies, E. B. *Time-dependent scattering theory.* Math. Ann. 210 (1974), 149–162.

[25] Deift, P., Trubowitz, E. *Inverse scattering on the line.* Comm. Pure Appl. Math. XXXII (1979), 121–251.

[26] Doi, S. *Smoothing effects of Schrödinger evolution groups on Riemannian manifolds.* Duke Math. J. 82 (1996), no. 3, 679–706.

[27] Erdogan, M. B., Killip, R., Schlag, W. *Energy growth in Schrödinger's equation with Markovian forcing.* Comm. Math. Phys. 240 (2003), no. 1-2, 1–29.

[28] Erdogan, B., Schlag, W. *Dispersive estimates for Schrödinger operators in the presence of a resonance and/or an eigenvalue at zero energy in dimension three,* Dynamics of PDE 1 (2004), 359–379.

[29] Erdogan, B., Schlag, W. *Dispersive estimates for Schrödinger operators in the presence of a resonance and/or an eigenvalue at zero energy in dimension three: II,* preprint 2005, to appear in Journal d'Analyse.

[30] Galtbayar, A., Jensen, A., Yajima, K. *Local time-decay of solutions to Schrödinger equations with time-periodic potentials.* J. Statist. Phys. 116 (2004), no. 1-4, 231–282.

[31] Georgiev, V., Visciglia, N. *Decay estimates for the wave equation with potential.* Comm. Partial Differential Equations 28 (2003), no. 7-8, 1325–1369.

[32] Ginibre, J. *Personal communication.*

[33] Goldberg, M. *Dispersive estimates for the three-dimensional Schrödinger equation with rough potential,* preprint 2004, to appear in Amer. J. Math.

[34] Goldberg, M. *Dispersive bounds for the three-dimensional Schrödinger equation with almost critical potential,* preprint 2004, to appear in Geom. and Funct. Anal.

[35] Goldberg, M., Schlag, W. *Dispersive estimates for Schrödinger operators in dimensions one and three.* Comm. Math. Phys. 251 (2004), no. 1, 157–178.

[36] Goldberg, M., Schlag, W. *A Limiting absorption principle for the three-dimensional Schrödinger equation with L^p potentials,* Intern. Math. Res. Notices, no. 75 (2004), 4049–4071.

[37] Goldberg, M., Visan, M. *A Counterexample to Dispersive Estimates for Schrödinger operators in Higher Dimensions,* preprint 2005, to appear in Comm. Math. Phys.

[38] Graf, J. M. *Phase space analysis of the charge transfer model.* Helv. Physica Acta 63 (1990), 107–138.

[39] Hassell, A., Tao, T., Wunsch, J. *A Strichartz inequality for the Schrödinger equation on non-trapping asymptotically conic manifolds,* Comm. Partial Differential Equations 30 (2005), no. 1-3, 157–205.

[40] Hassell, A., Tao, T., Wunsch, J. *Sharp Strichartz estimates on non-trapping asymptotically conic manifolds*, preprint 2004.

[41] Howland, J. S. *Born series and scattering by time-dependent potentials.* Rocky Mount. J. of Math. 10 (1980), no. 3, 521–531.

[42] Howland, J. S. *Stationary scattering theory for time-dependent Hamiltonians.* Math. Ann. 207 (1974), 315–335.

[43] Howland, J. S. *Scattering theory for Hamiltonians periodic in time.* Indiana Univ. Math. J. 28 (1979), no. 3, 471–494.

[44] Ionescu, A., Jerison, D. *On the absence of positive eigenvalues of Schrödinger operators with rough potentials.* Geom. Funct. Anal. 13 (2003), 1029–1081.

[45] Ionescu, A., Schlag, W. *Agmon–Kato–Kuroda theorems for a large class of perturbations*, preprint 2004, to appear in Duke Math. Journal.

[46] Jensen, A. *Spectral properties of Schrödinger operators and time-decay of the wave functions results in $L^2(R^m)$, $m \geq 5$.* Duke Math. J. 47 (1980), no. 1, 57–80.

[47] Jensen, A. *Spectral properties of Schrödinger operators and time-decay of the wave functions. Results in $L^2(R^4)$.* J. Math. Anal. Appl. 101 (1984), no. 2, 397–422.

[48] Jensen, A., Kato, T. *Spectral properties of Schrödinger operators and time-decay of the wave functions.* Duke Math. J. 46 (1979), no. 3, 583–611.

[49] Jensen, A., Nenciu, G. *A unified approach to resolvent expansions at thresholds.* Rev. Math. Phys. 13 (2001), no. 6, 717–754.

[50] Jensen, A., Yajima, K. *A remark on L^p-boundedness of wave operators for two-dimensional Schrödinger operators.* Comm. Math. Phys. 225 (2002), no. 3, 633–637.

[51] Journé, J.-L., Soffer, A., Sogge, C. D. *Decay estimates for Schrödinger operators.* Comm. Pure Appl. Math. 44 (1991), no. 5, 573–604.

[52] Kato, T. *Wave operators and similarity for some non-selfadjoint operators.* Math. Ann. 162 (1965/1966), 258–279.

[53] Keel, M., Tao, T. *Endpoint Strichartz estimates.* Amer. J. Math. 120 (1998), no. 5, 955–980.

[54] Kenig, C., Ruiz, A., Sogge, C. D. *Uniform Sobolev inequalities and unique continuation for second order constant coefficient differential operators.* Duke Math. J. 55 (1987), 329–347.

[55] Krieger, J., Schlag, W. *Stable manifolds for all monic supercritical NLS in one dimension*, preprint 2005, to appear in Journal AMS.

[56] Krieger, J., Schlag, W. *On the focusing critical semi-linear wave equation*, preprint 2005.

[57] Murata, M. *Asymptotic expansions in time for solutions of Schrödinger-type equations* J. Funct. Anal. 49 (1) (1982), 10–56.

[58] Naibo, V., Stepanov, A. *On some Schrödinger and wave equations with time dependent potentials*, preprint 2004. Math. Ann. 334 (2006), p. 325–338.

[59] Pierfelice, V. *Decay estimate for the wave equation with a small potential*, preprint 2003.

[60] Planchon, F., Stalker, J., Tahvildar-Zadeh, A. S. *Dispersive estimate for the wave equation with the inverse-square potential.* Discrete Contin. Dyn. Syst. 9 (2003), no. 6, 1387–1400.

[61] Planchon, F., Stalker, J., Tahvildar-Zadeh, A. S. L^p *estimates for the wave equation with the inverse-square potential.* Discrete Contin. Dyn. Syst. 9 (2003), no. 2, 427–442.

[62] Rauch, J. *Local decay of scattering solutions to Schrödinger's equation.* Comm. Math. Phys. 61 (1978), no. 2, 149–168.

[63] Reed, M., Simon, B. *Methods of modern mathematical physics. IV. Analysis of operators.* Academic Press [Harcourt Brace Jovanovich, Publishers], New York–London, 1978.

[64] Rodnianski, I., Schlag, W. *Time decay for solutions of Schrödinger equations with rough and time-dependent potentials.* Invent. Math. 155 (2004), 451–513.

[65] Rodnianski, I., Schlag, W., Soffer, A. *Dispersive analysis of charge transfer models.* Comm. Pure Appl. Math. 58 (2005), no. 2, 149–216.

[66] Rodnianski, I., Schlag, W., Soffer, A. *Asymptotic stability of N-soliton states of NLS*, preprint 2003.

[67] Rodnianski, I., Tao, T. *Quantitative limiting absorption principles on manifolds, and applications*, preprint 2005.

[68] Ruiz, A., Vega, L. *On local regularity of Schrödinger equations.* Int. Math. Res. Not. 1993 (1993), 13–27.

[69] Schlag, W. *Dispersive estimates for Schrödinger operators in dimension two,* Comm. Math. Phys. 257 (2005), 87–117.

[70] Schlag, W. *Stable manifolds for orbitally unstable NLS*, preprint 2004, to appear in Annals of Math.

[71] Simon, B. *Schrödinger semigroups.* Bull. AMS. vol. 7, 447–526.

[72] Smith, H., Sogge, C. *Global Strichartz estimates for nontrapping perturbations of the Laplacean.* Comm. Partial Differential Equations 25 (2000), no. 11-12, 2171–2183.

[73] Staffilani, G., Tataru, D. *Strichartz estimates for a Schrödinger operator with nonsmooth coefficients.* Comm. Partial Differential Equations 27 (2002), no. 7-8, 1337–1372.

[74] Stalker, J., Tavildar-Zadeh, S. *Scalar waves on a naked-singularity background,* Classical Quantum Gravity 21 (2004), no. 12, 2831–2848.

[75] Stein, E. *Beijing Lectures in Harmonic Analysis.* Princeton University Press, Princeton, NJ, 1986.

[76] Stoiciu, M. *An estimate for the number of bound states of the Schrödinger operator in two dimensions.* Proc. Amer. Math. Soc. 132 (2004), no. 4, 1143–1151.

[77] Strichartz, R. *Restrictions of Fourier transforms to quadratic surfaces and decay of solutions of wave equations.* Duke Math. J. 44 (1977), no. 3, 705–714.

[78] Weder, R. *The $W_{k,p}$-continuity of the Schrödinger wave operators on the line.* Comm. Math. Phys. 208 (1999), no. 2, 507–520.

[79] Weder, R. *Inverse scattering on the line for the nonlinear Klein-Gordon equation with a potential.* J. Math. Anal. Appl. 252 (2000), no. 1, 102–123.

[80] Weder, R. L^p-$L^{\tilde{p}}$ *estimates for the Schrödinger equation on the line and inverse scattering for the nonlinear Schrödinger equation with a potential.* J. Funct. Anal. 170 (2000), no. 1, 37–68.

[81] Weder, R. *The L^p-$L^{p'}$ estimate for the Schrödinger equation on the half-line.* J. Math. Anal. Appl. 281 (2003), no. 1, 233–243.

[82] Yajima, K. *A multichannel scattering theory for some time dependent Hamiltonians, charge transfer problem.* Comm. Math. Phys. 75 (1980), no. 2, 153–178.

[83] Yajima, K. *Scattering theory for Schrödinger equations with potentials periodic in time.* J. Math. Soc. Japan 29 (1977), no. 4, 729–743.

[84] Yajima, K. *The $W^{k,p}$-continuity of wave operators for Schrödinger operators.* J. Math. Soc. Japan 47 (1995), no. 3, 551–581.

[85] Yajima, K. *The $W^{k,p}$-continuity of wave operators for Schrödinger operators. II. Positive potentials in even dimensions $m \geq 4$.* Spectral and scattering theory (Sanda, 1992), 287–300, Lecture Notes in Pure and Appl. Math., 161, Dekker, New York, 1994.

[86] Yajima, K. *The $W^{k,p}$-continuity of wave operators for Schrödinger operators. III. Even-dimensional cases $m \geq 4$.* J. Math. Sci. Univ. Tokyo 2 (1995), no. 2, 311–346.

[87] Yajima, K. *L^p-boundedness of wave operators for two-dimensional Schrödinger operators.* Comm. Math. Phys. 208 (1999), no. 1, 125–152.

[88] Yajima, K. *Dispersive estimates for Schrödinger equations with threshold resonance and eigenvalue,* Comm. Math. Phys. 259 (2005), no. 2, 475–509.

Contributors

Paolo Baiti
Dipartimento di Matematica e Informatica
Università di Udine
Via delle Scienze, 206
Udine, Italia
(baiti@dimi.uniud.it)

Jean Bourgain
Institute for Advanced Study
Einstein Drive
Princeton, NJ 08540
(bourgain@math.ias.edu)

Alberto Bressan
Department of Mathematics
Pennsylvania State University
University Park, PA 16802
(bressan@math.psu.edu)

Haim Brezis
Department of Mathematics
Rutgers University, Hill Center, Busch Campus
110 Frelinghuysen Road
Piscataway, NJ 08854
(brezis@math.rutgers.edu).
Also, University P. et M. Curie
4 pl. Jussieu
75252 Paris Cedex 05, France
(brezis@ccr.jussieu.fr)

N. Burq
Laboratoire de Mathematiques
Université Paris-Sud—Bât 425
91405 Orsay Cedex, France
CNRS UMR 8628
(nicholas.burq@math.u-psud.fr)

Michael Christ
Department of Mathematics
University of California
Berkeley, CA 94720-3840
(mchrist@math.berkeley.edu)

Peter Constantin
Department of Mathematics
University of Chicago
5734 S. University Avenue
Chicago, Illinois 60637
(const@math.uchicago.edu)

J.-M. Delort
Départment de Mathématiques
Institut Galilée
Université Paris 13
99, avenue Jean-Baptiste Clément
93430 Villetaneuse, France
(delort@math.univ-paris13.fr)

P. Gerard
Laboratoire de Mathematiques
Université Paris-Sud—Bât 425
91405 Orsay Cedex, France
CNRS UMR 8628
(gerard@math.univ-paris13.fr)

Yoshikazu Giga
Department of Mathematics
Hokkaido University
Kita 10, Nishi 8, Kita-Ku
Sapporo, Hokkaido, 060-0810, Japan
(labgiga@ms.u-tolyo.ac.jp)

Alexandru D. Ionescu
Department of Mathematics
University of Wisconsin
Madison, WI 53706
(ionescu@math.wisc.edu)

Helge Kristian Jenssen
Department of Mathematics
North Carolina State University
Raleigh, NC 27695
(hkjensse@unity.ncsu.edu)

Carlos E. Kenig
Department of Mathematics
University of Chicago
5734 S. University Avenue
Chicago, Illinois 60637
(cek@math.uchicago.edu)

Alex Mathalov
Department of Mathematics
Arizona State University
Tempe, AZ 85287-1804
(mahalov@asu.edu)

Moshe Marcus
Department of Mathematics
Technion
Haifa 32000, Israel
(marcusm@tx.technion.ac.il)

Basil Nicolaenko
Department of Mathematics
Arizona State University
Tempe, AZ 85287-1804
(byn@stokes.la.asu.edu)

Augusto C. Ponce
Laboratoire de Mathematiques et Physique Theorique
Faculté des Sciences et Techniques
Université de Tours
Parc de Grandmont 37200, Tours, France
(ponce@lmpt.univ-tours.fr)

Igor Rodnianski
Department of Mathematics
Princeton University
Fine Hall, Washington Road
Princeton NJ 08544-1000
(irod@math.princeton.edu)

W. Schlag
Department of Mathematics
University of Chicago
5734 S. University Avenue
Chicago, Illinois 60637
(schlag@caltech.edu)

Jeremie Szeftel
Department of Mathematics
Princeton University
Fine Hall, Washington Road
Princeton NJ 08544-1000
(sxeftel@math.univ-paris13.fr)

Terence Tao
Department of Mathematics
University of California
Los Angeles, CA 90095-1596
(tao@math.ucla.edu)

N. Tzvetkov
Laboratory Paul Painlevé
University of Lille 1
U.F.R. de Mathématiques
59 655 Villeneuve d'Ascq Cédex, France
U.M.R. CNRS 8524
(nickolay.tzvetkov@math.p-sud.fr)

W.M. Wang
Department of Mathematics and Statistics
Lederle Graduate Research Tower, 16th Floor
University of Massachusetts
Amherst, MA 01003-9305
(weimin@math.umass.edu)

Index

active scalar system, 159–63
Agmon, S., 260, 268
Aizenman, M., 263
Akhiezer, N. I., 213–15, 219
Anderson localization, 23
Artbazar, G., 258, 268
Arzela-Ascoli diagonalization, 234–35
averaging property, 219–21

Babin, A., 218
Baiti, P., 43–53, 287
Banach spaces, 182, 185, 206, 213
Baras-Pierre, 55, 59, 61, 85
barrier zones, 24–25
Baxter, J. R., 87
Bénilan, Ph., 55, 84, 100
Bertozzi, A., 160
Besov space, 216
Biot-Savart law, 158
Birman-Schwinger type bounds, 277
blow-up problem, 157, 160
Boccardo, L., 59–60, 99
Bochner identity, 234
Bohr approach, 213, 215
Bollé, D., 257, 277
Bombieri, E., 2
Bona-Smith approximations, 199–200
Borel-Centelli theorem, 41
Borel sets: Newtonian capacity and, 58–59, 80, 88; nonlinear elliptic equations and, 56, 58–59, 63, 70, 80, 84, 88, 91, 102–4
Born series, 263, 265, 269–70
boundary conditions: barrier zones and, 24–25; basic multilinear operators and, 150–52; Borel-Cantelli theorem and, 41; decay and, 22, 24–25, 37; diffusion bound and, 21–42; Dirichlet conditions and, 21; dispersive estimates and, 268–79; dynamical localization and, 23; energy barriers and, 22, 26; finite difference schemes instability and, 43–52; Godunov scheme and, 44–45, 48–51; Hamiltonians and, 21–22; interaction amplitudes and, 142–44; Kadomstev-Petviashvili initial equations and, 193–208; kinematic viscosity and,

164–68; Klein-Gordon equations and, 176–77; monomials and, 22–23, 30–36; Navier-Stokes equations and, 213 (*see also* Navier-Stokes equations); nonlinear elliptic equations and, 55–106 (*see also* nonlinear elliptic equations); non-resonance condition and, 36; Poisson brackets and, 26–36; probability distribution and, 23; simplified multilinear operators and, 144–47; Sobolev norms and, 21; Strichartz's inequalities and, 1–20; symplectic transformations and, 24, 28–36; three-dimensional compact manifolds and, 111–27; time-classical solution and, 213; traveling profiles and, 44–48; Weber formula and, 158–59; zero energy and, 256–60
Bourgain, Jean, 181, 287; diffusion bound and, 21–42; dispersive estimates and, 260, 279; nonlinear Schrödinger equation and, 112, 117–118, 124, 127, 131, 135, 149; Strichartz's inequalities and, 1–20
Bressan, Alberto, 43–53, 287
Brezis, Haim, 55–109, 287
Brezis-Gallouët inequality, 114
Browder, F. E., 59, 98
Brownian motion, 162
bump function, 266
Burq, N., 3, 252, 287; dispersive estimates and, 258, 260; nonlinear Schrödinger equation and, 111–129, 131; wellposedness and, 182
Buslaev, V. S., 268

Cai, K., 276, 280
Cannon, J. R., 213
Cannone, M., 213
cardinalities, 138
Cauchy problem, 1, 3, 131; active scalar system and, 162–63; continuous mapping of, 132–33, 142; Dirac operator and, 132; Eulerian-Langrangian invariants and, 162–63; Fourier analysis and, 134–37; Galilean symmetries and, 132; Hölder inequality and, 142; inverse scattering theory and, 132; Kadomstev-Petviashvili initial equations and, 202; Laplacians and, 171; linear/nonlinear discrepancy and, 133–34;

www.ingramcontent.com/pod-product-compliance
Ingram Content Group UK Ltd.
Pitfield, Milton Keynes, MK11 3LW, UK
UKHW020237161224
452563UK00006B/211